CHROMOSPHERIC FINE STRUCTURE

INTERNATIONAL ASTRONOMICAL UNION
UNION ASTRONOMIQUE INTERNATIONALE

SYMPOSIUM No. 56
HELD AT SURFER'S PARADISE, QLD., AUSTRALIA,
3–7 SEPTEMBER 1973

CHROMOSPHERIC FINE STRUCTURE

EDITED BY

R. GRANT ATHAY

*High Altitude Observatory of the National Center for Atmospheric Research
(Sponsored by the National Science Foundation),
Boulder, Colo., U.S.A.*

D. REIDEL PUBLISHING COMPANY
DORDRECHT-HOLLAND / BOSTON-U.S.A.

1974

Published on behalf of
the International Astronomical Union
by
D. Reidel Publishing Company, P.O. Box 17, Dordrecht, Holland

Sold and distributed in the U.S.A., Canada, and Mexico
by D. Reidel Publishing Company, Inc.,
306 Dartmouth Street, Boston,
Mass. 02116, U.S.A.

Library of Congress Catalog Card Number 74–79568

ISBN 978-90-277-0288-3   ISBN 978-94-010-2103-6 (eBook)
DOI 10.1007/978-94-010-2103-6

TABLE OF CONTENTS

PART I
THE QUIET CHROMOSPHERE: LIMB PHENOMENA

PART II
THE QUIET CHROMOSPHERE: DISK PHENOMENA

PART III
THE UPPER CHROMOSPHERE

PART VII
ENERGY BALANCE, HEAT TRANSFER AND HEATING MECHANISMS IN CHROMOSPHERIC FINE STRUCTURES

PREFACE

The devotion of an IAU symposium entirely to the topic of chromospheric fine structure at a time when models of the spherically symmetric chromosphere are still evolving constitutes a valid recognition of the growing feeling among solar astronomers that the chromosphere cannot be understood independently of its discrete structural features. Network structure, which seemingly borders the photospheric supergranule cells, persists intact throughout the chromosphere and most of the chromosphere-corona transition region. The network is the locus of the bright coarse mottles, and the spicule bushes and is the terminus for one end of the quiet chromospheric fibrils as well. Additionally, it is the locus of most of the magnetic flux of the quiet chromosphere. It is not surprising, therefore, that current studies of the chromosphere tend to center around efforts to better describe the network phenomena and to ascertain the physical properties of the network features. Clearly, the supergranule cells and associated network structures constitute a fundamental and singularly important feature of solar structure in the boundary layers.

Just as it is now clear that much of the chromospheric fine structure is associated with the network bordering supergranule cells, it seems equally clear that structural features are almost universally associated with both fluid flow and magnetic geometry. Indeed, many observers claim that the brightness features faithfully map the magnetic lines of force while still others claim that associated with each class of brightness feature there is a more or less unique fluid flow.

Fluid streaming and wave phenomena associated with fine scale features of the solar atmosphere represent the transport of non-radiative energy that, in turn, heats the chromosphere and corona. Most of what astronomers now casually classify as 'micro-' and 'macro-turbulence' undoubtedly arises from these combined streaming and wave motions. Thus, an understanding of the fluid motions associated with chromospheric fine structure appears to be fundamental to both the interpretations of line broadening and mechanical energy transport in the Sun. Similarly, the concentration of magnetic flux in the network and the strong correlation between regions of enhanced magnetic flux, increased chromospheric brightness (in all spectral regions) and increased flow speeds suggest that the magnetic field is in some way important to the very existence of the chromosphere and corona.

Chromospheres undoubtedly exist in the majority of stars; and by implication, chromospheric fine structure exists in the majority of stars as well. Only in the case of the Sun, however, can we hope to isolate and identify the true nature of the fluid motions giving rise to spectral line broadening and to the transport and dissipation of mechanical energy.

R. GRANT ATHAY

ORGANIZING COMMITTEE

R. G. Athay, M. K. V. Bappu, J. M. Beckers, R. G. Giovanelli (Chairman),
K. O. Kiepenheuer, R. Michard, G. Righini, A. B. Severny, Z. Švestka.

LIST OF PARTICIPANTS

Acton, L. W., Lockheed Palo Alto Research Labs., Dept. 52-14, Bldg. 202, 3251 Hanover Street, Palo Alto, Calif. 94304

Altschuler, M. D., High Altitude Observatory, P. O. Box 3000, Boulder, Colo. 80303.

Athay, R. G., High Altitude Observatory, P. O. Box 3000, Boulder, Colo. 80303

Bappu, M. K. V., Indian Institute of Astrophysics, Kodaikanal 624103, India

Beckers, J. M., Sacramento Peak Observatory, AFCRL, Sunspot, N.M. 88349

Bhattacharyya, J. C., Indian Institute of Astrophysics, Hebbal, Bangalore 56006, India

Bhavilai, R., Department of Physics, Chulalongkorn University, Bangkok 5, Thailand

Boischot, A., Observatoire de Meudon, 92160 Meudon, France

Bonnet, R. M., Laboratoire de Physique Stellaire et Planetaire, B. P. 10, 91370 Verrieres-le-Buisson, France

Bracewell, R. N., Radio Astronomy Institute, Stanford University, Stanford, Calif. 94305

Brandt, J. C., NASA Goddard Space Flight Center, Code 680, Greenbelt, Md. 20771

Brault, J. W., Kitt Peak National Observatory, P. O. Box 26732, Tucson, Ariz. 85726

Bray, R. J., CSIRO Division of Physics, University Grounds, Chippendale, N.S.W., 2008, Australia

Brown, J. C., Department of Astronomy, University of Glasgow, Glasgow G12 80W, Scotland·

Brown, N., CSIRO Solar Observatory, P. O. Box 147, Narrabri, N.S.W., 2390, Australia

Brueckner, G. E., U.S. Naval Research Laboratory, Code 7142, Washington, D.C. 20375

Bruzek, A., Fraunhofer Institut, Schoneckstrasse 6, 7800 Freiburg i.Br., F.R.G.

Bumba, V., Astronomical Institute of the Czechoslovak Academy of Sciences, 251 65 Ondrejov, Czechoslovakia

Cannon, C. J., Department of Applied Mathematics, University of Sydney, Sydney, N.S.W., 2006, Australia

Caroubalos, C., Observatoire de Meudon, 92160 Meudon, France

Castelli, J. P., Air Force Cambridge Research Labs., L. G. Hanscom Field, Bedford, Mass. 01730

Catura, R. C., Lockheed Palo Alto Research Labs., Dept. 52–14, Bldg. 202, 3251 Hanover Street, Palo Alto, Calif. 94304

Cole, T. W., CSIRO Division of Radiophysics, P. O. Box 76, Epping, N.S.W., 2121, Australia

Corderroure, J. Casanovas, Instituto Universitario de Astrofisica, La Laguna, Tenerife, Spain

Cram, L. E., Department of Applied Mathematics, Sydney University, Sydney, N.S.W., 2006, Australia

Delache, P., Observatoire de Nice, Le Mont Gros, 06 Nice, France

Deubner, F.-L., Fraunhofer Institut, Schoneckstrasse 6, 7800 Freiburg i.Br., F.R.G.

Dravins, D., Lund Observatory, Svanegatan 9, S-222 24 Lund, Sweden

Elgaroy, O., Institute of Theoretical Astrophysics, Oslo University, P. O. Box 1029, Blindern, Oslo-3, Norway

Enome, S., Research Institute of Atmospherics, Nagoya University, Toyokawa 442, Japan

Fainberg, J., NASA Goddard Space Flight Center, Code 693, Greenbelt, Md. 20771

Frazier, E. N., Aerospace Corporation, P. O. Box 92957, Los Angeles, Calif. 90009

Fredga, K., Royal Institute of Technology, Department of Plasma Physics, S-10044 Stockholm, Sweden

Frost, K. J., NASA Goddard Space Flight Center, Code 682, Greenbelt, Md. 20771

Gabriel, A. H., Culham Laboratory, Abingdon, Berks., England

Gebbie, K. B., Joint Institute for Laboratory Astrophysics, University of Colorado, Boulder, Colo. 80302

Giovanelli, R. G., CSIRO Division of Physics, University Grounds, Chippendale, N.S.W., 2007, Australia

Godoli, G., Osservatorio Astrofisico, viale A. Doria, citta universit., 95125 Catania, Italy

Gotwols, B. L., The John Hopkins University, Applied Physics Laboratory, 8621 Georgia Avenue, Silver Spring, Md 20910

Grossmann-Doerth, U., Fraunhofer Institut, Schoneckstrasse 6, 7800 Freiburg i.Br., F.R.G.

Hachenberg, O., Max-Planck-Institut fur Radioastronomie, Auf dem Hugel 69, 53 Bonn 1, F.R.G.

Holt, J., Department of Applied Mathematics, University of Sydney, Sydney, N.S.W., 2006, Australia

Jefferies, J. T., Institute for Astronomy, University of Hawaii, 2840 Kolowalu Street, Honolulu, Hawaii 96822

Jensen, E., Institute of Theoretical Astrophysics, University of Oslo, P. O. Box 1029, Blindern, Norway

Jordan, C., SRC Astrophysics Research Division, Culham Laboratory, Abingdon, Berks., England

Kai, K., Tokyo Astronomical Observatory, Mitaka, Tokyo, Japan

Kiepenheuer, K. O., Fraunhofer Institut, Schoneckstrasse 6, 7800 Freiburg i.Br., F.R.G.

Kubota, J., Kwasan Observatory, Kyoto University, Kyoto City, Japan

Kundu, M. R., Astronomy Program, University of Maryland, College Park, Md 20742

Labrum, N. R., CSIRO Division of Radiophysics, P. O. Box 76, Epping, N.S.W., 2121, Australia

Lantos, P., Observatoire de Meudon, Departement Solaire, 92-Meudon, France

Leblanc, Y., Observatoire de Meudon, 92-Meudon, France

Leibacher, J. W., Laboratoire de Physique Stellaire et Planetaire, B.P. 10, 91370 Verrieres-le-Buisson, France

Liebenberg, D. H., Los Alamos Scientific Laboratory, University of California, P. O. Box 1663, Los Alamos, N.M. 87544

Loughead, R. E., CSIRO Division of Physics, University Grounds, Chippendale, N.S.W., 2008, Australia

McCabe, M., Institute for Astronomy, University of Hawaii, 2840 Kolowalu Street, Honolulu, Hawaii 96822

McKenna-Lawlor, S., Rooske Road, Dunboyne, Co. Meath, Ireland

McLean, D. J., CSIRO Division of Radiophysics, P. O. Box 76, Epping, N.S.W., 2121, Australia

Malitson, H. H., NASA Goddard Space Flight Center, Code 693, Greenbelt, Md. 20771

Mangeney, A., Observatoire de Meudon, 92-Meudon, France

Martin, S. F., Lockheed Solar Observatory, Plant 2, Bld. 243, Dept. 74-33, Lockheed California Company, Burbank, Calif. 91503

Martres, M. J., Observatoire de Meudon, D.A.S.O.P., 92-Meudon, France

Maxwell, A., Harvard College Observatory, 60 Garden Street, Cambridge, Mass. 02138

Mayfield, E. B., Aerospace Corporation, P.O. Box 92957, Los Angeles, Calif. 90009

Melrose, D. B., Department of Theoretical Physics, Australian National University, P. O. Box 4, Canberra, A.C.T., 2600, Australia

Meyer, F., Max-Planck-Institut fur Physik und Astrophysik, Fohringer Ring 6, 8 Munchen 40, F.R.G.

Molnar, H., Observatorio de Fisica Cosmica, de San Miguel, Avda. Mitre 3.100, San Miguel – FCGSM, Argentina

Mullaly, R. F., School of Electrical Engineering, University of Sydney, Sydney, N.S.W., 2006, Australia

Muller, E. A., Observatoire de Geneve, 1290 Sauverny, Geneva, Switzerland

Nakagawa, Y., High Altitude Observatory, P. O. Box 3000, Boulder, Colo. 80303

Newkirk, G. A., High Altitude Observatory, P. O. Box 3000, Boulder, Colo. 80303

Pande, M. C., Uttar Pradesh State Observatory, Manora Peak, Naini Tal, India

Pasachoff, J. M., Williams College Observatory, Williamstown, Mass. 01267

Payten, W. J., CSIRO Solar Observatory, P. O. Box 94, Narrabri, N.S.W., 2390, Australia

Pecker, J.-C., Institut d'Astrophysique de CNRS, 98 bis bd. Arago, Paris 75014, France

Pick, M., Observatoire de Meudon, D.A.S.O.P., 92-Meudon, France

Piddington, J. H., CSIRO Division of Physics, University Grounds, Chippendale, N.S.W., 2008, Australia

Pierce, A. K., Kitt Peak National Observatory, P. O. Box 26732, Tucson, Ariz. 85726

Pneuman, G. W., High Altitude Observatory, P. O. Box 3000, Boulder, Colo. 80303

Prokakis, T., National Observatory of Athens, Athens 306, Greece

Rees, D. E., Department of Applied Mathematics, University of Sydney, N.S.W., 2006, Australia

Ribes, E., Observatoire de Meudon, D.A.S.O.P., 92190-Meudon, France

Riddle, A. C., Department of Astro-Geophysics, University of Colorado, Boulder, Colo. 80302

Robinson, R., CSIRO Division of Radiophysics, P. O. Box 76, Epping, N.S.W., 2121, Australia

Roddier, F., Departement d'Astrophysique, IMSP, Universite de Nice, Parc-Valrose, 06034 Nice, France

Rosch, J., Observatoires du Pic-du-Midi et de Toulouse, 65200-Bagneres-de-Bigorre, France

Rosenberg, J., Observatory "Sonnenborgh", Zonnenburg 2, Utrecht, The Netherlands

Rutten, R. J., Observatory "Sonnenborgh", Zonnenburg 2, Utrecht, The Netherlands

Sakurai, K., NASA Goddard Space Flight Center, Code 693, Greenbelt, Md. 20771

Schatten, K. H., Physics Department, Victoria University, P. O. Box 196, Wellington, New Zealand

Schmidt, H. U., Max-Planck-Institut fur Physik und Astrophysik, Fohringer Ring 6, 8 Munchen 40, F.R.G.

Sheridan, K. V., CSIRO Division of Radiophysics, P. O. Box 76, Epping, N.S.W., 2121, Australia

Simon, P., Observatoire de Meudon, 92190-Meudon, France

Sivaraman, K. R., Indian Institute of Astrophysics, Hebbal, Bangalore 560006, India

Smerd, S. F., CSIRO Division of Radiophysics, P. O. Box 76, Epping, N.S.W., 2121, Australia

Smith, D. F., High Altitude Observatory, P. O. Box 3000, Boulder, Colo. 80303

Souffrin, P., Observatoire de Nice, 06300-Nice, France

Speer, R. J., Department of Physics, Imperial College of Science and Technology, Prince Consort Road, London, S.W. 7, England

Steinberg, J. L., Department de Recherches Spatiales, Observatoire de Meudon, 92190-Meudon, France

Stewart, R. T., CSIRO Division of Radiophysics, P. O. Box 76, Epping, N.S.W., 2121, Australia

Stix, M., Universitats-Sternwarte, Geismarlandstrasse 11, 34 Gottingen, F.R.G.

Sturrock, P. A., Institute for Plasma Research, Stanford University, Via Crespi, Stanford, Calif. 94305

Takakura, T., Department of Astronomy, University of Tokyo, Bunkyo-ku, Tokyo, Japan

Tanaka, K., Tokyo Astronomical Observatory, Mitaka, Tokyo, Japan

Thomas, R. N., Institut d'Astrophysique, 98 bis Blvd. Arago, Paris, France

Uchida, Y., Tokyo Astronomical Observatory, Mitaka, Tokyo, Japan

van Nieuwkoop, J., Sterrewacht "Sonnenborgh", Servaas Bolwerk 13, Utrecht, The Netherlands

Vardavas, I. M., Department of Applied Mathematics, University of Sydney, Sydney, N.S.W., 2006, Australia

Vorpahl, J., Sacramento City College, Physics and Astronomy, 3835 Freeport Boulevard, Sacramento, Calif. 95822

Vrabec, D., The Aerospace Corporation, P. O. Box 92957, Los Angeles, Calif. 90009

Waldmeier, M., Swiss Federal Observatory, Schmelzbergstrasse 25, 8006 Zurich, Switzerland

Wentzel, D. G., Astronomy Program, University of Maryland, College Park, Md. 20742

Wiehr, E., Universitats-Sternwarte, Geismarlandstrasse 10, 34 Gottingen, F.R.G.

Wild, J. P., CSIRO Division of Radiophysics, P. O. Box 76, Epping, N.S.W., 2121, Australia

Wilson, P. R., Department of Applied Mathematics, University of Sydney, Sydney, N.S.W., 2006, Australia

Wolf, B., European Southern Observatory, 131 Bergedorferstrasse, 205 Hamburg 80, F.R.G.

Yousef, S. M. A., Astronomy Department, Cairo University, Giza, Cairo, Egypt

Zirin, H., California Institute of Technology, Room 354-33, Pasadena, Calif. 91109

Zwaan, C., Observatory "Sonnenborgh", Zonnenburg 2, Utrecht, The Netherlands

PART I

THE QUIET CHROMOSPHERE: LIMB PHENOMENA

SPICULES AND THEIR SURROUNDINGS

R. MICHARD

Observatoire de Paris, France

(presented by J. Rösch)

I have taken the liberty to change the title of this review from the one announced in the programme, because I intend to look at the empirical knowledge about spicules both from limb and disk observations, also because it does not seem feasible to discuss the problem of *relations* between spicules and their surroundings without starting by what is known about spicules themselves.

This introductory lecture has of course no ambition to be exhaustive. For more thoroughness one should go to the excellent reviews by Giovanelli *et al.* (1967), and and Beckers (1968, 1972). The 1972 review paper by Beckers was a great help to me in preparing this contribution.

Spicules were first extensively observed and described by Secchi one hundred years ago. Their name was, I believe, invented by Roberts in 1945, and it has been a very successful word.

At present, one can distinguish between three definitions of spicules:

– *Definition 1:* Spicules are jet-like features which, under good seeing conditions, can be seen through a broad band Hα filter to extend *above* the more or less continuous chromospheric Hα fringe of roughly 4000 or 5000 km thickness (see Figures 1 and 2).

– *Definition 2:* Spicules are elongated features present at all heights in the chromosphere above 1000 km and made observable by a variety of techniques, but mainly by narrow band Hα filters used at various wave lengths inside the line, both on the disk and at the limb. (See Figure 3).

Some attention will be given later to the following question: are spicules according to Definitions 1 and 2 the same physical phenomena? or in other words do the high level, jet-like spicules above 5000 km represent only a selection of the tallest, most favourably placed relative to the limb, among a larger population? And does this larger population (spicules according to Definition 2) represent the only fine structure of the chromosphere?

– *Definition 3:* Spicules are a 'dense component', sometimes hot, sometimes cold, used to improve the fit of model chromospheres to various spectral data which refuse to conform to the analysis done under the hypothesis of sphericl symmetry.

1. Spicules in the High Chromosphere (above 6000 km)

1.1. SIZE

According to the best observations by Dunn (1960) or more recent data from the Sacramento Peak vacuum telescope, the average diameter of spicules is 900 km with a real dispersion between 400 and 1500 km.

R. Grant Athay (ed.), Chromospheric Fine Structure, 3–21. All Rights Reserved.
Copyright © 1974 by the IAU.

Fig. 1.   Chromosphere photographed 0.9 Å in the wing of Hα with a 0.25 Å bandwidth filter,
1970, Oct. 14, Vacuum Solar Telescope, Sacramento Peak Observatory,
Air Force Cambridge Research Laboratories.

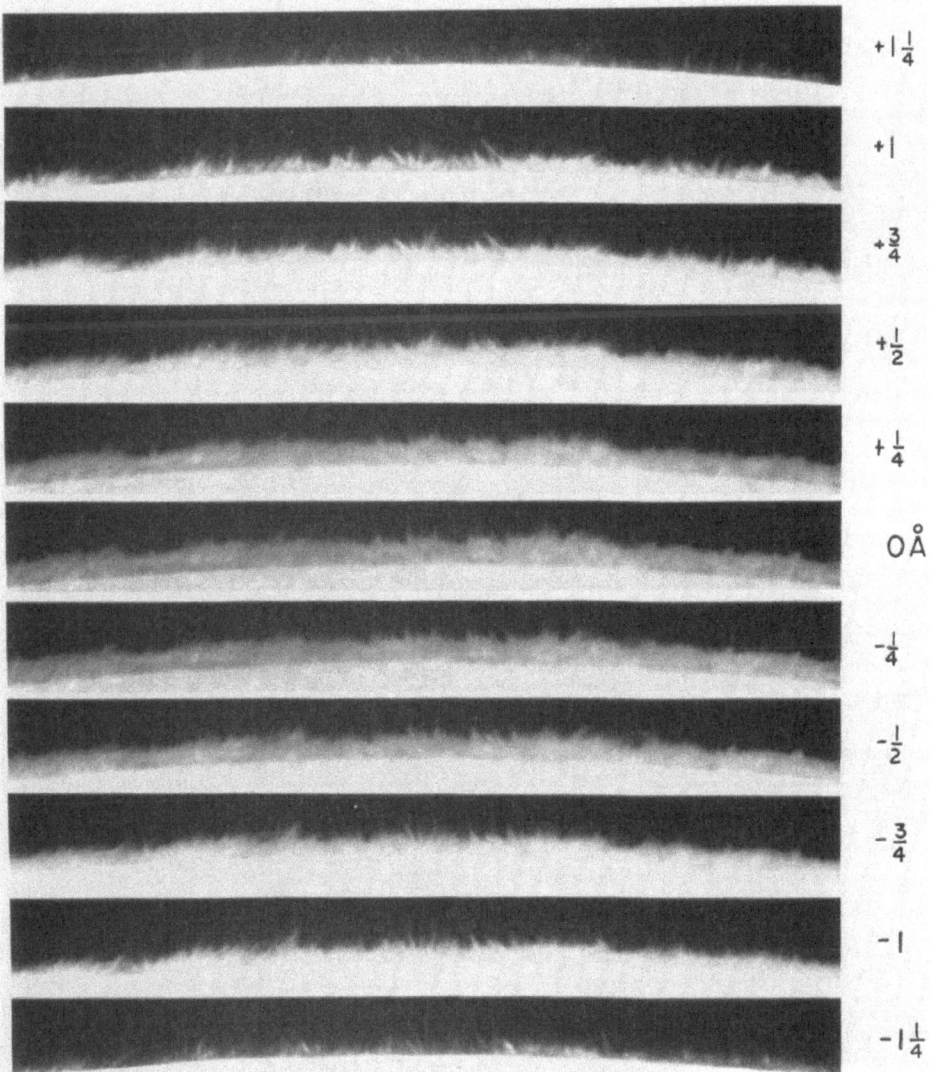

$+1\frac{1}{4}$

$+1$

$+\frac{3}{4}$

$+\frac{1}{2}$

$+\frac{1}{4}$

$0\,\overset{\circ}{A}$

$-\frac{1}{4}$

$-\frac{1}{2}$

$-\frac{3}{4}$

-1

$-1\frac{1}{4}$

Fig. 2. A wavelength scan through Hα of the chromosphere. The birefringent filter has a bandwidth of 0.25 Å. A small amount of continuum light leaks through the filter causing the solar disk to show at the 0 Å position, 1970, Oct. 14, Vacuum Solar Telescope, Sacramento Peak Observatory, Air Force Cambridge Research Laboratories.

1.2. COUNTS

Above 6000 km the spicules are essentially isolated, and the countings along a given arc of the limb approximately give their actual distribution $S(h)$ in apparent height. According to various authors, $S(h)$ has a maximum of about 40 per 12 deg between 5000 and 6000 km; it decreases exponentially at greater heights: it also decreases sharply towards the limb until, somewhat below 3000 km, it becomes difficult to

Fig. 3.   Spicules outline the edges of supergranule cells in this 0.25 Å bandwidth photograph taken 0.9 Å
to the red of the Hα line. Dark absorption in the tops of loop prominences appears over the active region,
1971, Feb. 13, Vacuum Solar Telescope, Sacramento Peak Observatory,
Air Force Cambridge Research Laboratories.

distinguish spicules (still with the broad band filter). Assuming this trend of the $S(h)$ function to be due to the mutual superposition of spicules, Athay (1959) corrected the counts to find the true number $N(h)$ of spicules intersecting a 12° arc of the limb; his corrections reach a factor of ≈ 3 at 5000 km and 15 at 3000 km and are quantitatively uncertain for many reasons.

In his review, Beckers concludes that the corrected counts can be fairly well represented by

$$N(h) = 10^3 \exp\left(-h/1950 \text{ km}\right) \text{ spicules per } 12°.$$

From this, assuming an even distribution of the spicules over the solar surface, we find their total number on the Sun to be

$$\Pi(h) = 10^6 \exp\left(-h/1750 \text{ km}\right).$$

These representations must not be used without caution in the ranges where they are mere extrapolations, in fact below 3000 km.

It should be noted that the observed function $S(h)$ depends on the distribution of spicules on the spherical solar surface, their distribution in lengths and in tilts to the vertical. Tilted spicules clustered in bushes, as shown by disk observations (Figure 3), are likely to overlap more often than vertical, evenly distributed, features.

Attempts have been made to evaluate the average proportion of spicule material along a given line of sight from spectral data, for instance by Michard (1959); these estimates very roughly agree with corrected counts, in showing that a tangential line of sight intersects in the mean more than one spicule below 3000 km.

1.3. APPARENT MOTIONS AND LIFETIMES

On successive Hα pictures, spicules appear to be 'ejected' from the Hα fringe, to ascend at an apparent velocity which is in the mean 25 km s^{-1}, to reach an average height of 9000 km; their top seems to remain stationary for a while; then the feature disappears, either by falling back along the same path (roughly half the cases) or by fading in brightness along its full length.

It is my opinion that the difficulty of observations are such, that the given description does not necessarily imply that two different mechanisms of disappearance are really at work. The observers note that the descent is much more poorly defined than the ascent.

On Dunn's films, Mouradian (1967) detected a tendency for many spicules to broaden or explode as they are fading. Although more recent data do not corroborate this result, according to Beckers (1972), it remains likely that spicules decay by a combined mechanism, part of the material falling back and part dissolving in the corona.

The 'lifetime' defined as the average duration of visibility above the inner Hα fringe is 5 min. There appears to be a genuine dispersion in the velocities, maximum heights and life-times. It is also interesting to note the reported positive correlation between velocity and maximum height and between life-time and maximum height: they are suggestive of a motion primarily governed by gravitational forces.

It is not yet certain that the apparent displacements of spicules tops are entirely due to material motions: we shall return to this in discussing observed Doppler effects.

1.4. GROUPING OF SPICULES

Large scale configurations have been recognized in the occurrence and orientation of spicules by Lippincott (1957) and more recently by Platov and Shilova (1969, 1971).

The name of 'porcupine' was given to a fan-like pattern of about 10 000 to 20 000 km extent along the limb, in which spicules seem to radiate outward from an origin buried inside the Sun, 10 000 to 30 000 km deep. These formations should be related with the 'bushes' or 'rosettes' of dark Hα features on the disk to be described later.

Other systematic groupings in inclination to the limb are observed. At high solar latitude, the spicules have systematic inclinations, paralleling those of the polar plumes of the corona.

The local magnetic fields seem to be the only possible agent likely to give a non-random orientation to spicules, whatever their physical mechanism; it is thus believed that spicules trace magnetic lines of force.

Systematic equator-pole variations in spicule properties have been reported: polar spicules are less inclined, somewhat taller and faster and less numerous.

2. Spicules in the Inner Chromosphere

2.1. EVIDENCE FOR AN 'INTERSPICULAR MEDIUM'

By using graded height sequences of high resolution spectrograms, Michard (1959) showed that all spicules seen above 5000 km can be traced much lower in the chromosphere in *the wings of the line* (Figure 2). This improved visibility arises from two reasons:

(1) Spicules with violet-shifted emissions are not masked by spicules with redshifted profiles and conversely,

(2) While at the line center most of the spicules are likely to be obscured by an absorbing *interspicular* medium, this absorption is much reduced in the wings.

Michard developed a model for the interpretation of spicule spectra in the presence of a postulated interspicular medium. The local profile on a spicule should be represented by

$$\Delta I_{s,\lambda} = B_s [1 - \exp(-\tau_{s,\lambda})] \exp(-\tau_{is,\lambda}), \tag{1}$$

where $\Delta I_{s,\lambda}$ represents the difference of emission on a line of sight crossing the spicule and a nearby line of sight 'between' spicules, B_s is the source function and $\tau_{s,\lambda}$ the optical thickness of the spicule and $\tau_{is,\lambda}$ is the optical depth of the interspicular medium along the line of sight to the spicule.

A coherent interpretation of the data about spicules and the averaged spectrum of the chromosphere, is obtained by assuming that the source-function and the line-broadening parameter are both larger in the measurable spicules than in the interspicular medium. The main proof for the existence of an interspicular medium of significant optical thickness is the presence of a central reversal in the profiles of spicules below 5000 km; and the fact that this central reversal remains at the average wavelength of Hα, even when the spicule profiles are Doppler shifted.

According to Michard the total optical thickness of the chromosphere *between spicules* for a tangential line of sight and for the center of Hα increases from zero at 6000 km to $\simeq 10$ at 2000 km; below it increases to large and poorly determined values. The visibility of spicules is due to their larger broadening parameters and to their Doppler effects.

The arguments for an interpsicular medium of significant opacity in very strong lines have been further checked by Pasachoff *et al.* (1968) using high quality spectrograms in various lines.

Filtergrams of superior quality have been obtained by Dunn at the Sacramento Peak tower, using a filter of pass-band 0.25 Å shifted throughout the Hα line. A first analysis was recently reported by Lynch *et al.* (1973). An enormous number of spicules (according to definition 2!) are seen in the low chromosphere, sufficiently away from the line center: direct counts on the original negatives give $S(h) = 110$ per 12° at Hα ± 0.75 Å and ± 1 Å, for $h \simeq 4000$ km. Since the structures are very different in opposite wings, the true number should be increased by a factor of nearly 2, and still larger to take account of the chance overlaps of features with similar Doppler shifts.

There is some inconsistency between these very large counts and the value of 950 km for the true spicule diameter, reported in the same paper. An arc of 12° at the limb contains only 155 times this spicule diameter: according to the Athay and Thomas (1961) analysis, if any number of spicules are projected on an arc of 155 times the spicule diameter, the maximum number of isolated features is $155/2e \simeq 28$.

A more quantitative, and perhaps more objective study of spicule abundances is presented in the same paper from counts of intensity enhancements on microphotograms at a given intensity threshold. The numbers of features at $I = 2\%$ (in percentage of the Hα continuum at the disk center) reach maximum values $S \simeq 30$, at lower and lower heights when observing farther and farther from the center of Hα (Figure 4).

Although a quantitative interpretation of these fine data is still lacking, we suggest that the very fast increase of the number of features when going from the center of the line towards the wings is due to the decrease of interspicular absorption.

2.2. DOPPLER SHIFTS AND VELOCITY FIELD OF SPICULES

As shown by spectrograms and by the recent narrow band filtergrams, many spicules show significant Doppler shifts. Among others Mouradian (1965) obtained extensive observations of spicules emission profiles carefully corrected for the influence of the 'interspicule medium'. His data are shown in Figure 5. *Above 6000 km the rms line of sight velocities is about 10 km s^{-1}*; the average velocity being zero. These results are confirmed by others, such as Nikolsky and Sasanov (1966), Athay and Bessey (1964) who also find a decrease of spicule velocities near the pole, and Pasachoff *et al.* (1968). As noted by Beckers, the Doppler data are consistent with the idea that the apparent displacements of spicule tops, are indeed due to genuine motions at the same

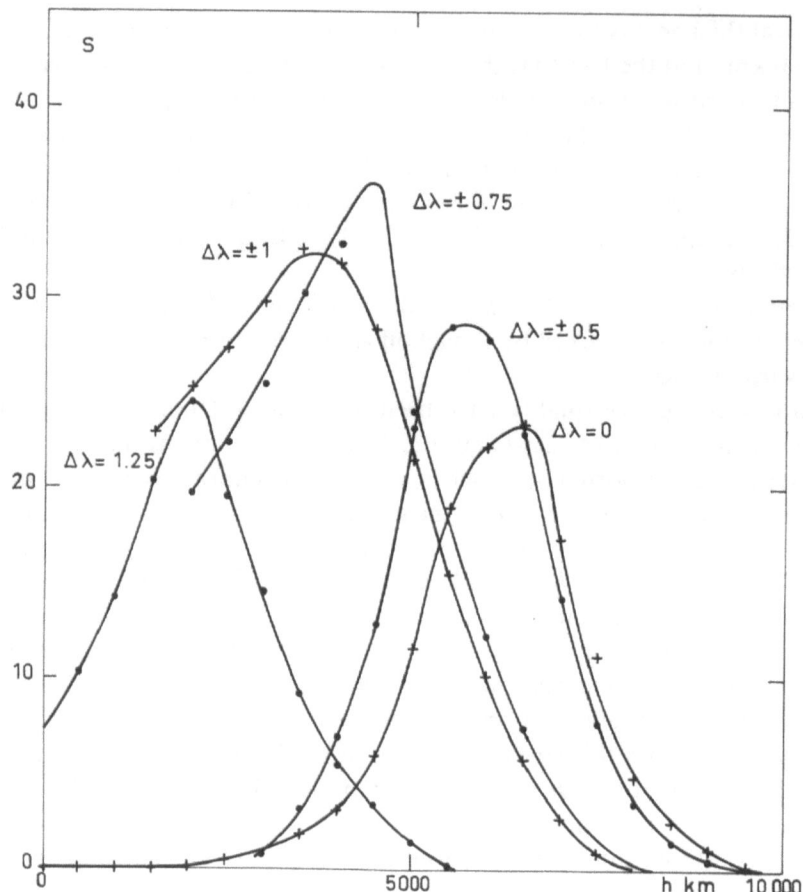

Fig. 4. Counts of spicules of treshold intensity 2% of the Hα continuum at the center of the disk, for various wavelengths $\Delta\lambda$ throughout the Hα line. The figure is drawn from the results of Lynch *et al.* (1973).

velocities: assuming a constant velocity of 25 km the rms line of sight somponent of 10 km s^{-1} would correspond to a rms tilt of 24° to the vertical. The rms tilt observed by Beckers is 19° and he assumes this to be a lower limit.

While many spicules, particularly those with the largest Doppler shifts, show an increase of Doppler shift with height, there appears to be no significant systematic variation of the rms Doppler shift with altitude above the limb.

During the time of spicule visibility their Doppler shift increases progressively, goes through a maximum and decrease rather abruptly, as first reported by Mouradian. This author did not find spicules reversing their Doppler motion, and suggested that the material does not fall back in the physical state and trajectory of its ascent. However Nikolsky *et al.* at Izmiran, Pasachoff *et al.* at Sacraments Peak, find a number of spicules actually reversing their Doppler motion before disappearing. However a majority of 80% conform with Mouradian's description, according to Pasachoff *et al.*

As already found for the apparent motions, the Doppler motions seem to indicate

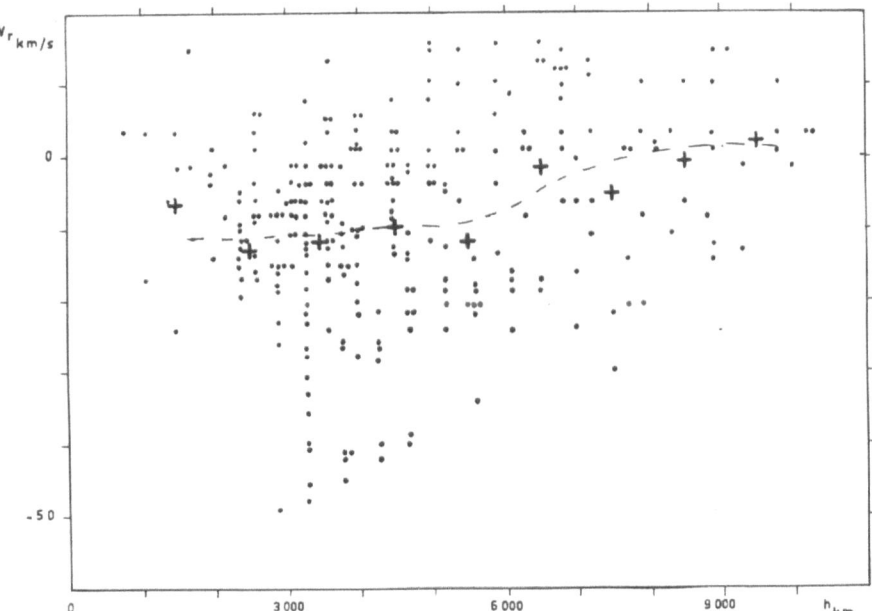

Fig. 5. Line-of-sight velocities of spicules as a function of height, from the analysis of Pic-du-Midi spectrograms (Mouradian, 1965). In the lower chromosphere, the Doppler shifts are systematically towards the violet, a result questioned by other observers. (Reprinted from *Ann. Astrophys.* 1965).

that the spicule material falls back in part in spicular form, and in part after 'dissolution' in the corona. It is possible that the two mecanisms are simultaneously at work, with a different relative importance in each spicule. Although direct correlations between the Doppler curve and brightness curve of spicules have not been presented, it seems likely that the large velocities occur during the youth and ascending phase of spicules; therefore ascending spicules should be brighter.

In the *low chromosphere*, Mouradian found the Doppler effects of spicule to have an rms value of the same order of magnitude as above, but to scatter around a mean of -10 km s^{-1}, i.e. spicules have systematic violet shift. This systematic effect was qualitatively explained by assuming (see Figure 4):

– that the sample at greater depths contains a large number of spicules situated towards the observer, the spicules rooted beyond the limb being masked by the interspicular medium;

– that these more easily visible spicules are predominantly tilted towards the observer, and ascending, according to the remark above.

Observations by Pasachoff *et al.* (1968), which refer to two levels, 5000 and 7000 km, do not confirm this curious result, although there is no real contradiction because of the different ranges of heights best covered by the two sets of data.

We should finally mention a result of great physical significance obtained by Pasachoff *et al.* (1968) from their simultaneous time sequences of Hα spectrograms at two heights, separated by 1800 km; the time variations of the Doppler shift occur nearly simultaneously at the two heights, the average time delay being only 2.5 ± 2 s.

2.3. Spicules in other lines than Hα

Spicules have been observed in practically all strong chromospheric lines of H, He and Ca$^+$. Here I shall summarize some results relevant to spicule morphology and perhaps in relation with the interspicule medium. Later in this symposium Dr Jefferies will review the physical conditions in spicules as derived from their radiative properties, and Dr Frazier will talk about the mysterious internal motions inferred from peculiarities in the profiles of spectral lines.

Profiles in Hβ and Hγ and also in the H and K lines are qualitatively similar to those of Hα. However in Hγ and Hβ the characteristic self-reversals that we have attributed to the interspicular medium occur only at lower apparent heights.

Mouradian (1965) measured simultaneous height sequences of spectrograms in Hα and Hγ. At great heights a number of faint Hα spicules cannot be detected in Hγ due to reduced emissivity. At low heights, however, the number of spicules seen at the center of Hγ is larger by 30% than can be distinguished in Hα, although the resolution of the Hγ spectra is less good. This is attributed to the reduced opacity of the interspicular medium in Hγ.

Incidentally, the broadening parameters in Hα and Hγ for a given spicule, assumed to be purely a Doppler broadening, are in the ratio of the wavelength of the two lines, as expected.

The He lines are of special interest; simultaneous observations in the D$_3$ line and Hβ were examined by Giovanelli *et al.* (1965). There is a one-to-one correspondence between the features seen in both lines and they coincide in position within the accuracy of the comparison, i.e. better than 1″. On the other hand, the shape of the D$_3$ line is gaussian, showing that both the spicules and the interspicular medium are optically thin in this line.

These results were confirmed, refined and extended to the comparison of other lines, by Pasachoff *et al.* (1968). These authors conclude

that a given spicule emits simultaneously in lines of hydrogen, helium and calcium. This does not mean that the emission comes from the same volume within that spicule; for instance sheathed models are not ruled out. We have found further that although the ratios of intensities in the lines of these elements are often constant, there are certainly features that are relatively brighter in one or another of these lines.

3. Chromospheric Fine Structures on the Disk

The description of the chromospheric fine structure on the disk has been rather a controversial subject, notwithstanding the good observations obtained by several observers. In such lines as Hα or K the chromospheric structure appears on the disk with rather complex changes of morphology depending on wavelength inside the line, the considered line, and position on the disk, Also the question of whether a given observed pattern should be described in terms of 'dark features' or 'bright features' or a convenient mixture of both dark and bright things, has always been a troublesome one with solar observers.

3.1. SUMMARY DESCRIPTION

Let us first look at the image in Hα somewhere towards the limb, and rather far off the center of the line, say 0.75 Å, for instance from the beautiful observations by Dunn shown in Figure 3.

One see rows (or 'bushes') of elongated *dark fine mottles* of lengths 5–10000 km. The center-limb variations of the appearance of the dark mottles indicates that they are moderately inclined to the vertical. Nearer the disk center the grouping of dark mottles takes the appearances described by Cragg *et al.* (1963), Beckers (1963) and

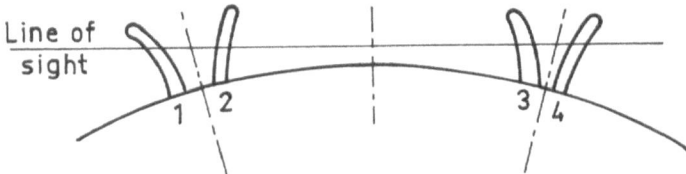

Fig. 6.   Tentative explanation of Figure 5 results by a bias favoring the visibility of spicules tilted towards the observer and generally ascending.

others, under the names of 'rosettes' and 'chains' (or more accurately 'double chains').

Shifting to the center of the line, one also sees *bright fine mottles* which occur in close association with the dark ones. When looking away from the disk's center, the bright fine mottles are situated farther from the limb than the dark ones, which seem to be rooted in a row of bright features. Banos and Macris (1970) claim that there is a one-to-one correspondence, each bright mottle being the base of a dark elongated one. Other observers, such as Bhavilai (1965) and Bray (1969), do not agree with this simple relationship, and give to bright mottles the full status of independent features. Near the center of the disk, bright mottles are more concentrated towards the center of the rosettes than the dark ones. Both the appearances, at the center of the disk or near the limb, suggest that bright fine mottles do not reach as high in the chromosphere as do the dark ones.

When following individual mottles throughout the Hα line either on spectrograms as done by Grossman-Doerth and von Uexküll (1971, 1973) or on filtergrams by various authors, such as recently Bray (1973) and Loughhead (1973), it is found that most bright fine mottles become dark in the wings, while dark fine mottles remain dark in the whole profile.

The rosettes, chains and bushes of dark and bright fine mottles form under low resolution, the *coarse mottles*; these are either bright or dark depending on the wavelength: bright at the center of Hα or in the strongest lines of Ca^+, dark in Hβ, Hγ, ... and He 10830. The coarse mottles trace (in rather 'dotted lines') the contours of large cells; these were shown by George Simon and Leighton to correspond to the cells of the supergranulation and their contours to the regions of locally enhanced magnetic field. It is thus assumed that the occurrence of the fine structures described above are associated with the geometry of the solar fields. These relationships are illustrated, qualitatively, in Figure 7.

It is important to note that the most prominent and most studied fine features of the

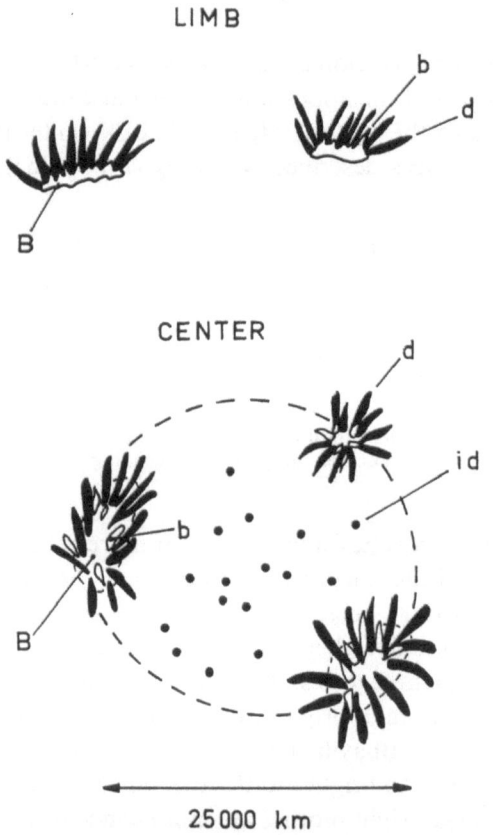

25 000 km

Fig. 7.   Scheme of Hα fine structures after Beckers (1968). Bright and dark fine mottles (resp. *b* and *d*) cluster in 'rosettes' or 'double chains' at the center of the disk; near the limb their projected appearance is termed 'bushes'. Bright coarse mottle (*B*) occur in the central area of rosettes. Inner cell dots (*id*) are scattered between chains and rosettes.

chromosphere seem to form a *special class of structures*, associated with the presence of enhanced magnetic field. They are referred by Beckers to BSG structures (for Boundaries of Supergranules).

I am not aware of very systematic attempts to describe the structures inside the cells (or ISG regions for inner supergranules). Near the center of Hα or slightly redwards they look striated by faint contrast threads (bright or dark?) which seem to prolongate the fine elongated mottles of chains and rosettes, and are called fibriles. These disappear in the wings while the photospheric granulation appears. On the other hand Beckers emphasizes the presence of numerous dark point-like mottles at Hα − 0.5 Å, in possible correspondence with similar bright grains in the K line.

3.2. SOME PROPERTIES OF FINE MOTTLES IN BSG REGIONS

The most easily observed and best studied features are the *dark fine mottles* in Hα.
   - Their average size, about 800×6000 km;
   - Their total number on the Sun, about 300000;
   - Their average duration, 10 min or somewhat smaller;
   - Their average inclination to the vertical, 21°, according to Beckers (1963).

As regards *bright fine mottles*, the situation is less satisfactory because there are discrepancies between the results reported by different observers. It seems, however, that their properties are practically the same as for dark mottles, except that they are shorter and perhaps less numerous. The morphological properties listed so far, show that both types of fine mottles are good candidates for indentification with spicules. The case is most clear for the *dark* fine mottles in view of the divergence remaining between observers in the description of the bright mottles. We consider as quite certain that the *dark fine mottles as observed in the wings of Hα* form a large proportion and perhaps the bulk of the spicule population (according to our Definition 2).

There occurred a controversial point in this problem because Bhavilai (1965) and Mouradian (1965), when trying to connect elongated structures across the solar limb, found correspondence between spicules and *bright* features *just inside* the limb. Accordingly, Bhavilai (1965) and Giovanelli (1967) favored the identification of spicules with bright fine mottles only.

This argument is not however decisive, because the same feature in the outer solar atmosphere may very well look dark at the center of the disk and bright at 1000 km or 2000 km inside the limb, due to the change of the effective depth of origin of the radiation in both the feature and the 'background'.

Beckers (1968) carefully discussed the problem of identification of spicules with disk details. He suggested going further than the simple alternative of always dark against always bright: spicules might be dark or bright on the disk depending on wavelength in Hα, position on the disk, position within the spicule (root or top) and, eventually, phase in the evolution of the feature. From a physical model of spicules (diameter, Te, Ne as a function of height) he derived the emergent radiation of the feature seen on the disk as a function of μ and λ in a few lines, thus predicting the visibility and contrast.

In the most recent calculations of Beckers, theoretical spicules should be dark throughout Hα at the disk center (except that their root below 2000 km could easily be bright!); towards the limb the spicules are bright for the low part of their length and dark for the upper part at the center of Hα; they are dark in the wings.

In the K_3 line their lower part only is bright at the disk center: they are bright all along near the limb. In the He 10830 line spicules should be dark everywhere.

The matter is still complicated because spicules have a *genuine dispersion* in physical properties, probably in correlation with phase in their evolution. We must therefore conclude that *both* dark and bright elongated fine mottles of the disk Hα picture appear as spicules in the low chromosphere and contribute to the various aspects of the limb image. It is very likely that there exist 'connecting' bright and dark features that are part of the same spicule, as observed by Banos and Macris.

3.3. VELOCITY FIELD OF THE DISK CHROMOSPHERIC FINE STRUCTURE

The disk observations of Doppler effects are potentially suitable to extend our present incomplete information on the velocities of spicules. However, the problem is difficult and the present situation rather confused, with discrepancies reported by different observers. This is not surprising in view of the intrinsic difficulties of the observations themselves, and of the problems outlined above in associating with spicules some easily identified class of disk structures.

One should not also underestimate the difficulty of recovering information on the actual velocities, from the casual observation of the profiles, when one has to deal with moving material embedded in an absorbing medium at rest. This was pointed out by Beckers (1968) and others.

Indications on the velocities have been deduced by Beckers (1963) and Bhavilai (1965) from the differences of appearance of dark fine mottles in both wings of Hα. Title (1966, 1967) made velocity records by photographic subtraction of the blue and red images. These authors agree in the fact that downflow occurs predominantly towards the centers of rosettes. Beckers finds dark mottles reversing their velocities during their lives, but Title does not.

Recently Bray and Grossman-Doerth *et al.* derived the Doppler effects of fine mottles near the disk center from spectrophotometry of the profiles: these were represented by the so-called 'cloud model', that is with the Equation (1) given above (but for the factor representing absorption by the interspicular medium). The results fall in the range ± 9 km s^{-1} and the rms value is much smaller than found from limb spicules, indeed by a factor of 2 or 3.

It is the opinion of the reviewer that this discrepancy does not cast doubt on the identity of spicules and the fine structures of the BSG regions. Its explanation must be sought in systematic effects arising from the fact that the structures are observed under very different conditions: in one case (limb), at great heights and along an horizontal line of sight; in the other case (disk center), at low heights and along the vertical.

The systematic effects will include the genuine properties of the velocity field in spicules, various selection effects arising from different visibilities of the features in the two geometrical situations, and also eventual bias in the data arising from the methods of analysis.

In attempting to explain the discrepancy between line-of-sight velocities at the disk center and apparent velocities on the plane of the sky, or line-of-sight velocities at the limb, one should first consider as likely, systematic changes of the bulk velocities with height.

In several sets of observations there is evidence for an increse of velocities with height. This increase is large and evident for some spicules with large Doppler effects

(Michard, 1959). Beckers in his 1968 review suggested that perhaps these form a special class; they may also be cases where a general phenomenon is most readily seen.

One should also consider the probable distribution of spicule inclinations to the vertical which is far from being definitely known. For instance one could perhaps assume that most spicules are far from vertical, with a distribution of inclinations θ peaked at say 60° but extending towards small values.

Then above 5000 km at the limb only the tail of the distribution with small θ would show up, and provide for the large apparent velocities; near the center, the bulk of the population with large θ would provide most of the elongated mottles in rosettes, with velocities reduced by $\cos \theta$.

A situation with a wide variety of spicules inclinations, and therefore different visibilities near the disk center and at the limb, is in line with the ideas of Bhavilai and Giovanelli. However, there is presently no very strong argument to assume that the high standing, near vertical objects, and the very inclined ones, are of a different physical nature.

In summary I would say that careful and extended studies of the velocities of the fine structure of the chromosphere all across the disk and at the limb are still needed.

4. Summary

(1) The most conspicuous fine structures of the solar chromosphere under various appearances can be brought under the general name of spicules because they share the fundamental properties of size and forms, enhanced opacity (i.e. density), lifetimes and numbers.

(2) They show an intrinsic dispersion in all parameters that can be individually measured, notably in the source functions for Balmer lines emission. These fluctuations are related to height in the chromosphere, and most likely also to phase in the evolution of these features.

(3) In the spicules there is a predominant bulk motion along the axis, which is ascending during the first phase in the life of each feature, then stops and eventually reverses.

Important aspects in the kinematics of spicules are still controversial or unknown, although there are indications of changes with height, which need not be of the same sense for ascending and descending motions.

(4) Spicules are denser *than their immediate surroundings* at all heights. They are also most likely colder at heights greater than some unknown level, and possibly hotter below this level. However the height where the spicules start is uncertain between 0 and 2000 km. Models in which spicules start where the chromosphere ceases to be roughly isothermal are not excluded.

(This point 4 is a personal remark from data not reviewed above!).

(5) Spicules cluster in 'coarse mottles' forming a 'coarse network', delineating the boundaries of supergranules (BSG regions). The surroundings of spicules are thus somewhat special regions of the chromosphere with enhanced emission in many lines

such as center of Hα K line, Ly-α, and XUV lines characteristics of transition to coronal temperatures, as shown by OSO-4 and -6 data.

This shows enhanced density and perhaps also enhanced temperature; it is not clear how much of these emissions come from spicules, transition sheaths around spicules or the general coarse mottle areas.

The coarse mottles are coincident with enhanced magnetic fields and there is a nearly continuous sequence of situations between the quiet sun coarse mottles, and the plages of active regions.

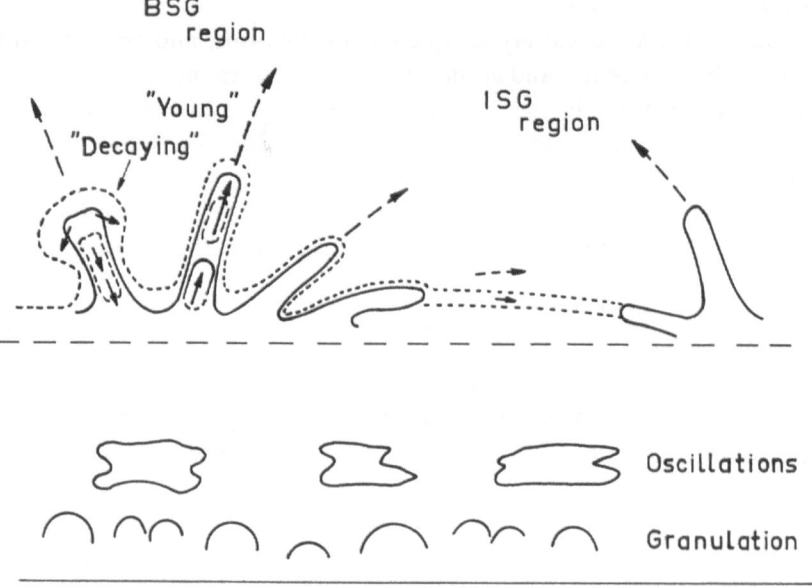

Fig. 8.   Tentative summary picture of the chromospheric structure, shown along a vertical cut in the solar atmosphere. A distinction has been made between 'young' and 'decaying' spicules, assuming that the two classes have systematically different velocities and brightnesses.
– – → magnetic field; → velocity; – – – limit of roughly isothermal chromosphere; ------ transition layer $T \simeq 10^5$ K; == Spicular material with $S(\text{H}\alpha) > 0.2$; and ----- spicular material with $S(\text{H}) \simeq 0.1$.

(6) Away from the BSG, the chromosphere still has fine structures of at least two different types near the disk center:

– very elongated 'fibrils' near the Hα center with lower opacities than spicules, which are horizontal formations probably oriented by faint magnetic fields;

– point-like grains, prominent in the K line.

(7) The characteristic structures of the *photosphere* and *low chromosphere* (granulation, oscillations) are essentially unchaged between BSG and ISG, except perhaps for a tendency of oscillations to have a broader spectrum in BSG.

A concluding Figure (8) tentatively summarizes the possible situation as seen by the reviewer.

References

1. *Reviews about spicules*
Athay, R. G. and Thomas, R. N.: 1961, in *Physics of the Solar Chromosphere*, Interscience Publ., New York and London.
Beckers, J. M.: 1968, *Solar Phys.* **3**, 367.
Beckers, J. M.: 1972, *Ann. Rev. Astron. Astrophys.* **10**, 73.
Giovanelli, R. G.: 1967, in J. N. Xanthakis (ed.), Proc. of NATO Advanced Study Institute at Lagonissi, Sept. 1965, Interscience Publ., New York and London.

2. *Other papers quoted in the present review*
Athay, R. G. and Bessey, R. J.: 1964, *Astrophys. J.* **140**, 1174.
Banos, G. J. and Macris, C. J.: 1970, *Solar Phys.* **12**, 106.
Beckers, J. M.: 1963, *Astrophys. J.* **138**, 648.
Bhavilai, R.: 1965, *Monthly Notices Roy. Astron. Soc.* **130**, 411.
Bray, R. J.: 1969, *Solar Phys.* **10**, 63.
Bray, R. J.: 1973, *Solar Phys.* **29**, 317.
Cragg, T., Howard, R., and Zirin, H.: 1963, *Astrophys. J.* **138**, 303.
Giovanelli, R. G., Michard, R., and Mouradian, Z.: 1965, *Ann. Astrophys.* **28**, 871.
Grossmann-Doerth, U. and von Uexküll, M.: 1971, *Solar Phys.* **20**, 31.
Grossmann-Doerth, U. and von Uexküll, M.: 1973, *Solar Phys.* **28**, 319.
Loughhead, R. E.: 1973, *Solar Phys.* **29**, 327.
Lynch, D. K., Beckers, J. M., and Dunn, R. B.: 1973, *Solar Phys.* **30**, 63.
Michard, R.: 1959, *Ann. Astrophys.* **22**, 547.
Mouradian, Z.: 1965, *Ann. Astrophys.* **28**, 805.
Mouradian, Z.: 1967, *Solar Phys.* **2**, 258.
Nikolsky, G. M. and Sasanov, A. A.: 1966, *Astron. J. SSSR* **43**, 928.
Pasachoff, J. M., Noyes, R. W., and Beckers, J. M.: 1968, *Solar Phys.* **5**, 131.
Simon, G. W. and Leighton, R. B.: 1964, *Astrophys. J.* **140**, 1120.
Title, A. M.: 1966, Thesis, Caltech.
Title, A. M.: 1967, *Astron. J.* **72**, 323.

DISCUSSION

The discussion was summarized by the chairman J. M. Beckers and divided according to topics:

(i) *Relation of Spicules and Polar Plumes*

Athay questioned the observational support for the association of spicules with polar plumes in the corona. An association could possibly exist (*Athay*) or would almost certainly exist (*Zirin*) but would be hard to establish observationally since coronal observations include many plumes well in front and behind the limb (*Beckers*).

(ii) *Height of Origin of Spicules*

Souffrin raised the question of the height of origin of spicules. It is very hard to answer this question observationally because of the merging of spicules with the background, or with other spicules, at low heights above the limb (*Athay*). In the wing of the Hα line one can see spicules well down because one selects only the few spicules with large Doppler shifts and/or line width. Indeed one sees there that some spicules start near ∼1500 km. This may however be explained by a sharp increase with height of the spicule Doppler shift or line width (*Beckers*). *Beckers* suggested to look in optically thinner lines like Hγ or Hδ where one might see spicules extending down even in the line center. Schmidt suggested to study the spicule apparent or Doppler velocity with height and/or time and to extrapolate downward to find the height with zero velocity which could be considered to be the height of origin. This seems however a very difficult and dangerous task (*Beckers, Thomas*).

(iii) *Spectral Tilts of Spicules*

If a correlation exists between spectral tilts of spicules and their Doppler shift, this may have some implications about their kinetics, the magnetic field, for example, both accelerating and twisting motion (*Altschuler*). No such correlation is however clearly established (*Pasachoff*). What causes the spectral tilt (*Giovanelli*)? Two interpretations were offered: blending of two spicules (*Rösch*) and/or rotating spicules (*Pasachoff*).

(iv) *Height and Abundance of Spicules*

The average height and abundance of spicules is derived from spicule count statistics made at the solar limb. *Schatten* pointed out that counts should be made per unit area and not per unit angle at the limb. There is however no way known to us to do this and one will have to rely on limb counts per unit angle and convert these to numbers per unit area (*Athay, Beckers*). These counts will depend on the threshold intensity adopted for the spicule visibility (*Beckers*). *Zirin* suggested that threshold effects may be responsible for the differences in height of the spicule bushes as seen on the disk near the limb and at the limb.

(v) *Spicule Velocities and Mass Flux*

The apparent motions of Hα spicules is predominantly upward; 75 to 80% of the spicules show an upward motion (*Athay* quoting Rush and Roberts). Apparent velocities are near 25 km s^{-1}. Doppler shift statistics confirm the magnitude of this velocity provided that the shifts are caused by the motion along the spicule. This results in an upward mass flux sufficient to replace the solar corona in less than an hour (*Pasachoff*) or to give one hundred solar winds (*Beckers*). Since this is clearly not possible the spicule matter has to return in some less visible way (*Beckers*). *Souffrin* asked about the motion of the interspicular medium. Nothing is known about it. At the limb it shows a zero Doppler shift in contrast to the spicular motions (*Pasachoff*). Spicule-like structures seen on the disk show both up and downward motions (*Grossman-Doerth*) although some observers claim that in the wings of Hα (e.g. $\Delta\lambda \gtrsim 0.6$ Å) one sees a predominant downward motion at the supergranule boundaries (*Giovanelli*). Doppler velocities of the disk structures are however small (~ 5 km s^{-1}). *Grossman-Doerth* believes that this small velocity is caused by seeing effects. The real Doppler shifts may well be of the order of the 25 km s^{-1} spicule velocity. It could be that the predominantly dark, downward moving, mottles at the supergranule boundaries represent the returning spicule matter (*Beckers*).

(vi) *Observations in λ10830 Helium Line*

Giovanelli stresses the importance of observations in the λ10830 line. On the disk some 30–50% of the total absorption in this line comes from the network boundaries. The absorption structures at the boundaries correspond to both Hα bright and dark mottles lending support to the notion that both dark and bright Hα mottles should be identified with spicules. The λ10830 absorption at the centers of the supergranule cells has no pronounced structure.

(vii) *Spicule Temperature*

Beckers raised the question of spicule temperatures. For the last 15 yr our estimates of the temperature has dropped from 50000 K to 15000 K, and according to some estimates, even to 6000 K. These low estimates (6000–8000 K) are the results of observations which do not resolve spicules but which rely on center-limb variation of EUV and radio data. The temperature is a very hard quantity to measure. There is probably not a unique temperature for all spicules and at all heights within a spicule (*Zirin*). *Giovanelli* reports some low spatial resolution observations of the limb chromosphere in λ10830, *Paschen β* and λ8542 (Ca$^+$). If the emission lines originate in spicules one derives $T \leqslant 7500$ K when one assumes that the lines originate in regions with the same nonthermal motion and temperature.

(viii) *Significance of Spicules for their Surroundings*

If spicules are removed from the Sun ('shaved off') how will the corona and solar wind be affected? There was no agreement on this question. They are probably important in the mass balance and in the energy flux (*Athay*). If spicules are absent in plage regions their absence and the behaviour of the corona and the solar wind above the plage may indicate the effect of spicules (*Wilson*). It is not clear whether spicules are absent in plages (*Beckers*). *Pasachoff* suggested that the solar wind expands mainly from the 'coronal holes'. *Brandt* responded that we really don't know the solar origin of the solar wind. Other things are different in plage regions apart for spicules so that even when differences are found in the corona and solar wind one cannot necessarily derive information on spicules (*Beckers, Jordan*). *Zirin* points out that independent of the question if spicules are significant for the solar corona and wind, spicules are important for our interpretation of observations of the chromosphere. *Sturrock* and *Schmidt* consider the existence of spicules quite unessential as far as the solar wind per se is concerned. *Schmidt* believes, however, that the understanding of spicules, and specifically of their mechanical energy flux, is important since it should be related to the as yet poorly known and understood mechanical energy flux which is responsible for the coronal heating.

(ix) *Role of Spicules in the EUV Line Intensities*

The combined area of the spicule surface is only approximately 5% of the total solar surface. At supergranule boundaries spicules contain however ten times the area of the underlying region (*Beckers*). *Pasachoff* suggests even higher values. It is therefore likely that spicules are a significant contributor to the enhanced EUV emission at the supergranule boundary. *Schmidt* pointed out that dT/dx may be very high across a spicule because of the insulation by the magnetic field. This would tend to decrease the spicule EUV emission measure. This topic will be further discussed by *Jordan* at another session.

(x) *Spatial Orientation of Spicules*

Vrabec commented on the appearance of the dark mottle (= spicule?) rosette structure when viewed at the disk center. Very few mottles are visible at the center of the rosette, most mottles seem to make rather large angles with the vertical forming what he calls "a flowering horn of a trumpet."

RADIATION PRESSURE IN STELLAR ATMOSPHERES
WITH APPLICATION TO SOLAR SPICULES

R. GRANT ATHAY

High Altitude Observatory, National Center for Atmospheric Research*, Boulder, Colo., U.S.A.

Abstract. Radiation pressure due to absorption in spectral lines can play an important role in the equilibrium conditions in the outer portions of the stellar atmosphere. In particular, it is suggested that spicules ejected from the upper solar chromosphere as well as other fine structural features of the chromosphere may very likely be a result of radiation pressure from Ly-α. Numerical computation of the radial radiation pressure due to Ly-α in the chromospheric model of Vernazza *et al.* (1973) gives a maximum outward pressure gradient due to radiation that is 0.06 of the gravitational pressure gradient. The model chromosphere of Vernazza *et al.* is a spherically symmetric model that reproduces the average Sun Ly-α intensity and profile. Supergranule borders (network), where spicules arise, are known to have a Ly-α brightness some ten times the average Sun brightness. It is suggested, therefore, that in the network areas the pressure gradient due to radiation is comparable to (and oppositely directed to) that due to gravitation and that in local areas of unusual brightness the net pressure gradient is outwards. An estimate of the expected outwards velocity of spicule material driven by radiation pressure was made by equating the kinetic energy flux ($\frac{1}{2}\varrho V^3$) to the Ly-α energy flux. This gives values of v that are comparable to observed spicule velocities. Also, the maximum outwards radiation pressure occurs in the upper chromosphere where the material density in the model of Vernazza *et al.* is close to the value 4×10^{-14} g cm^{-3}, which is about a factor of two less than the currently accepted value for spicules. Increased density in the network areas could easily remove this minor difference.

A downward radiation pressure gradient due to Ly-α occurs in the model of Vernazza *et al.* at heights below 2250 km. The maximum downward radiation pressure gradient in the spherically symmetric model is 0.006 of the gravitational pressure gradient. Again, this value may be increased in the network areas by well over a factor of ten and, together with the lateral gradient in radiation pressure in the network, may help give rise to much of the chromospheric fine structure associated with the network.

The author is indebted to J. E. Vernazza, E. H. Avrett and R. Loeser for supplying details of their Ly-α computations that were used for computing the radiation pressure.

Reference

Vernazza, J. E., Avrett, E. H., and Loeser, R.: 1973, *Astrophys. J.* **184**, 605.

* The National Center for Atmospheric Research is sponsored by the National Science Foundation.

DISCUSSION

There is some disagreement as to the significance of the Ly-α radiation pressure as an important factor in the dynamics of spicules (*Beckers, Pecker*). The following items were discussed in particular:

(i) *To What Extent does this Model Reproduce the Observed Spicule Properties?*

Athay's model refers to a one-dimensional, spherically symmetric atmosphere. To what extent can its properties be used to preduct spicule properties? It seems that one of the first requirements is the presence of small bright Ly-α points with 100 × the average brightness (*Delache, Beckers, Giovanelli*). These bright points may be caused by conduction downward from the corona along a magnetic flux tube (*Athay*). Given these small bright points, why does radiation pressure cause the sudden beginning of a spicule and why does it cause its very elongated cylindrical shape (*Sturrock*)? The properties can really only be predicted after the full dynamic model has been worked out (*Athay*). The absence of this full dynamic calculation led to other unanswered questions like: (a) how long does it take to build up a spicule (*Delache*), (b) what is the influehce of drag forces (*Pecker*), and (c) how much energy is needed to overcome the gravitational potential (*Schmidt*)? (See Section (iv).)

(ii) *Coupling of Neutral Hydrogen Atoms to the Gas*

The radiation pressure of Ly-α works in the first instance only on the neutral hydrogen atoms. Are there enough collisions to transfer their momentum to the plasma as a whole (*Brandt*)? The answer seems affirmative (*Athay*).

(iii) *Radiation Pressure vs Electron Pressure*

Giovanelli pointed out that radiation pressure and electron pressure are related concepts. Photon generation results from electron collisions, so that for a given total energy an increase in radiation pressure must be accompanied by a reduction in electron pressure. Their combined effects should be more or less equivalent except for numerical differences between photon and electron properties such as the mean free path. Radiation pressure at small Ly-α optical depths ($\tau \leqslant 3$ according to *Athay*) is very anisotropic in contrast to the electron pressure (*Gabriel*). Therefore it can be used as an acceleration mechanism.

(iv) *The Effect of Gravity on the Model*

Schmidt pointed out that you have to include the work done against gravity in lifting the spicule material in order to compute a realistic expected velocity. This was already pointed out in a review paper by Pikel'ner. Most of the energy flux is actually used this way rather than as a kinetic energy flux. The resultant spicule velocity becomes consequently much smaller. *Athay* agreed with the importance of this effect, which he had inadvertently overlooked, and noted that it will change the conclusions substantially.

CHROMOSPHERIC FINE STRUCTURES
NEAR THE SOLAR LIMB IN Hα

RAWI BHAVILAI

Dept. of Physics, Faculty of Science, Chulalongkorn University, Bangkok 5, Thailand

Abstract. The solar limb has been studied visually and photographically in the light of Hα and the combined results are reported. The appearance of a spicule changes with wavelength settings. Effects due to changes of spicule velocities with height and opacity-wavelength relations are indicated. A number of bright spicules seen in the wings are identified with bright mottles seen in the line core. The dark band appears between $\Delta\lambda = \pm 1.0$ Å and $\Delta\lambda = \pm 0.70$ Å. Examination of a small mottle cluster in a series of photographs taken in steps from Hα + 0.85 Å to Hα + 0.10 Å leads to a three-dimensional model of a typical mottle cluster.

DISCUSSION

(i) *On the Magnetic Field Configuration in the Dark Mottles*

Meyer remarked that the configuration of the magnetic field associated with the structures shown in the figures seems a little bit strange. Can most dark mottles be described as such a curved, prominence type phenomena (*Beckers*)? Yes, this is the common, not the exceptional behaviour (*Bhavilai*).

(ii) *The Inclination of the Feet of the Dark Mottles*

The feet of the dark mottles, or the parts closest to the rosette center, generally appear straight since the region where the mottle curves around coincides with the region where it becomes diffuse and therefore transparent (*Bhavilai*).

Schmidt asked about the inclination, or steepness, of the feet of the dark mottles. The inclination is more or less the same as that for the bright mottles (*Bhavilai*).

RECENT DEVELOPMENTS IN IMPROVING DAYTIME
ANGULAR RESOLUTION FROM THE GROUND

K. O. KIEPENHEUER

Fraunhofer Institut, Freiburg, F.R.G.

The subject of this symposium is the fine structure of the solar chromosphere. Progress in this field of reserach will depend to a high degree of the quality of seeing, resp. on the effective angular resolution available on the ground. Today's situation of solar ground seeing has changed distinctly in the last years. I would like to report here a few new aspects, which could be condensed into 3 questions:

(1) Are there on the ground 'good seeing windows', comparable in quality with stratosphere results obtained from balloon borne equipment?

(2) Is there a chance to resolve from the ground the solar scale height, corresponding to about 0.1″.

(3) Is there a residual fundamental atmospheric seeing noise resp. a basic limit to the atmospheric Modulation Transfer Function (MTF)?

Within the European JOSO project (Joint Organization for Solar Observation) we have tried hard to answer these questions. Quite on purpose we deferred for the present the construction of a joint observatory and concentrated on the systematic investigation of our own atmosphere and the search for an optimal site. It is clear: as soon as an observatory site has been agreed upon this kind of activity, somewhat unusual for an astronomer, will stop.

I will leave out here the mistakes and detours which we made in this field which was new to us. One thing which we have learned is that to move around on the ground with an appropriate telescope is the most uneconomic way to find a site. This telescope would have to be put at each testing point on a very rigid and expensive tower of some 20 m height. And this telescope would give only the integrated effect of the atmospheric disturbances all along the line of sight. Instead we aimed first to investigate the atmosphere above promising areas directly by recording the inhomogeneities 'in situ' in the form of temperature (resp. density) fluctuations.

The measurements have been made from high masts, from tethered and from free flying balloons (with radiosonde up to 15000 m) and from slow flying aircrafts (up to 5000 m). Temperature fluctuations down to 0.01 K with a time resolution of better than 0.002 s could be recorded. This corresponds for an aircraft flying with 150 km h^{-1} to a spatial resolution of better than 10 cm.

Here are our main results:

(1) We could prove quantitatively the high degree of homogeneity of the atlantic airmass as compared to the airmass over the continent by aircraft scans. The difference of the two is most striking in altitudes below 3000 m. By means of balloon-borne radiosondes for recording temperature fluctuations up to about 18000 m altitude, it was

R. Grant Athay (ed.), Chromospheric Fine Structure, 27–30. All Rights Reserved.

furthermore shown that the contribution of the upper troposphere and of the tropo-pause is relativily small or even negligible, at least above the southern atlantic ocean within the high pressure system of the Azores and under fairly stable weather conditions.

Under certain assumptions (isotropic turbulence, validity of the Kolmogoroff law) one can – following J. E. Coulman (1969) – calculate the Modulation Transfer Function (MTF) of an atmosphere with given temperature fluctuations. Based on the aircraft and balloon records obtained over the Canary islands we found quite often that the *free* airmass above ca. 3000 m (troposphere and tropopause) would yield an MTF of ca. 40% for an angular separation of 0.1"? This implies of course that the telescope can be brought in direct contact with this airmass.

(2) The direct optical access to this exceptional airmass resp. to its MTF is the more difficult problem: Usually there will be a convective or a turbulent layer of several hundred up to about 2000 m thickness above ground. The only places which we found and proved to be suitable were certain mountain sites in the atlantic, which because of their specific topography showed a laminar flow of air down to < 30 m above ground, at least for a certain range of windspeed. It is this laminar flow which brings the undisturbed airmass close to the site. In addition it wipes off the local turbulence and keeps it away from the line of sight. This situation we found on the Pico de Teide (3717 m) on the island of Tenerife, some 600 km away from the African W-coast, with its very regular conical shape, and on the Muchachos (2430 m), highest point of the island of La Palma, showing very smooth slopes to NW and W, which are the prevailing wind directions at this altitude. Figures 1 and 2 give some typical records of temperature fluctuations obtained above the two sites for various windspeeds. They show clearly, that already 30 m above the sites the fluctuations can be as small as a few hundredths of a degree, if the wind speed exceeds ca. 6 m s^{-1}, while for calm days with wind speed < 4 m s^{-1} the local convection forms a turbulent layer of a few hundred meters thickness above the site.

No doubt, there are many other mountains of this height, possibly even with better aerodynamic properties, but obviously there are extremely few being exposed continuously to this kind of stable and homogeneous airmass as in the case of the Canary islands. The Teide site will certainly give more hours of excellent seeing for a given height of the telescope above ground, than the Muchachos site. But it will produce more technical and logistic problems because of its height and its small size. It will be the task of JOSO – after having conducted the final optical test – to come to a reasonable compromise between these two sites.

(3) Altogether we have learned that the absolute altitude of a site is more important than we had expected. Apart from disturbances near to the ground tropospheric turbulence seems to occur mostly in the form of relatively thin layers, most of them appearing below 3000 m altitude. Sea level or low level sites, whether above the ocean or above the continent, will therefore probably reach the quality of a site above 2000 or 3000 m, only for short moments. We are nevertheless still investigating a sea level site in southern Portugal (Utrecht Group).

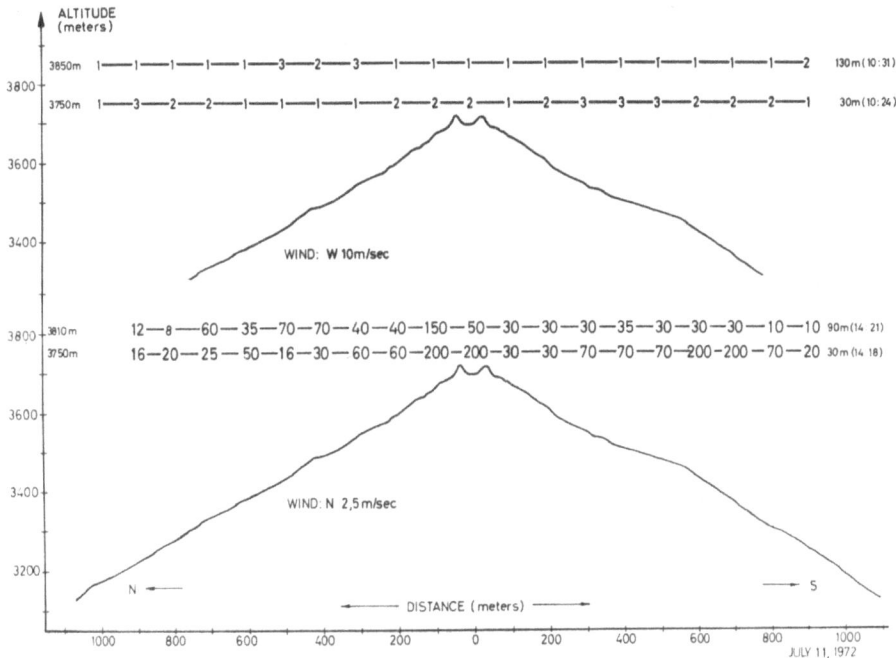

Fig. 1.   Temperature fluctuations recorded by aircraft above the Pico de Teide (3718 m) in units of 0.01 K.
Upper record: 1972, July 15, windspeed 12.5 m s^{-1}; below 1972, July 11, windspeed 5 m s^{-1}.

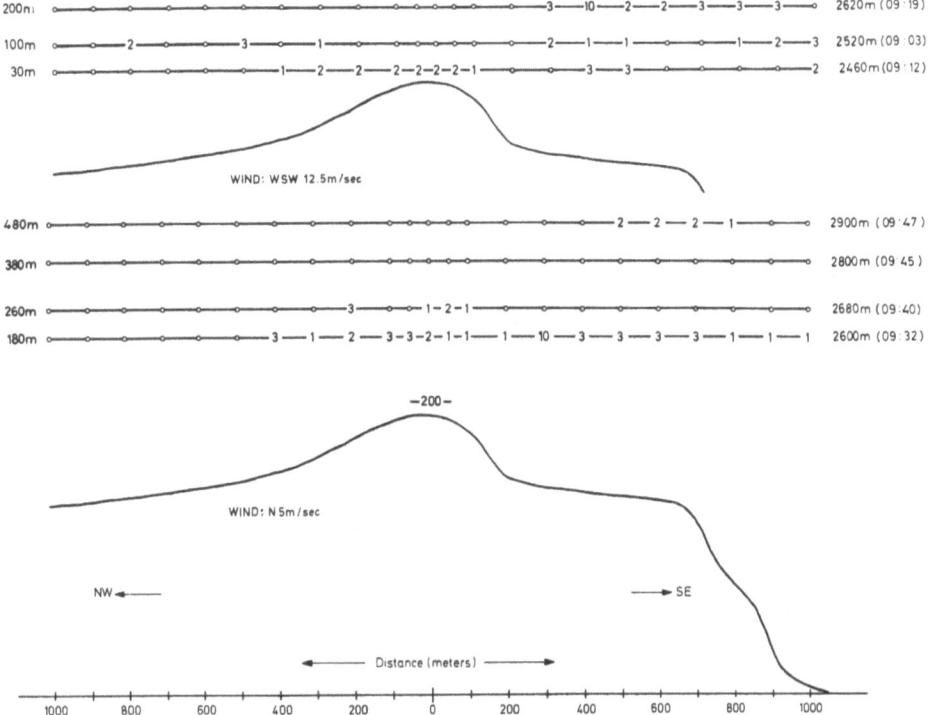

Fig. 2.   Temperature fluctuation recorded by aircraft above the Roque de los Muchachos (2430 m) in units
of 0.01 K. Upper record: 1973, June 20, windspeed 10 m s^{-1}; below 1973, June 20, windspeed 5 m s^{-1}.
The value of 200 ($=2$ K) just above the summit was obtained on a flight perpendicular
to the drawing plane.

With this information in hand, we can now try to answer provisionally the three questions which I presented initially:

There are indeed large 'windows' of excellent seeing in our atmosphere. They occur above 2000 m altitude above the ocean and will guarantee for long periods an image quality as obtained with balloon-borne equipment in the stratosphere, probably even better than this, if larger instruments can be used. These large 'windows', however, shrink to bull's eyes if a direct optical access to the stable airmass on the ground or in buildable heights above the ground is wanted. Because then the complicated condition of laminar airflow and absence (or wiping off) of the thermal convection has to be fullfilled. Indeed, such windows are extremely rare. But they *do* exist and the corresponding sites will very probably enable us to resolve the solar scale height from the ground in pictures as well as in long exposure spectrograms during long periods of time.

The residual inhomogeneities which we could trace even high above the ocean in a stable airmass (Clear Air Turbulence) by aircraft and by free flying balloons turned out to be mostly so small that their optical effects along the line of sight will on the average be important only for very large solar telescopes (aperture $\gtrsim 150$ cm).

These results could not have been obtained without the fortunate and stimulating cooperation of solar astronomers of all European countries which devoted a considerable part of their time to the JOSO project. The two German aircraft campaigns (Fraunhofer Institut, Freiburg) were kindly sponsored by the Deutsche Forschungsgemeinschaft in Bonn and by the Institut National d'Astronomie et de Géophysique in Paris. The Italian radio sonde campaign in Tenerife (Arcetri and Catania observatories) was sponsored by the Consiglio Nacionale delle Richerche, Rome.

Further details can be found in the JOSO Reports SIT 15 and SIT 17 as well as in the Annual JOSO Reports for 1970, 1971 and 1972 and in the forthcoming Italian JOSO Report on the radiosonde campaign 1973.

Reference

Coulman, J. E.: 1969, *Solar Phys.* **7**, 122.

DISCUSSION

Jefferies commented that the same favorable conditions apply to the mountain observatories in the Hawaiian Islands. *Kiepenheuer* agrees, except that the conditions there are not quite as stable because of the larger extent of the mountains. The critical point is therefore higher up so that one probably has to go higher above the ground to reach the steady flow.

The sunshine record is 95% for the Teide, a little less at the site on La Palma (this in reply to a question by *Giovanelli*).

SPATIAL AND SPECTRAL STRUCTURE
OF CHROMOSPHERIC LINES

JAY M. PASACHOFF and FREDERICK S. HARRIS

Williams College-Hopkins Observatory, Williamstown, Mass. 01267, U.S.A.

and

JACQUES M. BECKERS

Sacramento Peak Observatory, AFCRL, Sunspot, N.M. 88349, U.S.A.

Abstract. We are reducing a set of spectra covering the region from 3400 to 4330 Å that show both spatial and spectral structure in chromospheric emission lines from many elements and ions. The spectra were taken with the vacuum tower telescope at the Sacramento Peak Observatory with the spectrograph slit tangent to and touching the limb. Thus some height resolution in the chromosphere is present. Many of the lines show doubly-reversed emission, often asymmetric either for the whole line or for fine structure. The quality of the data is such as to improve our understanding of line blends. The spectral structure can be compared with the well-known structure in the H and K lines of Ca II and in the resonance lines of Mg II in order to deduce a model for the lower chromosphere.

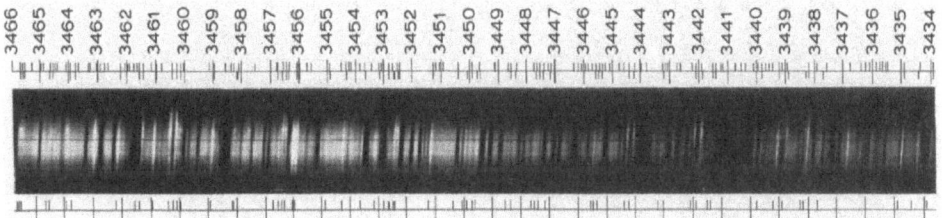

Fig. 1. Part of one of the spectra that are being analyzed for the new chromospheric atlas. Tick marks facing inwards towards the spectra on the top and bottom are lines identified on this print; tick marks facing outward represent lines reported in the atlas of Pierce. This sample covers the region from 3434 to 3465. As the center of the slit is slightly lower down than the edges, it shows photospheric continuum in addition to the emission lines.

Many rare-Earth lines are also visible in emission on the spectra and, contrary to the other lines, show no spatial or spectral structure even under this high resolution.

The wavelength region is covered with 13 separate spectra, each covering about 70 Å at dispersions ranging from 9 to 13 mm Å^{-1}. About 200 emission lines are visible on the best of the plates.

An atlas is being compiled from these spectra that will comment on the spatial and spectral structure, for comparison with the atlas of Pierce.

R. Grant Athay (ed.), Chromospheric Fine Structure, 31–32. All Rights Reserved.

DISCUSSION

(i) *On the Reality of the Central Reversal*

Beckers questioned the origin of the self reversed profiles. In similar spectra taken at the solar limb the central reversal disappeared after the scattered light was subtracted. Are all central reversals due to scattered light (*Zirin*)? No, especially eclipse spectra show many self reversed lines which must be real (*Jefferies, Pierce, Pasachoff* and *Rutten*). *Beckers* persists in questioning the reality of many of the reversals in the spectra presented by *Pasachoff*.

PART II

THE QUIET CHROMOSPHERE: DISK PHENOMENA

WHY THE CHROMOSPHERE HAS ITS DISCRETE FINE STRUCTURE

HERMANN U. SCHMIDT

Max-Planck-Institut für Physik und Astrophysik, München, Germany

The fine structure in the chromosphere is probably controlled by magnetic flux concentrations which cellular convection produces in the photosphere. We will first deal with this flux concentration by the convection and its constraints. Then we will discuss the chromospheric network and its main constituent, the spicules. Following that we will look at the inside of the network, i.e. the fibrils.

1. Convection and Magnetic Flux

Cellular convection in supergranules and granules is well observed in the chromosphere and photosphere respectively. In these thin regions we know its sizes, l, velocities, v, and life times, t, as well as the mass density, ϱ. Therefore we can estimate the local dynamic stresses, ϱv^2, which may counter-balance magnetic stresses, and a quantity with the dimension of a viscosity, vl, which may act on flows of larger scale as a turbulent viscosity or on the large scale magnetic flux as a diffusivity. But it is the ability of cellular convection to expel and to concentrate the magnetic field, which is of more importance for the understanding of our subject matter. More than 90% of the field is concentrated into small flux units at the border between adjacent supergranules (Frazier and Stenflo, 1972).

The transport of flux by cellular convection, its accumulation and its dissipation has been described by Parker (1963), Weiss (1966), and Clark and Johnson (1967). Clark and Johnston estimated the effect of pure horizontal transport at the top of an idealized hexagonal supergranular cell with reasonable extension and velocity. They found that if Lorentz forces would not interfer, an initially homogeneous vertical field at the top of this cell would in 14 h grow by a factor 55 at midboundary points, by a factor 400 at the vertices and decay to 6×10^{-6} of its original value at the center. In reality the Lorentz forces stop this explosive accumulation much earlier since the dynamic pressure $\frac{1}{2}\varrho v^2$ in the photosphere would be balanced by the magnetic pressure $B^2/8\pi$ for a field of 50 G. But much larger fields, > 500 G, are observed. How can they be formed? There may be about 7 such units of 3×10^{18} Mx per supergranule, or about twice that number around its borderline. Therefore they are not too closely related to the less numerous vertices in the corners common to several supergranules. Instead their small size points to granules which happen to occur on the borderline and produce dynamic pressures equivalent to about 250 G in the photosphere. The most thorough theoretical investigations relevant to this problem have been made by N. O. Weiss and his coworkers (e.g. Peckover and Weiss, 1972). They have computed the effects of a thermally driven two-dimensional cellular Boussinesq convection on

R. Grant Athay (ed.), Chromospheric Fine Structure, 35–44. All Rights Reserved.

the magnetic field with full account of the non-linear Lorentz forces. The most important result is that the simple concept of equipartition between kinetic and magnetic energy holds only under special circumstances. Instead their Figure 5 shows flux concentrations in which the magnetic energy is about ten times stronger than the total kinetic energy of the subdued convective motion whenever the effects of ohmic dissipation can be neglected against those of the Lorentz forces and for a wide range of parameters. The surplus of magnetic pressure over dynamic pressure is balanced by gas pressure. The magnetic field reaches a level (cf. their Figure 6) that suppresses the convection by balancing the differences of buoyancy in the temperature field, i.e. the magnetic pressure balances the full dynamic pressure of a cell without a field. Such effects are certainly operative at a level in the hydrogen convection zone where the supergranules are effective in transporting the solar heat flux. There the full dynamic pressure of the convection will balance magnetic pressures for 5000 G. Near the surface this effect seems to be inoperative, since the supergranulation has not been recognized in white light. However, it may be that the flux concentration can be localized in dark intergranular space by exact comparison. Therefore once more we must expect the dynamic pressure of the granules to be operative in the compression of the magnetic flux concentrations at the border of supergranules. It seems to be the combined action of supergranules and granules which produces these concentrations, within hours, at the borderline of a supergranule which confines new erupting flux e.g. in an arch filamentary system. Thereafter the flux concentrations forming a stretch of supergranular boundary may well live for a long time before they are dissipated, e.g. much longer than a single supergranule.

2. The Ultraviolet Network

The chromospheric network outlining the borders of supergranules is clearly recognized in Hα, Ca$^+$ and in the UV up to a temperature of 1×10^6 K for the line formation (cf. Reeves and Parkinson, 1972). This was not expected, as Kopp and Kuperus (1968) predicted an anticorrelation between the Ca network and the UV emission pattern for the transition zone. Their Figure 1 indicated a weaker temperature gradient and thereby a stronger emission measure for those lines from the interior of the network cells. Here a major revision of this plausible picture is needed. An important step in this direction was done by Kopp (1972). He recognized another discrepancy which exists for the spherical symmetric model of the transition zone derived from the UV emission measures. These emission measures are inversely proportional to the temperature gradient at the temperature of line formation, so that a spherical symmetric temperature profile can be calculated. The conductive heat flow thus derived has a peak in divergence which would deposit more energy in the gas than what actually can be radiated at this temperature and density by a factor of about five, as can be seen from Kopp's Figure 2. It shows the calculated sink of the conductive flux and the source of the radiative flux. The sink peak is actually larger than the radiative losses by a factor of five for the appropriate pressure at the transition zone. Since mechanical wave energy sinks can only aggravate the problem, the conductive sink cannot be correct, but must be reduced by at least a factor of 5. If this were to be an overall reduction of the temperature gradient, the observed emission measures could not be matched unless one assumes the observed radiation to originate from a network which covers on the average only one fifth of the visible Sun. If this is done the discrepancy is resolved and the total conductive flux of 6×10^5 erg cm^{-2} s^{-1} is reduced by a factor 25 to 2.5×10^4, since both the gradient and the area through which the conductive flux crosses back is reduced by a factor 5. This is a drastic tentative change, but it is in line with all observations, especially of the continuation of the network into the corona. It implies that also the input of mechanical energy should be enhanced in the network or rather reduced in the interior of the supergranules. Such a relative enhancement of the mechanical input in the of the super-granules. Such a relative enhancement of the mechanical input in the stronger magnetic field connected with the network is entirely reasonable, as the excitation of soundwaves is more effective under these conditions (cf. Parker, 1964). Over the interior of the supergranule we can assume that the field spreads out from the network very efficiently such that the vertical temperature gradient can be rather large. Also the density will be lower than in the network. Hence the total UV emission from the transition layer can be low, too, because the horizontal field insulates the overlying corona from the chromosphere against thermal contact even for a very large temperature gradient.

3. Spicules

We now turn to the most interesting and most puzzling inhabitants of the solar chromosphere, the spicules. They constitute the network in the chromosphere. The processes which have been considered to produce them are shockwaves guided by the enhanced field, thermal instability, diamagnetic accelerations, annihilation of opposite fluxes. In order to judge their applicability I will try to recall and to extrapolate some of the features of spicules reported by Beckers (1972). There seem to be about 80 spicules at any one time per supergranule, i.e. about an order of magnitude more than the number of underlying photospheric flux concentrations with which they certainly are correlated. They form groups or bushes around them, not simply a thin sheet, they do not follow all the directions of the field lines spreading out from the flux concentration but they seem to prefer the vertical direction slightly. This is certainly an extrapolation, as I assume that there must be a sizable amount of very low lying flux which insulates the cooler interior of the chromospheric supergranule from the corona, but does not guide spicules. The question whether the spicules can heat the corona has been answered negatively by Beckers because of the average upward flow of kinetic energy in the spicules is less than 10^4 erg cm^{-2} s^{-1}. Beckers also stresses the similarity of size, abundance and lifetime for granules and spicules. The spicules apparently rise with velocities near 20 km s^{-1}, but may be less (cf. Grossmann-Doerth and von Uexküll, 1971). They reach nearly 10^4 km with densities near

10^{11} cm^{-3} and temperatures near 15 000 K. Shockwaves as spicules have been proposed by Parker (1964), and Lüst and Scholer (1966) have done computations. But the results of these were not conclusive, as large amplitudes could not be reached. Since the material of the real spicule is rising with less than 25 km s^{-1} it cannot coast to the observed altitudes but only to less than 1/10 of it if no external forces are applied to it during its rise. This simple argument of Pikelner (1971a) rules out many models in the literature.

Thermal instability in the chromosphere was first discussed by Athay and Thomas (1956) and by Kuperus and Athay (1967). The most thorough investigation of a linearized set of model equations for a pure one level hydrogen chromosphere was put forward by Defouw (1970a, b). This local instability is set up, whenever the total change of the net energy loss of a chromospheric plasma parcel with temperature under realistic conditions becomes negative. The most simple approach is to investigate this change statically under constant pressure. But also for more refined methods there seems to appear consistently near 17 000 K an instability. With a positive temperature perturbation the increase of ionization diminishes the number of atoms that can cool the electrons in collisional ionizations and as the pressure nearly stays constant the number density itself must decrease so that the rate of cooling by collisions must decrease. The instability is especially sensitive in a magnetic field strong enough to align the motion in the perturbation. A quantitative discussion of the growth rates and the dependence on temperature, field and density looks quite good. But this does not mean too much, because once more after the onset of this instability the matter cannot rise very far unless its thermal velocity would reach 100 km s^{-1}, which are not observed. It is also not clear whether the quiet chromosphere does not manage to hold the geometrical region of instability sufficiently small by a steep jump to higher temperatures, where other atoms can take over the cooling task. A small region which is only locally unstable may easily be stabilized by conduction and generation of waves at the boundary towards stable regions. There is also a model by Hollweg (1972). His Figure 7 shows supergranular flow towards the supergranular border which creeps downwards to avoid the magnetic field there. Above this flow spicules rise, also obeying alignment with the field. The balances of energy and momentum are not discussed. Instead, his Figure 8 illustrates the model in which mass conservation alone explains the spicule as the velocity v moves in over $2L \cdot z_0$ and the spicule moves out with v_s over $L \cdot d$. There is even an increase in v_s over that because the inward moving matter happens to obey a barometric formula which brings in a lot of mass at the bottom which has no choice but to move out at the top with the local density there:

$$v_s = v \cdot 2 \frac{h}{d} (e^{z_0/h} - 1)$$

which brings 40 km s^{-1} at $z_0 \approx 1000$ km out of 0.4 km s^{-1} at the bottom. In my opinion the model does not discuss the dynamics and therefore does not explain how a spicule overcomes gravity.

Now we come to those models which involve external forces, which under the circumstances can be only magnetic, as the dynamic pressures above the photosphere are certainly insufficient. There are really two possibilities and they may be combined: dissipation or 'annihilation' of flux and/or diamagnetic acceleration. Schlüter (1957) put forward the latter process which is now often referred to as the 'melon seed mechanism'. It seems to me that this process probably is the only cause of the spicule. For the spicules, it was used, together with annihilation of opposite flux, by Pikelner (1969, 1971a) and simultaneously by Uchida (1969). The diamagnetic process produces in an inhomogeneous magnetic field a net volume force on an imbedded body which is field free or has its own disconnected field. This force is proportional to ∇B^2 and its strength depends on the surface and will change as the diamagnetic body expands in the process. Livshits and Pikelner (1964) have done calculations for ejecta from active regions, readily applicable to surges. The process basically does not run at the expense of magnetic energy but of the internal energy of the diamagnetic body. By its pressure it must push on the field, which is restored after the body has moved past. Therefore this is an ideal mechanism for recurrent phenomena. It may also act on bodies, which after an acceleration and some cooling in the expanded phase in a weaker field fall back by gravity, get new heat in the compressed phase, e.g., by radiation and then start on the same path again. The problem is always to store sufficient internal energy in the body and to get it disconnected from its magnetic environment. Therefore both authors quite naturally tried to use dissipation of magnetic flux as a source of internal energy. The statement that at the supergranular boundaries opposite polarities are mixed was once fashionable but it certainly is not nowadays. However, this may change again with better resolution. In any case, once you have opposite polarities you can easily get reconnection of field lines by Petschek's mechanism and you also get the needed increase in internal energy more easily (see Figure 1 of Pikelner). But this reconnection may be rather unimportant for the spicules, because there is probably not sufficient flux available in the solar cycle to supply all the spicules with their share. My estimate of the average production of surface flux by the solar dynamo is $10^{16.9}$ Mx s^{-1} and for the solar wind losses I get with a lifetime of 3 rotations $10^{16.3}$ Mx s^{-1}. On the other hand for the spicules I get a conservative estimate of $10^{18.9}$ Mx s^{-1}. Therefore I am afraid, we have to look for deeper sources of energy than reconnection. Certainly the dynamo production has to be reconnected too, but we may simply say that this phenomenon should be harder to find for the eager observer, as it is probably hidden among more than a hundred spicules. We do not know yet where the spicules get there heat. But we should remember Beckers statement, that we do not need very much, some 10^3 erg cm^2 s^{-1}. My tip would be the same as the one of Beckers, i.e. to look into the layers underneath. There is a downstream flow towards the supergranule proper in the hydrogen convection zone. With the granulation this downstream pattern changes a bit to and forming a new azimuth with respect to an irregular field concentrations. Thereby, diamagnetic matter gets trapped near to the photosphere and in the field, is heated up and shot out. Anyway, it seems conceivable that observers may soon find a nice new 'moving diamagnetic body' which produces the spicule.

4. Inside the Network

Here we find the fibrils. One might ask though: do fibrils appear in the quiet chromosphere? I did not check whether there is a clean-cut answer to this question. In any case I shall discuss them briefly. They seem to outline a roughly continuous field of two-dimensional streamlines which apparently is confined to the chromosphere and mostly to parts of the active chromosphere. I do not now discuss the question whether this is the magnetic field, I simply assume it. They are probably confined to the active chromosphere for a simple reason, which is not so much the amount and the concentration of the flux but rather the fact that in the quiet chromosphere there is a much smaller chance to find a field line which does not leave the particular supergranular cell and therefore the chromosphere and even the corona for interplanetary space. Now why do we see this field? Meyer and Schmidt (1968a, b) argued that low lying fluxtubes connect opposite magnetic polarities of slightly different fieldstrength and, therefore, slightly different pressure. The high pressure will cause subsonic upword flow to the crest and supersonic downward flow which is dissipated in a shock into a subsonic inflow toward low pressure. These differences in density and pressure may become visible as contrast between fibrils.

Pikelner (1971b), on the other hand, has constructed a good quantitative model for fibrils, which assumes that they constitute the middle chromosphere between 1500 and 4000 km, but that they are imbedded into a hot corona. This is possible because the plasma tube aligned by the magnetic field of several Gauss is almost perfectly insulated against heat conduction from the corona. Further it is assumed that the maximum height is limited because at this height the chromospheric thermal instability limit of 18 000 K determined by Defouw is reached. The pressure distribution in the fibrils has to be hydrostatic, at least in its subsonic parts. In its supersonic parts the pressure will be even lower. Now the roughly known temperature profile in the lower chromosphere with fixed points 6000 K at 1000 km and 10 000 K at 2000 km can be extrapolated for the fibrils so that at their visible crest at 4000 km the stability limit of Defouw mentioned above is reached, i.e. 18 000 K. Now the density can be determined from hydrostatic equilibrium and one fixed value, say 8×10^{10} cm^{-3} at 2000 km. Once that is known the ionization of hydrogen and the optical depth of the fibril in Hα at each height can be calculated from a detailed non-LTE model. It is found to be larger than 0.25 for the temperature range from 13 000 to 18 000 K in contrast to the transparent corona. Thus an Hα filtergram should show the fibril structure as it does.

Quite naturally the optical depth in Hα drops rapidly at the upper edge for the same reason that causes the thermal instability. The gas pressure at this level in the fibril is below the pressure of the ambient surrounding corona by a factor 5. This does not cause a difficulty for the model as the magnetic field in the fibril has to compensate the total coronal pressure.

5. Bright Points

I want to add a few words about the bright points in the wings of Hα as they are also an important fine structure: For a shortlived very local heating in a strong magnetic field concentration in the lowest chromosphere one would expect the same characteristics for three rather different processes, which all can occur under proper circumstances. They are ohmic dissipation of magnetic energy, viscous dissipation of kinetic energy and diamagnetic buoyancy of matter heated beneath the photosphere. In the first case the field concentration contains opposite fluxes and as it is formed it is annihilated. This would be the process for spicule formation described by Pikelner and Uchida, but which will probably form only one spicule out of many. It would be identified by resolution of the flux polarities as they move together. In the second case matter falls back down from sufficient height into the flux concentration so that its energy produces sufficient dynamic pressure to balance the gas pressure in the lower chromosphere or more. In the third case the internal energy is provided to a diamagnetic body below the chromosphere, so that buoyancy brings it up to the chromosphere.

In *conclusion* I list some open questions.

(1) From the numerical studies of the Cambridge group we have seen that in thermally driven cellular convection the enhancement in the magnetic energy can surpass the kinetic energy in the cell by factors up to 10. The local magnetic pressure may then still be larger compared to the actual dynamic stresses, since it is balanced in this case by pressure gradients. Here is certainly a new good tool invented and applied by N. O. Weiss and his coworkers which will in the future tell us more about the effects of solar surface convection on the magnetic field. The questions one might ask here are:

Can one measure the pressure gradients in the photosphere which balance the magnetic pressure of the field concentrations?

What are the exact phenomenological correlations between the field concentrations and the granules? e.g. do the concentrations fall into intergranular regions, is there a measurable low in the pressure and in the temperature? How do the granules apply their dynamical pressure?

(2) What are the correlations between granules and supergranules? Do the granules move towards supergranular boundaries or vertices? Is the chromospheric supergranule driven locally by thermal gradients or by some viscous coupling to deeper layers? Simon (1967) answered the second question positively. This thorough and labourious work was never repeated, but it concerns a fundamental question.

(3) What is the exact variation of the input in mechanical energy with the vertical magnetic field in the photosphere and with height in the chromosphere? We may infer from Parker's discussion of the influence of the magnetic field on the production

of acoustic energy and from the XUV observations and their discussion by Kopp that the input is concentrated in the field underlying the network.

(4) What is the height, the thickness, and the pressure of the transition zone as the field varies? One might tentatively answer these questions in the quiet chromosphere using the assumption that the mechanical energy moves straight up. i.e., exactly radially outward, whereas the conduction follows a force free field configuration and hydrostatic equilibrium holds along those field lines. There is a danger that due to the widespread fear to make any inferences about the magnetic field we do not get out the fantastic amount of information contained in the basic observations. Important contributions to this symposion concerning these questions have been made by Jordan (1974) and by Gabriel (1974).

(5) What is the cause of the spicule? Is it diamagnetic acceleration driven by internal energy with input from annihilation of magnetic flux or by mechanical heating? The latter may occur during the time between the rise of repetitive spicules at the same stable location. What are the correlations between spicules and granules? Is magnetic annihilation ruled out as a cause by unipolarity on the small scale?

(6) Are the fibrils imbedded into the corona and down to which level is this possible? Being horizontal are the fibrils magnetically aligned? Is there an observable discontinuity of the magnetic field and its direction somewhere between the fibrils and the spicules?

References

Athay, R. G. and Thomas, R. N.: 1965, *Astrophys. J.* **123**, 299.
Beckers, J. M.: 1972, *Ann. Rev. Astron. Astrophys.* **10**, 73.
Burton, W. M., Jordan, C., Ridgeley, A., and Wilson, R.: 1974, this volume p. 89.
Clark Jr., A. and Johnson, A. C.: 1967, *Solar Phys.* **2**, 432.
Defouw, R. J.: 1970a, *Astrophys. J.* **160**, 659, **161**, 55.
Defouw, R. J.: 1970b, *Solar Phys.* **14**, 42.
Frazier, E. N. and Stenflo, J. O.: 1972, *Solar Phys.* **27**, 330.
Gabriel, N. H.: 1974, this volume, p. 295.
Grossmann-Doerth, U. and von Uexküll, M.: 1971, *Solar Phys.* **20**, 31.
Hollweg, J. V.: 1972, *Cosmic Electrodyn.* **2**, 423.
Kopp, R. A.: 1972, *Solar Phys.* **27**, 373.
Kopp, R. A. and Kuperus, M.: 1968, *Solar Phys.* **4**, 212.
Kuperus, M. and Athay, R. G.: 1967, *Solar Phys.* **3/4**, 361.
Livshits, M. A. and Pikelner, S. B.: 1964, *Astron. Zh.* **41**, 464.
Lüst, R. and Scholer, M.: 1966, *Z. Naturf.* **21a**, 1098.
Meyer, F. and Schmidt, H. U.: 1968a, *Z. angew. Math. Mech.* **48**, 218.
Meyer, F. and Schmidt, H. U.: 1968b, *Astron. J.* **73**, 72.
Parker, E. N.: 1963, *Astrophys. J.* **138**, 226, 552.
Parker, E. N.: 1964, *Astrophys. J.* **140**, 1170.
Peckover, R. S. and Weiss, N. O.: 1972, *Comp. Phys. Commun.* **4**, 339.
Pikelner, S. B.: 1969, *Astron. Zh.* **46**, 328.
Pikelner, S. B.: 1971a, *Commun. Astrophys. Space Phys.* **3**, 33.
Pikelner, S. B.: 1971b, *Solar Phys.* **20**, 286.
Reeves, E. M. and Parkinson, W. H.: 1972, *Solar Phys.* **24**, 113.
Schlüter, A.: 1957, in H. C. van de Hulst (ed.), 'Radio Astronomy', *IAU Symp.* **4**, 356.
Simon, G. W.: 1967, *Proc. Capri Sympos. Chromospheric Fine Structure.*
Uchida, Y.: 1969, *Publ. Astron. Soc. Japan* **21**, 128.
Weiss, N. O.: 1966, *Proc. Roy. Soc.* **A293**, 310.

DISCUSSION

Pecker first raised the question of the relationship between granulation and magnetic fields. He referred to a study initiated by his group in which some correlation appeared to exist between granules and magnetic fields measured on a granula scale. *Kiepenheuer* reported that, in well-resolved Hα line wing filtergrams showing mottles and granulation, there are cases where the foot points of the mottle (or spicule) seem to be well correlated with the intergranular region. *Deubner* emphasized the importance of sequential series of spectra in different spectral regions covering the whole extent of the atmosphere. He mentioned that some observations of this kind also point to a close relationship between spicules and granules. *Grossmann-Doerth* pointed out that, if fibrils are the elongated small features which stretch out over distances at least as far as the diameter of supergranular cells, they do not appear in the quiet chromosphere but only in the active chromosphere. He then went on:

Grossmann-Doerth: May I ask a question concerning the driving mechanism of spicules: do you mean to say that Pikel'ner has shown that no thermal instability can be the cause of spicules? Did I understand this correctly?

Schmidt: Yes, Pikel'ner argues that you have to supply sufficient energy to overcome the difference in gravitational potential. Now, you can take any model, even constant temperature gradient through the unstable layer and if you now accelerate an element from the bottom to the top of this unstable layer by this thermal instability, you would get only a few kilometers per second. Thus you may reach an elevation of 10 km or so.

Sturrock: It is my impression that spicules and surges represent the same phenomenon occurring on different scales. In each case, a mass of cool gas is shot up; sometimes it disperses and sometimes it falls back to the surface. No observer has pointed out to me any way of distinguishing between spicules and surges (other than length and time scales). If there is indeed a common mechanism for driving spicules and surges, this points towards a magnetic model such as Schluter's or Pikel'ner's, and probably rules out a radiation-driven model such as Athay's.

Schmidt: I certainly agree with you, and I put the very same question to Le Roy and Rust when they were working on surges and they agreed. There is a paper by Lifshits and Pikel'ner (*Soviet Astron. AJ* **8** (1964), 368) which describes the diamagnetic acceleration of surges with surprising quantitative detail. The same description may be adequate for spicules when scaled down.

Most of the remaining discussion was concerned with the interaction between magnetic and velocity fields in the Sun. Because of its basic importance it is reported here in some detail. The question was first raised by:

Stix: You mentioned that in the photosphere the supergranulation has not enough kinetic energy to produce flux densities of 1000 G but that the granulation possibly *has* enough. How can the granulation concentrate flux on a much larger scale than its own? Do you mean flux transport by turbulent diffusion?

Schmidt: I don't think so. The supergranulation cannot concentrate the flux up to 1000 G, but it can transport it. Weiss and Peckover have shown that cellular convection can concentrate the flux to a field strength above the equipartition value but for supergranules this would only give field strengths of order 60 G. The granules, on the other hand, cannot transport flux very effectively but they can concentrate it, probably up to field intensities of 1000 G. Later Meyer returned to the problem:

Meyer: According to computations by Weiss, which were carried out using the Bousinesq approximation, one gets a pressure decrease in the concentrated magnetic field region. In the non-Bousinesq case one must also take the vertical temperature structure into account and it may be that the reduced gas pressure produced by the supergranules in the lower regions also occurs at higher levels. If so, this pressure difference may assist in concentrating the magnetic fields.

Schmidt: I think you are right. In two recent papers Parker has argued that the granules impose a depression on the flux tube by two processes: turbulent pumping producing downward flow and the Bernoulli effect. I do not know whether we need these processes. The model calculations of Weiss and Peckover point in a somewhat different direction. The influence of the magnetic field on the thermal diffusion may well be such that the temperature depression does not spread out away from the magnetic column.

Vrabec: My comment is in response to that of Dr Stix. We generally believe that supergranules are responsible for sweeping any fields occurring in the cell interiors out to the cell boundaries, where the fields become concentrated into the supergranular network. Stix has expressed concern, because the supergranules cannot produce the field strengths of 1500 G or more, presently believed to occur in this network. Let me point out that Dr Schmidt has just told us that normal granulation, on the other hand, can produce these high field strengths. Can we then not think of normal granulation producing and maintaining the high, localized fields, and supergranulation carrying the granulation and the field concentrations continually being produced by it, out to the cell boundary? The field structure is then the result of two processes acting independently. Granulation produces the high field strengths and fine structure, while supergranulation arranges these fields into the familiar network.

An important distinction between model calculations and the solar problem was pointed out by Gabriel.

Gabriel: Models such as you describe for the time taken for a convection cell to sweep the field to its edges are all based upon the starting point of a uniform field distribution. This does not occur in reality. It is more realistic to consider the starting point as the field concentrations remaining from the old decaying network. I suggest that this picture may significantly shorten the time required to set up the new field concentrations.

Schmidt: Yes, I agree. This general description of the changes in the supergranule network is supported by the Aerospace movie taken at Thule.

These views regarding the roles played by granule and supergranule velocity fields in producing the observed network fields appeared to gain some degree of general support in this and other discussions during the Symposium. However, Zirin added a word of caution.

Zirin: I am still worried about the business of convection concentrating fields, for several reasons. The observations themselves are extremely subtle, mostly based on the work of Leighton and his co-authors years ago and on this rather subtle work of Simon. I am impressed myself with the possibility from the limited amount of white-light films that we have that granules in fact represent an oscillation with a 5-min period rather like the calcium network and not convection. Furthermore, even if the flow is observed there it is possible that this is simply part of the spicule outflow being supplied from the centre of the network and not necessarily convection. We know that the network itself is a very stable feature and we would expect stronger convection there to balance this magnetic pressure. If this were not so, the stronger the field the less long it would live. So, I'm afraid we've built a house of cards on not a very strong foundation.

Schmidt: Well, I think the foundation is not too bad because we can base it on numerical simulations of stationary convection. I placed the migration of granules on my list only as a very important question in its own right. But I think it's important to know whether the supergranulation is moving the photosphere or not. The concentration of flux by the granules is effective everywhere because they cover the whole photosphere. They do not have to move. They simply expel the flux and provides sufficient buoyancy stresses to balance the field up to 1000 G.

Zirin: Would the convection have to be stronger where the field is stronger?

Schmidt: No. It may actually be weaker. This prediction of these calculations would be very interesting to test. As you saw in the figure, the convective velocity goes slightly down when the field strength goes up. Nevertheless, the field energy can surpass even the level of the total kinetic energy calculated for the undisturbed cell without any magnetic field.

Wilson: I think Zirin's question also raises the very important point that the surface velocity fields which we call the supergranulation are only the surface effects of velocity fields which, if they are convective in origin, must be most efficient at depths of order 10 000 km. We know this because attempts to find temperature correlations with the velocity fields have failed significantly. This then supports the picture that the deep supergranule velocity fields are quite adequate to transport magnetic flux to the cell boundaries. While we seem to agree that the granules cannot transport fields, the assertion that they can concentrate fields into small flux units of strengths ~ 1000 G is based at present on the numerical simulations of Weiss and Peckover and to a lesser extent on arguments presented here by Meyer and by Schmidt. It will be particularly important to further probe this question for if the overall picture as outlined in this discussion can be confirmed, it represents an important advance in our understanding of the relation between solar velocity and magnetic fields.

PROPERTIES OF THE SOLAR FILIGREE STRUCTURE

R. B. DUNN, J. B. ZIRKER*, and J. M. BECKERS

Sacramento Peak Observatory, AFCRL, Sunspot, N.M. 88349, U.S.A.

Abstract. A number of observers have noted the presence of bright structures near the cores of the chromospheric rosettes when observed in the far wings of the Hα line (eg Hα ± 7/8 Å). Dunn and Zirker observed these bright structures with the highest possible resolution using the Sacramento Peak vacuum solar telescope. They find that these bright regions exhibit a very intricate fine structure which can be followed out much further into the Hα line wing (eg Hα + 2 Å) and even into the continuum. They called this fine structure 'solar filigree', the name referring mainly to the collective appearance of the fine structure elements. The elements themselves appear as dot-like structures and frequently also as small wiggly structures called 'crinkles'. The properties of the filigree structure are summarized as follows:

(i) *Size:* Measured diameter of the crinkles and dots equals 0.25, 0.40 and 0.60″ at Hα + 2 Å, Hα ± 7/8 Å and Hα ± 5/8 Å respectively. The telescope resolution equals 0.22″ so that at Hα + 2 Å the structure is extremely small. The drawings in Figure 1 show typical sizes of the crinkles and network patterns in the filigree.

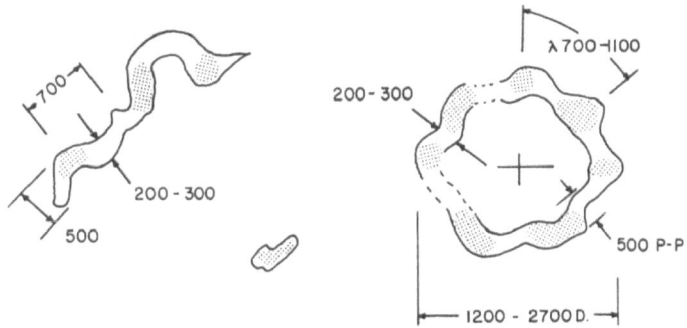

DIMENSIONS IN KILOMETERS

Fig. 1.   Typical size of filigree patterns.

(ii) *Contrast:* Filigree is enhanced in the blue wing of the Hα line. Measured contrast, uncorrected for seeing, equals 5–10%.

(iii) *Relation to the Granulation:* The filigree structures tend to lie between the granules. This is, however, not a strict rule. It seems that in the course of their lifetime the granules move the filigree structures around with velocities of about 1.5 km s⁻¹. Some of the crinkles also seem to wash out temporarily until compressed again by a

_____

* On leave from the University of Hawaii, Astronomy Department.

R. Grant Athay (ed.), Chromospheric Fine Structure, 45–47. All Rights Reserved.
Copyright © 1974 by the IAU.

new granule. The detailed structure of the filigree, therefore, changes significantly over times comparable to the granule lifetime. The overall structure is, however, preserved over much longer periods of time. The granulation pattern when observed in the continuum well outside the Hα line appears very peculiar in that it has substantially decreased in contrast. It appears 'soft' similar to granulation washed-out by seeing. This abnormal granulation can be traced over long times (> 30 min) and coincides in location to the filigree location. It is, therefore, definitely real.

(iv) *Relation to the spicules:* The filigree structure falls near the center of the Hα chromospheric rosettes. These rosettes consist of dark elongated mottles which

should probably be identified with spicules. There is, therefore, at least a coarse relation between the occurrence of spicules and the filigree. There is no clear evidence that variations in the filigree pattern are related to the generation of spicules. Some spicules seem to originate from the spaces between the crinkles. Too few, however, to conclude a definite relation.

(v) *Relation to the magnetic field:* Beckers studied the filigree with the Universal Birefringent Filter in the magnesium b_1 and b_2 lines. It is very well visible in the far wing of the lines (eg. $b_1 \pm 0.8$ Å). When traced into the line core the structures increase somewhat in size, as they do in Hα, and form structures similar to, and perhaps identical with, the so-called photospheric network. In the magnetically sensitive b_2 line one sees a one-to-one correspondence between these network structures and the magnetic field so that, at least in the layers seen near the core of the b_2 line, there is a one-to-one correspondence between the filigree structures and the enhancements in the magnetic field. Simon and Zirker (*Solar Physics*, submitted for publication) using a spectrograph also found that the filigree occurs in regions of enhanced magnetic field. However, in contrast to the filter observations, they found the magnetic field regions to be much more diffuse (2–3″) so that there is not a one-to-one spatial correspondence between filigree and magnetic field structure.

The filigree patterns have the appearance of a scaled-down supergranule network, a suggestion made also by others (e.g. Vrabec). Similar to the structuring of the magnetic fields by the supergranules (causing the magnetic/chromospheric network), there appears to be a structuring by the granulation at a ∼20 times smaller spatial scale and ∼100 times smaller time scale. Beckers' observations suggest a strong compact concentration of the magnetic field on the scale of the filigree whereas Simon and Zirker's observations indicate an enhancement which is less concentrated. The field strength will, of course, depend on the scale of the magnetic structures. It is at least a few hundred gauss. If concentrated on the scale of the filigree its strength may be much higher although it would be hard to see how granules could concentrate fields (and move flux tubes around) with strength as large as a few thousand gauss.

DISCUSSION

Acton: Do you feel that in the very quiet regions the filigree will be related to the network boundaries.
Beckers: In the very quiet regions the magnesium network is easily visible and is concentrated in very

small elements that are very stable. They have the same appearance as filigree features but we have not proven yet that the two are identical.

Meyer: Is there any chance to see a velocity pattern? It would help enormously for models of these features.

Beckers: The network is much more visible in the blue wing of the magnesium line than in the red wing and I therefore suspect that there is upward motion associated with these features.

Zirin: I am worried about the inference of big fields. There was another paper published recently in which someone measured fields of 5 G and from them inferred fields of thousands of gauss. I really wonder if we should go so far with our estimates of fields. If they are this strong why don't they become sunspots?

Beckers: I share your worries. I am just reporting results that others have obtained. I personally have a hard time accepting fields as large as 3000 G; I am willing to consider fields near 1500 G.

Deubner: The structure of the filigree very much resembles that of the inverted granular intensity as, for example, measured by Evans and Catalano. In the case of the inverted photospheric intensity you can see in all metal lines bright structures corresponding to the intergranular lanes that are associated with downward flows, so maybe we need to look again at the motion of filigree.

Beckers: There is a tendency for the filigree to occur in the intergranular lanes but the filigree are bright features observed against the dark intergranular lanes. When you observe in the intergranular lanes it will depend entirely on your resolution whether you see something bright or something dark. If your resolution is poor and the filigree contrast is not too great you may still see just the intergranular lanes and the downward movement associated with them.

SPICULES ARE BRIGHT AND DARK

HAROLD ZIRIN

*Big Bear Solar Observatory, Hale Observatories, Carnegie Institution of Washington,
California Institute of Technology, Pasadena, Calif., U.S.A.*

Abstract. I have tried to understand whether spicules appear dark or bright against the disk in Hα by comparing high resolution pictures in Hα and Ca II K. Since the chromospheric structure is more apparent near active regions, I have first compared the Hα and K fibrils near an active sunspot; these are illustrated in my review paper (this Symposium, p. 161). The elongated fibrils in the active center appear similar to spicules, but much larger and more horizontal. K line pictures show bright features better, while Hα shows dark features better.

Careful comparison of the K and Hα pictures shows a complete correspondence. Bright fibrils are bright in both lines and dark features, dark. The only difference is in the relative contrast. As one can judge from these pictures, there seem to be an equal number of bright and dark fibrils in the active region, the former being more prominent on the K-line flares and the latter, on the Hα. It is not surprising that this agreement exists. Both bright and dark fibrils seem to emanate from bright plages or plagettes.

In Figure 1, we see an Hα: K pair taken near the limb. In this case the K-line frame is unfortunately taken with 0.6 Å bandpass, which penetrates a little lower into the chromosphere, and does not show spicules very well. We have marked several pairs of bushes by corresponding numbers. The following points appear:

(1) The bases of the bushes show bright vertical structure in Hα (bush 2 is typical) with dark spicules above. However, there is not a clear correspondence between the bright jet at the base and the vertical absorption above.

(2) The bright bases in K correspond to the bright bases in Hα, but near the limb (3, 4, 5) much more vertical bright structure appears in K, partly because of the lower absorption by foreground objects.

(3) Some bright Hα spicules are seen, for example in the bushes 6 and 7, but they are in general few.

Thus the bush structure near the limb in quiet regions is different from that in active regions; there is a clear bright-dark asymmetry, and only a few spicules are hot enough or optically deep enough to go into emission in Hα.

DISCUSSION

Athay: I would like to make one comment, but I don't mean it to be as bad as it sounds. I did not see one bright mottle that you convinced me was a spicule.

Zirin: You're the one guy that escaped seeing the enlargements so far.

Wilson: You must have a certain amount of faith to see spicules – if Hal says he sees spicules I'll believe him.

R. Grant Athay (ed.), Chromospheric Fine Structure, 49–50. All Rights Reserved.

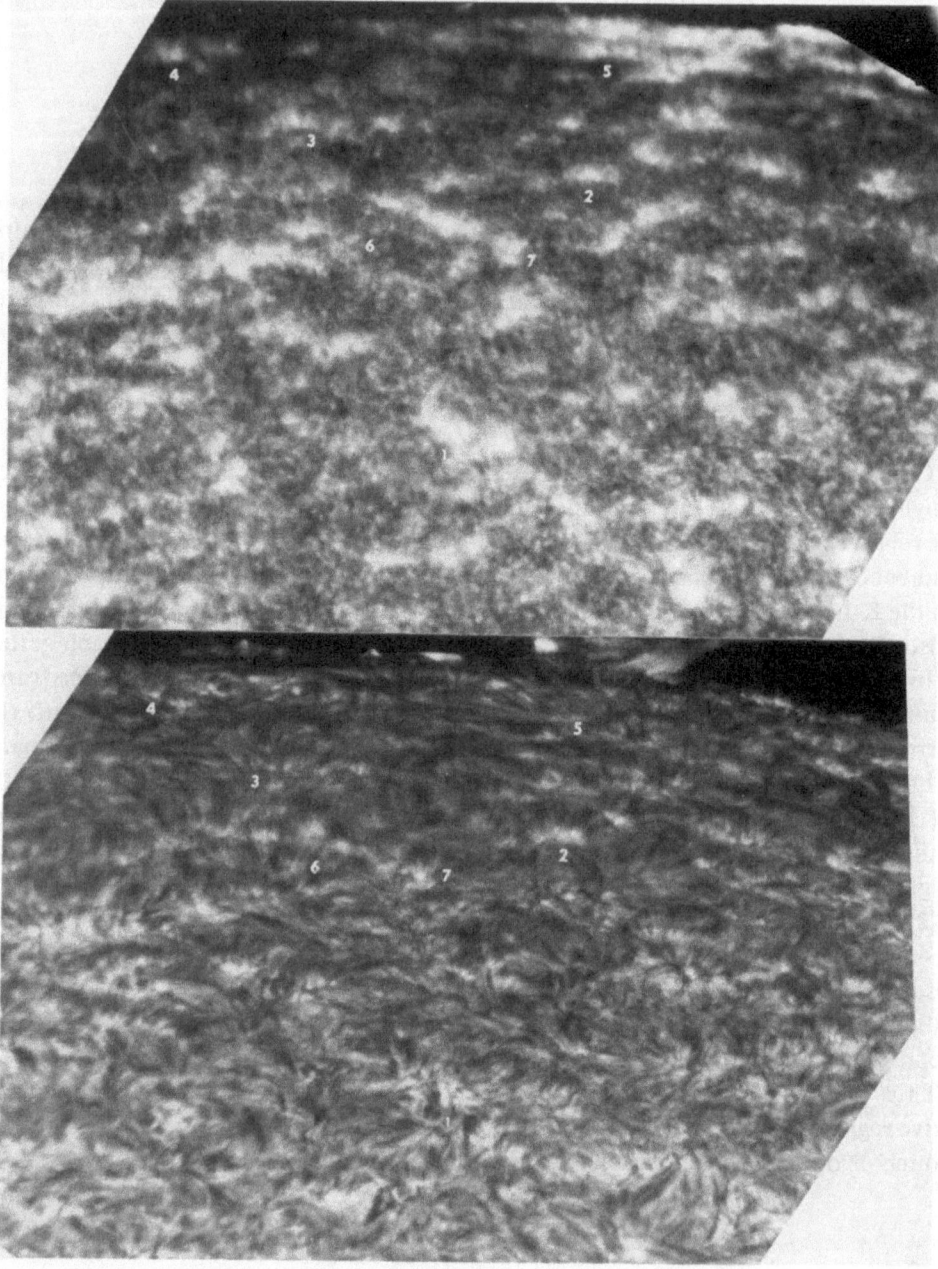

Fig. 1.   Filtergrams near the limb taken 1971, Sept. 23. K line 0.6 Å wide, above, Hα, 0.25 Å wide, below.
The granular K-line structure inside the network cells corresponds to a lower height. These pictures were
made 5 s apart.

HIGH RESOLUTION SPECTROSCOPY OF THE DISK CHROMOSPHERE

III: *Evidence for the Propagation and Dissipation of Mechanical Energy in the Chromosphere*

L. CRAM

Dept. of Applied Mathematics, University of Sydney, Sydney, Australia

Abstract. We describe time-series observations of small-scale Ca II emission features located outside the network in the quiet chromosphere. Simultaneous spectra in K and $\lambda 8542$ show unambiguously that the evolutionary behaviour of the K-line profile is due to an outwardly propagating velocity pulse. Assuming that this pulse is a progressive acoustic wave, as suggested by the inferred flow parameters, we show that the wave loses mechanical energy in traversing the chromosphere. This implies that the bright Ca II features (K grains) are the manifestation of local heating in the chromosphere, possibly by shock waves.

DISCUSSION

At first the discussion was concerned with clarification of the details of the evolution of these features and whether they occur in the interior of supergranule cells or in the network or both.

Grossmann-Doerth: I would like to know how many examples of this kind of evolving feature you have found.

Cram: This one was selected because it is a very clear example. Liu (1972) observed the evolution of many K_{2v} 'points' which were not followed by K_{2r} emission, and there are features on these spectra which do the same. I think this implies the wave motion in many of these points stops before the wave reaches the K_3 layers, and this is also a possible explanation of why the spatially-averaged K-line profile exhibits stronger K_{2v} emission than K_{2r}.

Wilson: I would just like to add that, although this is by far the best example I have seen, I have seen several different sequences of this sort of evolution in observations at Kitt Peak and with both the Big Dome and Tower Telescope at Sac Peak.

Athay: I would like to comment that there are many different examples. Liu has observed the K line out to about 7 Å in the wings, and he has followed the brightening from the wings right into the K_2 regions, progressively in time. So, that's a good indication that they're moving upwards.

Zwaan: I am interested in whether these features usually occur in the interior of a supergranular cell. Have the other observations been made together with filtergrams and slit-jaw pictures?

Wilson: In one case we had slit-jaw pictures and it was clear that the features were interior to the supergranule cell. In the others we didn't have pictures, but we had the impression by studying the spectra and noting where the really bright emission features occurred that the evolving features were indeed interior to the supergranule cells. You don't really see the same sort of evolution occurring in the emission in the network.

Beckers: These are the features which I called 'grains' about 10 yr ago, and which were also pointed out by Dr Rosch this morning on the drawing made by Dr Michard. The features were called 'dots' and on the drawing they were shown in the supergranule cells – in quiet regions. How do these features show in Hα?

Wilson: I don't know but that might be very interesting to find out.

Tanaka: I will show on Thursday that the grains in Hα show a very clear oscillation with 180-s period. Sou-Yang Liu has found similar oscillations in the K_{2v} cell points.

R. Grant Athay (ed.), Chromospheric Fine Structure, 51–53. All Rights Reserved.
Copyright © 1974 by the IAU.

Cram: Liu did not claim that every point was oscillating. He exhibited one grain that had something like 9 oscillations but other grains did not oscillate at all.

Tanaka: I think Liu claims most of them – 90% – oscillate.

Cram: That's not the impression I got from his thesis.

Schmidt: Do your results imply that the same physical processes are absent in the network, or can't they be seen?

Cram: I would say that the evolutionary behaviour of the network component is quite different from the K grains inside the cells. I think the K grains are shock waves which heat the chromosphere, but I don't think that the same interpretation applies in the network.

Schmidt: The spicules of course are something different, but would you be able to see the same processes in the network if they are there by this method, or are there other features which might prevent detection there?

Cram: One sees time-dependent behaviour in the network region, but I have certainly never seen singly-peaked features in these regions on the disk.

Wilson: I think the point is that, although there are time variations in the network structures, the predominant features in the network appear to be doubly peaked when they are resolved. There are so many extraordinary things going on that it's hard to be absolutely sure. The results of statistical counts which I think some people are doing should be very interesting.

Meyer: What is the spatial extent of these features?

Cram: They are about 1000–2000 km in diameter and they are separated by something like 6000 km.

The discussion then turned to the types of models required to interpret these observations and in particular to the role of velocity fields in these models.

Pasachoff: I'm very interested by your calculations. How large are the velocities you need for the evolution you've been discussing?

Cram: In the case where there is upward motion only in the K_3 layer, so that the K_{2v} peak is removed, the velocity field is about 6 or 8 km s^{-1}. That's smaller than the wavelength displacement from line center to K_2, but the absorption is still enough to obscure the violet peak.

Athay: When you extract the velocity from the asymmetric profile you have to know the thickness of the layer that's moving. What is it in this case?

Cram: The upwardly moving K_3 slab is about 300 km thick.

Athay: Isn't that ambiguous? If you increase the slab thickness you can get the same effects with a smaller velocity.

Cram: Not when the slab is so thick that the K_2 layer moves as well, because then the whole line core shifts.

At this point Thomas raised the important question of time-dependence in the radiative transfer theory used in the model calculations.

Thomas: This is a question to both Cram and Cannon. You have both made calculations of time-dependent behaviour, but have you correctly included time-dependent distributions of atmospheric thermodynamic parameters?

Cram: In order to understand the rather complex effects of velocity fields on the K line, I have performed model calculations which postulate ad hoc source functions and velocity fields. I am now combining my non-LTE CaII models with Leibacher's solution of the time-dependent continuum hydrodynamic equations in order to obtain self-consistent solutions.

Cannon: My main concern has been the development of a method of solution to the aerodynamic equation of radiative transfer, i.e. the macroscopic equations specifying conservation of mass, energy and linear momentum coupled to the non-LTE equation of radiative transfer. One of the problems studied, however, involved the propagation of some initial periodic disturbances through a medium of finite optical thickness with the subsequent development of this disturbance into a shock. It was found that once every period of the disturbance, the red emission peak quickly doubled in intensity and then equally quickly disappeared over a time of order 1/20th of the cycle. This was *not* a result of mechanical energy dissipation. Although the atmosphere chosen was somewhat pathological in structure and not meant to relate to any specific problem, the results strongly suggest that the time development of the CaII K peaks mentioned by Cram may possibly be explained solely by radiative transfer effects.

Thomas: I made this point because if you look at a lot of the earlier work on cepheid spectra, each phase mimics a supergiant with a different effective gravity, so that you apparently get instantaneous adjustment all the way through. I think it's very interesting to do the self-consistent time-dependent calculations. Do you include the correct time-dependence in the transfer equation?

Cram: Cannon and I have shown that it is important to take into account the fact that a moving atom can excite and de-excite in different physical conditions. The expression for the line source function includes the Lagrangian derivative of population ratios, in the form $S_L = (1 - \varepsilon) \, J + \varepsilon B + \alpha (dS/dt)$.

Delache: To be more specific on this point, you have not taken into account the $(1/c) \, (\partial I/\partial t)$ term in the radiation transfer equation, and I think that that can be important in some cases where a photon scatters many times before destruction.

Cram: Do you really believe that the relative difference between the velocity of light and of the material could be important?

Delache: I think that the time-dependent term has to be retained but this I cannot explain briefly (see paper by Delache and Fróeschle (*Astron. Astrophys.* **16** (1972), 356).

COMPARISON OF Hα AND CaII H AND K SPECTRO-HELIOGRAMS AS A DIAGNOSTIC PROBE

K. B. GEBBIE* and R. STEINITZ**

Joint Institute for Laboratory Astrophysics, University of Colorado, Boulder, Colo. 80302, U.S.A.

Abstract. The line formations of Hα and Ca II H and K are compared in order to differentiate the various mechanisms giving rise to observable contrasts in the emergent intensities. Table II summarizes the criteria for distinguishing between horizontal spatial variations in temperature, density, and turbulent velocity.

1. Introduction

Spectroheliograms taken in the light of Hα are readily distinguished from those in Ca II H and K both by their overall appearance and by many of their specific details. Figures 1a and 1b are spectroheliograms taken simultaneously in the line centers of Hα and Ca II K. Comparison of these figures shows that whereas some features, such as filaments and plages, appear in both spectroheliograms, others, like the dark fibrils seen in Hα, do not appear in Ca II K, while still others are seen in Ca II K and not in Hα. These features, or contrasts in the emergent intensity, may arise directly from lateral changes in electron temperature, T_e, or in electron density, N_e, or from changes in the shape of the absorption profile such as could result from mass motions or from changes in T_e or in turbulent velocity. In this paper, we suggest how the differences between the Hα and Ca II H and K spectroheliograms may be used, together with a knowledge of the physical processes by which each line is formed, to distinguish the various mechanisms giving rise to the observed features. Here, however, we exclude from our discussion features such as prominences that arise from systematic mass motions (Hyder and Lites, 1970). Our theory is based on the supposition that Hα and Ca II are formed in roughly the same regions of the chromosphere (Vernazza *et al.*, 1973).

2. The Line Formation

The physics involved in the formation of any given line depends on the dominant processes by which line photons are created and destroyed (Thomas, 1965). A new line photon is created in the radiation field when the upper level of the transition is excited from the lower level either by direct collisions or by indirect processes, which are usually photoionizations from the lower level followed by photorecombinations to the upper level. Similarly, line photons in the radiation field are destroyed either by direct collisional de-excitations or by indirect processes. Whether direct collisions or photoionizations dominate these *source* and *sink* processes depends, *for a given*

* Staff Member, Laboratory Astrophysics Division, National Bureau of Standards.
** On leave of absence from Tel-Aviv University, Department of Physics and Astronomy.

R. Grant Athay (ed.), Chromospheric Fine Structure, 55–63. All Rights Reserved.

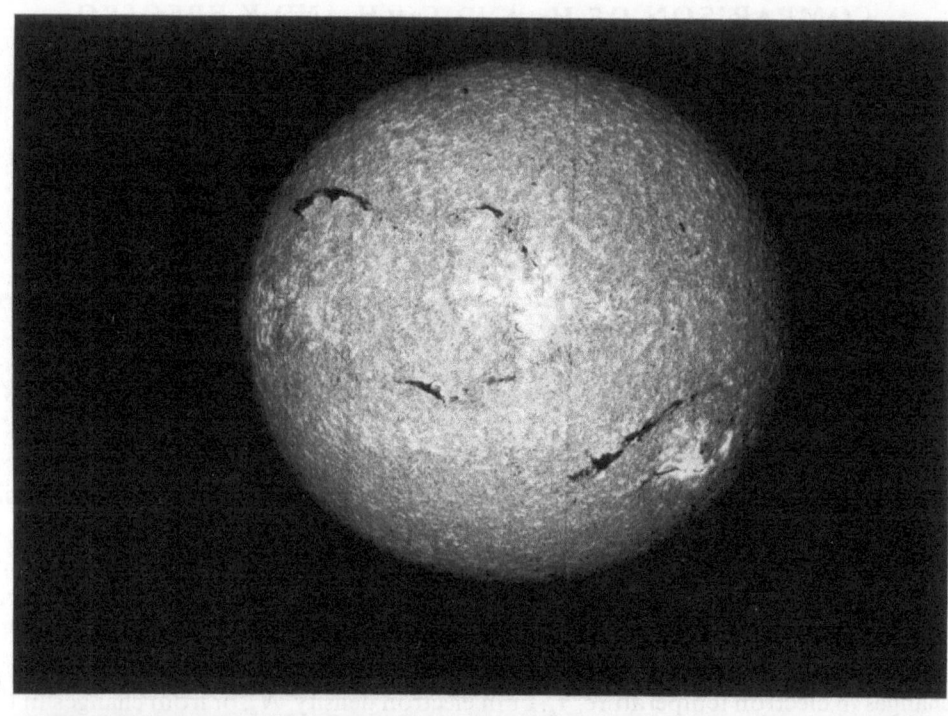

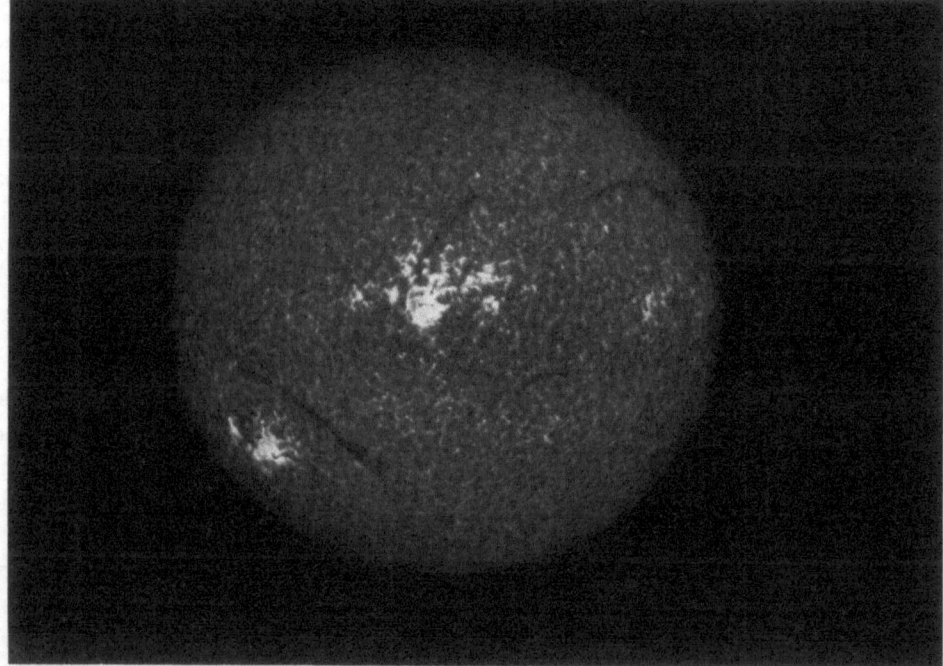

Fig. 1.   Spectroheliograms in (a) Hα and (b) Ca ɪɪ K, taken at line centers on 1973, May 25 at 13:22: UT.
Note the fine detail that appears in Hα but not in Ca ɪɪ K. (Sacramento Peak Observatory, Air Force
Cambridge Research Laboratory.)

transition, on the electron density, the collision cross sections, and the intensity of the ionizing radiation fields.

The essential difference between the formation of Hα and that of Ca II H and K in the solar chromosphere may be understood from a comparison of their energy level diagrams (Figure 2). For Hα, the upper and lower levels of the transition are only 1.5 eV and 3.4 eV, respectively, below the continuum. Thus the solar radiation fields in the Balmer and Paschen continua suffice to insure that photoionizations and photo-recombinations dominate over direct collisions as sources and sinks of line photons. Since the Balmer and Paschen continua are fixed in the photosphere, well below the region of Hα line formation, we would not expect any horizontal spatial variation in these source and sink terms. For Ca II H and K, on the other hand, the upper and lower levels of the transition are 8.7 eV and 11.9 eV below the continuum. Because the flux of solar radiation at these energies is low, direct collisions control the creation and destruction of line photons. Since the rate of these collisional transitions depends on electron temperature and density, lateral changes in these parameters will affect the source and sink terms.

Actually, there has been some controversy in the past over whether Hα is collision or photoionization controlled in the solar chromosphere. Using the most recent atomic data to compute and compare the relevant rates, we feel that we have now resolved this controversy in favor of photoionization control. This result is displayed in the form of a source-sink-control diagram, Figure 3 (Gebbie and Steinitz, 1974).

3. The Features

Having distinguished between collision and photoionization controlled lines, we now make the distinction between optically thick and optically thin features.

The emergent intensity at a frequency v is given by

$$I(v, 0) = \int_0^\infty S(\tau)\, e^{-\phi(v)\tau}\, \phi(v)\, d\tau, \tag{1}$$

where τ is the optical depth in the line center, $\phi(v)$ is the normalized absorption profile *assumed independent of depth*, and $S(\tau)$ is the frequency independent line source function, which may be written

$$S(\tau) = \frac{\int_0^\infty J(v, \tau)\, \phi(v)\, dv + \text{sources}}{1 + \text{sinks}}. \tag{2}$$

Here the first term in the numerator is the so-called scattering term, and $J(v, \tau)$ is the mean radiation field given by

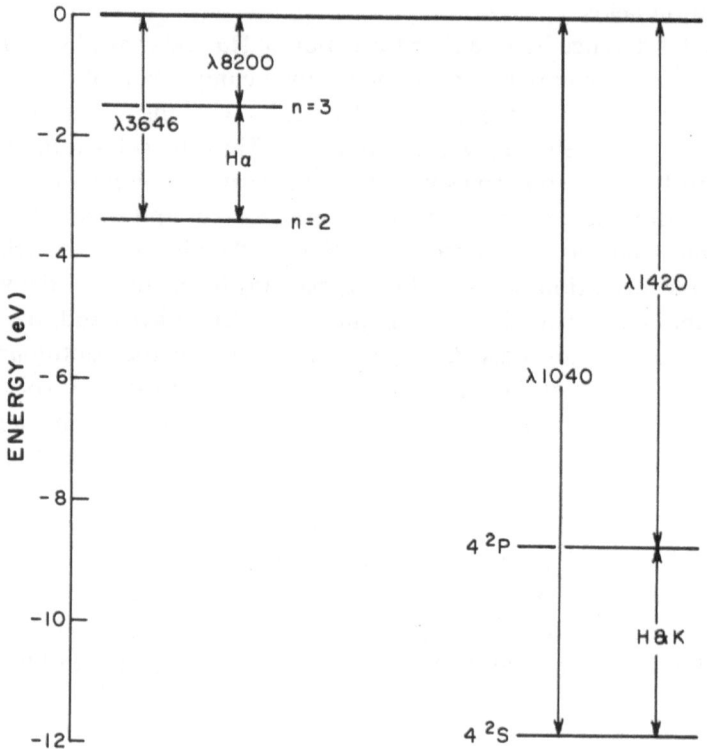

Fig. 2.   A comparison of Hα and Ca II H and K energy level diagrams. The Hα transition is 2→3, and the Ca II H and K is $4^2S_{1/2}\rightarrow4^2P_{1/2,\,3/2}$.

$$J(v,\,\tau)=\tfrac{1}{2}\int_0^\infty S(t)\,E_1\,|\phi(v)\,(\tau-t)|\,\mathrm{d}t.\tag{3}$$

Because most 'photon encounters' are in fact scatterings, $S(\tau)$ is numerically equal – to within about one percent – to the value of the scattering term, which is itself determined by the ambient radiation field in the line.

We define a feature as a lateral inhomogeneity that gives rise to an observable contrast, $C(v)$, defined as follows:

$$C(v)=\frac{I_f(v,\,0)}{I_0(v,\,0)}-1,\tag{4}$$

where $I_f(v,0)$ and $I_0(v,0)$ are, respectively, the observed intensities of the feature and the featureless regions.

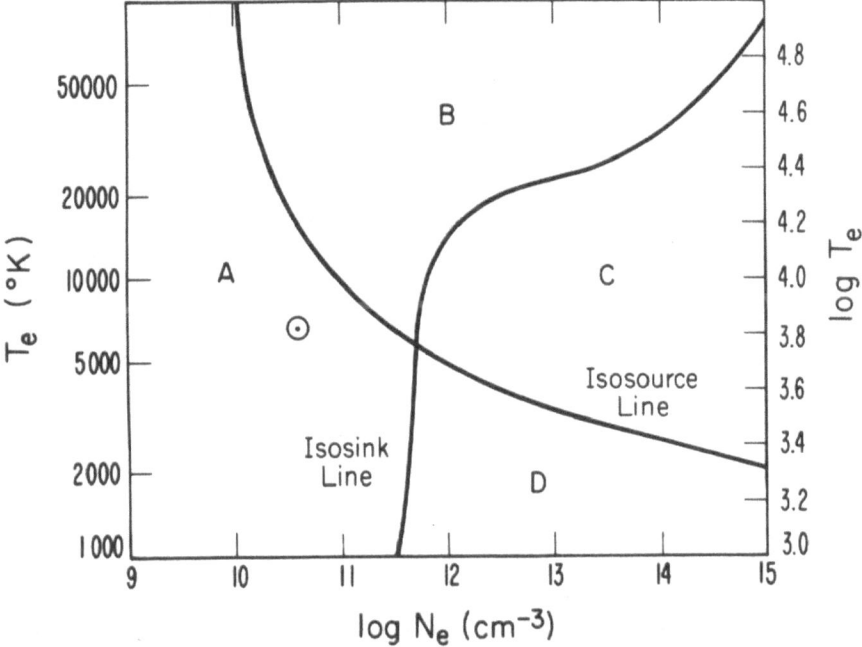

Fig. 3. Source-sink-control diagram for Hα and a radiation temperature of 5800 K. The isosource line delineates those values of T_e and N_e for which the rate of creation of line photons by direct collisional excitations equals that by indirect processes; the isosink line is that on which the rate of destruction of line photons by direct collisional de-excitations equals the rate by indirect processes. *Zone A:* sources and sinks indirectly controlled; *Zone B:* sources controlled by direct collisions; sinks indirectly controlled; *Zone C:* sources and sinks controlled by direct collisions; and *Zone D:* sources indirectly controlled; sinks controlled by direct collisions.

The Sun has been placed at $T_e = 6500$ K and $N_e = 4 \times 10^{10}$ cm^{-3} in accordance with the model of Vernazza *et al.* (1973).

3.1. OPTICALLY THICK FEATURES

In an optically thick feature, the source function builds up its own self-consistent radiation field, in the sense that Equations (2) and (3) are coupled to yield an equation of the form

$$S(\tau) = \frac{1}{1 + \text{sinks}} \int_0^\infty K(t, \tau; \phi) S(t) \, dt + \frac{\text{sources}}{1 + \text{sinks}}, \tag{5}$$

where $K(t, \tau; \phi)$ is the kernel. As long as the absorption profile is fixed and independent of depth, the solution, $S(\tau)$, depends only on the τ dependence of the source and sink terms.

Thus in a collisionally controlled line, where the sources and sinks are affected by lateral differences in the depth distribution of T_e and N_e, we would expect such differences to be reflected as contrasts in the emergent intensity. In a photoionization controlled line, however, such differences in T_e and N_e will not be observed, provided the ionizing radiation fields are fixed outside the region of line formation.

3.2. OPTICALLY THIN FEATURES

In an optically thin feature, Equations (2) and (3) are *uncoupled*: instead of building up its own radiation field, such a feature 'sees' the radiation field of the surrounding, featureless region. Thus the scattering term, and hence the source function itself, are determined not by the distribution of source and sink terms inside the feature but by those in the surrounding region. In contrast to the optically thick case, therefore, the source and sink terms inside the feature have almost no effect on the emergent intensity, regardless of whether the line is collision or photoionization controlled. Thus, in an optically thin feature, the source function is a function of geometrical rather than of optical depth.

There are, however, two mechanisms that may, in optically thin features, give rise to contrasts in the observed intensity.

(1) A change in the density will introduce a shift in the optical depth scale. The effect of such a shift on the emergent intensity will then depend on the behavior of S as a function of τ. If, for example, the density is lower in the feature than in the surrounding region, the optical depth at a given geometrical level of the atmosphere will be lower; the main contribution to the integral $I(v, 0)$, Equation (1), will then come from deeper levels of the atmosphere, where the source function has, in general, a higher value. Thus the feature will appear brighter than its surroundings at *all frequencies* in the line profile. If, on the other hand, the density is larger in the feature than in the featureless region, we will 'see' values of the source function higher in the atmosphere, and the feature will appear darker at all frequencies. In neither case will there be a reversal of sign in the contrast profile. That is, a feature that, as a result of a change in density, appears bright at line center will appear bright in the wings, and correspondingly, a feature that appears dark at line center will appear dark in the wings.

(2) A change in the shape of the *normalized* absorption profile, $\phi(v)$, will in general affect the emergent intensity in two ways: (i) by changing the value of the scattering term in the source function, and (ii) by introducing a frequency dependent shift in the optical depth scale (Gebbie and Steinitz, 1973 a, b). The effect on the scattering term will depend on (a) the relative shapes of the absorption profiles inside and outside the feature and (b) the frequency dependence of $J(v, \tau)$ at the depth of the feature. The effect of the shift in the optical depth scale will, as when resulting from a change in density, depend on the shape of the source function. However, since the profile is normalized, there will be some frequencies at which the profile has a larger value in the feature than in the non-feature and other frequencies at which it has a lower value. Thus we predict a reversal of sign in a contrast profile produced by this mechanism. Such reversals have in fact been observed by Grossmann-Doerth and Von-Uexküll (1971, 1973), by Bray (1973), and by Bar *et al.* (1973).

This discussion clearly includes such special cases as Becker's (1964) cloud model and de Jager's (1957) mottle models.

3.3. General Classification

In the previous sections we have classified line formation according to the processes controlling the source and sink terms and according to the optical thickness of the feature. This general scheme is summarized in Table I, which is self-explanatory. In

TABLE I

Mechanisms for producing observed contrasts
(excluding systematic mass motions)

	Optically thick feature		Optically thin feature
Primary change	$S(\tau) = \dfrac{1}{1+\text{sinks}} \displaystyle\int_0^\infty K(t, \tau; \phi)\, S(t)\, dt + \dfrac{\text{sources}}{1+\text{sinks}}$		$S(\tau) \sim \displaystyle\int_0^\infty J(v)\, \phi(v)\, dv$

	Collision control	Photoionization control	Control irrelevant
Density, N_e, T_e	changes in source and sink terms	no changes	change in τ scale
$\phi(T_e, v_t)$	(possible change of kernel)	(possible change of kernel)	(1) change of τ scale (2) change of scattering integral

the following section, we apply this scheme to the formulation of criteria by which to distinguish the various mechanisms giving rise to the observed contrasts.

4. The Shape of the Line Profile

Changes in the shape of the line profile may result from changes in temperature or in turbulent velocity, or from mass motions, or from any combination of these mechanisms. In this paper, however, we deal only with stationary features, excluding any discussion of macroscopic velocities.

The effect of changes in temperature and turbulent velocity on the line width may be combined in the following manner. The Doppler width of the line is given by

$$\Delta\lambda_D = \frac{\lambda}{c}\left(\frac{2kT}{m_A} + v_t^2\right)^{1/2}, \tag{6}$$

where m_A is the mass of the atom and v_t is the ambient turbulent velocity. For a given spectral line, the Doppler width in a feature of temperature T' and turbulent velocity v_t', relative to that in the featureless region, may conveniently be expressed as

$$\frac{\Delta\lambda_D'}{\Delta\lambda_D} = \left(\frac{\theta + \mu\xi'}{1 + \mu\xi}\right)^{1/2}, \tag{7}$$

where all velocities have been normalized with respect to the thermal velocity of hydrogen at the temperature T in the featureless region. Thus $\theta' = T'/T$, $\mu = m_A/m_H$,

$\zeta = v_t^2/(2kT/m_H)$, and $\zeta' = v_t'^2/(2kT/m_H)$. Here it is the mass of the atoms ($\mu_H = 1$, $\mu_{ca} = 40$) that will distinguish the effects of changes in T and v_t on Hα and Ca II H and K. For turbulent velocities of the order of the thermal velocity of hydrogen, changes in temperature will have negligible effect on the calcium absorption profile but can be significant for hydrogen. Changes in the turbulent velocities, on the other hand, will tend to have a greater effect on calcium than on hydrogen.

One may therefore expect changes in temperature to be reflected in contrasts observed in Hα but not in Ca II H and K, whereas changes in turbulence should be observed in both lines.

5. Summary: The Criteria

We now classify the observed contrasts according to two criteria: (1) Is the feature seen in Hα but not in Ca II H and K, or is it seen in both lines, or is it seen only in Ca II H and K? (2) Is there a reversal of sign in the contrast profile? These criteria are applied in Table II to distinguish between changes in electron temperature, density,

TABLE II

Criteria for distinguishing the mechanisms
(excluding systematic mass motions)

Contrast seen?		Sign reversal in contrast profile?	Optical thickness	Operative mechanism
Hα	H and K			
yes	no	yes	thin	ΔT_e
yes	yes	yes	thin	Δv_t
yes	yes	no	thin	Δ (density)
no	yes	[a]	thick	$\Delta T_e \, \Delta N_e \, \Delta v_t$

[a] depends on the details of the particular processes.

and turbulent velocity. We distinguish systematic mass motions by asymmetric and shifted profiles.

Thus by comparing Hα and Ca II H and K spectroheliograms, and by observing the behavior of the contrast as a function of frequency in the line profile, it is possible to probe the conditions giving rise to the observed contrasts. We therefore encourage observers to take spectroheliograms or spectrograms simultaneously in Hα and in Ca II H and K. This could provide data for a more reliable interpretation of the physical conditions prevailing in the solar chromosphere.

Finally, it should be pointed out that conditions in flares and plages may shift the control of Hα formation from photoionizations to collisions; if so, the criteria given in Table II would not be valid.

It is a pleasure to thank L. Gilliam of Sacramento Peak Observatory for providing us with the spectroheliograms published here. We acknowledge support by the National

Bureau of Standards grant NBS(G)-164-IS-TA-A10 to the University of Tel-Aviv, by Sacramento Peak Observatory Contract Y73-802, and by NASA Contract NGR 06-003-057.

References

Bar, V., Steinitz, R., and Gebbie, K. B.: 1973 (in preparation).
Beckers, J. M.: 1964, Thesis, Utrecht.
Bray, R. J.: 1973, *Solar Phys.* **29**, 317.
de Jager, C.: 1957, *Bull. Astron. Inst. Neth.* **13**, 149 (No. 474).
Gebbie, K. B. and Steinitz, R.: 1973a, *Solar Phys.* **29**, 3.
Gebbie, K. B. and Steinitz, R.: 1974, *Astrophys. J.* **188**, 399.
Grossmann-Doerth, U. and von Uexküll, M.: 1971, *Solar Phys.* **20**, 31.
Grossmann-Doerth, U. and von Uexküll, M.: 1973, *Solar Phys.* **28**, 319.
Hyder, C. L. and Lites, B. W.: 1970, *Solar Phys.* **14**, 147.
Thomas, R. N.: 1965, *Some Aspects of Non-Equilibrium Thermodynamics in the Presence of a Radiation Field*, University of Colorado Press, Boulder.
Vernazza, J. E., Avrett, E. H., and Loeser, R.: 1973, *Astrophys. J.* **184**, 605.

DISCUSSION

Wilson: Would you explain exactly what you mean by the reversal in sign?

Gebbie: If, for example, the temperature or turbulent velocity were lower in the feature than in the surrounding region, the normalized absorption profile will be narrower; that is, the profile will have a higher value in the center of the line and a lower value in the wings. Thus provided the feature is optically thin and the source function decreases monotonically outwards, we would expect the feature to appear dark in the center of the line and bright in the wings. This is what we mean by a reversal in the contrast profile.

Athay: Could we see the first slide again? Dr Gebbie points out that she has plotted the sun on her source-sink-control diagram at an electron density of about 4×10^{10} cm^{-3}. If you go down a little deeper in the chromosphere, the electron density increases to something like 2×10^{11} cm^{-3}. You need an electron density definitely above 10^{11} cm^{-3} to get the maximum in the Balmer continuum emission as observed at eclipse. You see the region just above the temperature minimum at about half an angstrom from line center in Hα. Also, you expect to see regions of a little higher density when you go to the network regions. The effect of the higher densities will move your point toward the intersection between photoionization and collision control, and Hα will show some effect of collisions.

Gebbie: Yes, certainly an increase in density will 'move the Sun' *toward* the intersection, but Hα will still be photoionization dominated at densities of up to about 5×10^{11} cm^{-3}. According to the *one-component* model of Vernazza, Avrett and Loeser, this value is not reached until well below the temperature minimum. However, in denser chromospheric regions, such as perhaps flares and plages, collisions will certainly begin to dominate. Also, if, as has been suggested by Milkey, there is a photospheric contribution to the emergent intensity in Hα, this will be affected by collisions.

Giovanelli: Where we see bright features in Hα, we see these usually fairly small points, and when we look in the K line we will find that surrounding these points, we will certainly find it brighter than average in the K line. But surrounding this there will be a region where it is bright in the K line and dark in Hα. Now, would you like to interpret this?

Gebbie: I would interpret the feature that you see in Ca II K and not in Hα as being optically thick. Without further information, I could not then say whether the increase was due to an increase in temperature, density, or turbulent velocity.

Zirin: I guess I disagree with Ron Giovanelli just a little bit. Most of the time when you find the diffuse calcium brightening bigger than in Hα, it's because the calcium filter isn't well aligned. The only difference that I've ever been able to find is close to sunspots where there is overlying absorbing material. This material may be dark in Hα and somewhat transparent in calcium K, and I would say that from every observation I have made that there is only a quantitative and not a qualitative difference between K and Hα except close to sunspots. In the network I've never seen any difference. Even in flares the two are identical, and that's quite a departure in temperature and density from your critical value.

FINE STRUCTURE OF THE SUN AT CENTIMETER WAVELENGTHS

M. R. KUNDU, T. VELUSAMY, and R. H. BECKER

Astronomy Program, University of Maryland, Md., U.S.A.

Abstract. The Hat Creek Observatory's two-element interferometer and the NRAO 3-element interferometer have been used at 1.3, 3.7- and 11.1-cm wavelengths respectively, to study the fine structure of the radio emissive regions on the Sun. Observations of the quiet Sun at 1.3 cm show sudden increases followed by a gradual decrease in the fringe amplitude lasting for typically about 5–8 min. Assuming these events are identical in nature, a plot of peak amplitude against the projected baseline at the time of the event suggests emission from a region of angular size of about 10″. The corresponding brightness temperature is 50 000 K. It is possible that these events may be related to the appearance and disappearance of groups of spicules or mottles.

Observations at 3.7 and 11.1 cm were used to synthesize maps of the active region located at N14 W19 on 1973, June 8. The 11.1-cm map (Figure 1) exhibits three distinct peaks. The brightest components at both wavelengths do not seem to be completely

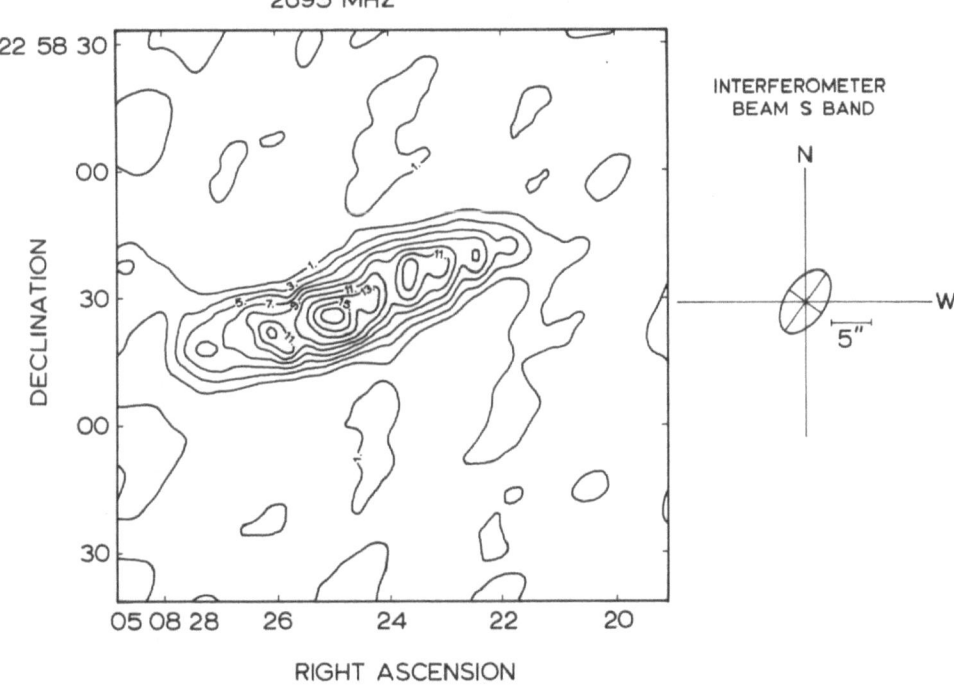

Fig. 1. Synthesized map at 11.1 cm of the region located at N14 W19 on 1973, June 8.

R. Grant Athay (ed.), Chromospheric Fine Structure, 65–68. All Rights Reserved.

resolved by the synthesized beams. The synthesized cleaned beams at 3.7 and 11.1 cm
are elliptical gaussians of dimensions 3″.23 × 1″.62 and 9″.70 × 4″.82. There appears to
be general agreement between the optical structures of the plage region and those of
the synthesized maps. The maps imply that the plage regions contain bright points or
cores along with diffuse areas or halos. The brightness temperatures of the strongest
components are 8.4×10^5 K and 7×10^4 K at 11.1 and 3.7 cm, respectively.

On 1973, June 9, a flare associated burst was observed at 3.7 and 11.1 cm. The burst
lasted for over 25 min. Figure 2 shows the fringe amplitudes during the flare at 3.7-cm

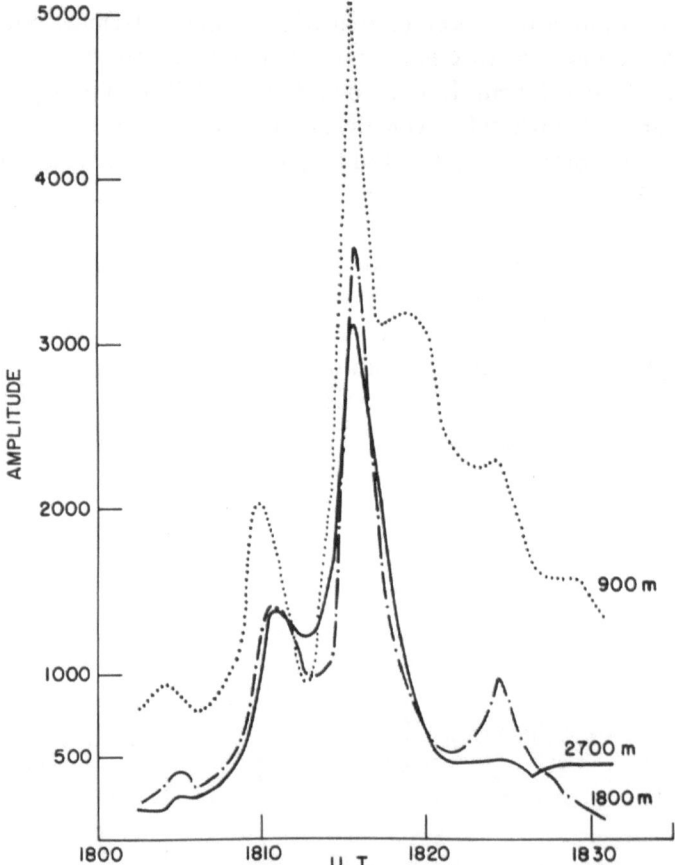

Fig. 2.   Fringe amplitude in arbitrary units of the interferometer at 3.7 cm at all three spacings during
the burst of 1973, June 9.

wavelength as a function of time. The interferometric records show a precursor from
18:05:5 to 18:12:5 UT followed by an impulsive phase with maximum at 18:15:5 UT.
Comparing the fringe amplitudes at 3.7 cm to the visibility computed for model flare
regions we found that the precursor data are best fitted by a region at 4″ in size while
at the time of the peak, the flare appears to have a size of 2″. During the post-maximum

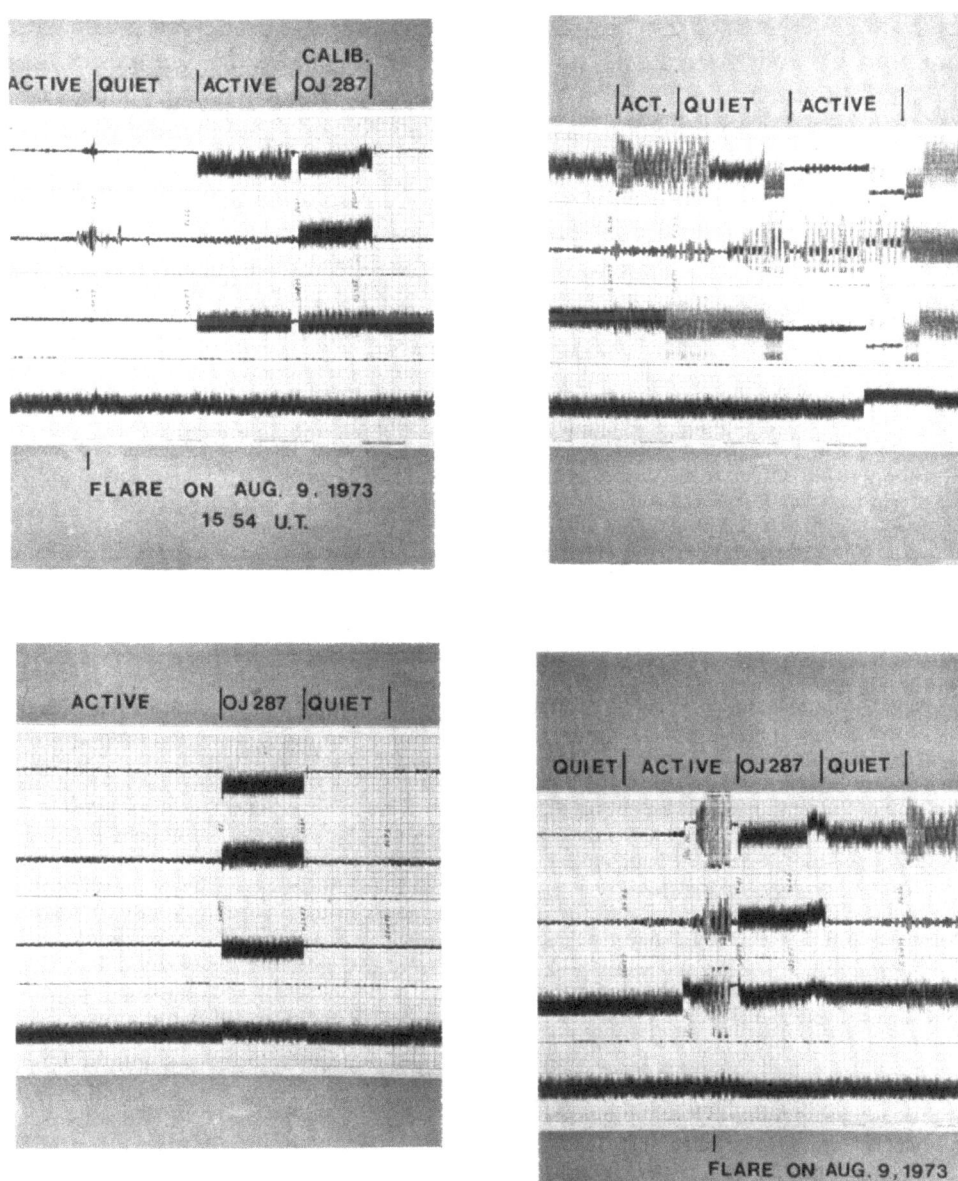

Fig. 3. Fringe amplitude output of the NRAO interferometer. Wavelength alternates every 30 s between 3.7 and 11.1 cm. Baselines displayed in each set are 1.8, 0.9, 2.7 and 35 km from top to bottom. Labels quiet and active refer to area of Sun producing fringes.

phase a size of 5″ is the best estimate. On the basis of the size of the flare as calculated from both 3.7- and 11.1-cm data, we compute peak brightness temperatures of 1.2×10^9 K and 1.65×10^8 K at 3.7 and 11.1 cm respectively. These high values imply that a significant fraction of radiation has a nonthermal origin. These results have been submitted for publication to *Solar Physics*.

Observations have also been made with angular resolution of 0″.2 and 0″.6 at 3.7- and 11.1-cm wavelengths. These observations indicate that there is fine structure on the scale of 0″.2 and 0″.6 at these wavelengths in the quiet, active and flare regions (see Figure 3).

DISCUSSION

Brandt: Our group (Hobbs, Jordan, Maran and Webster) has used the NRAO 3-element interferometer in the 2.7–1.8–0.9 configuration to obtain solar data at 3.7 and 11.1 cm. The results show the existence of structures in active regions which are not resolved by the instrument, i.e. less than 8.5″ at 11.1 cm and 2.8″ (~ 2000 km) at 3.7 cm. At the time of an optical flare on 1972, November 2 and 3 the 3.7- and 11.1-cm fluxes from these small scale structures increased significantly.

Schmidt: Could you comment on the time correlation with the optical flare?

Kundu: This flare which I observed with the longest base line was associated with an Hα flare. Regarding the X-rays, I don't have much detail except to say that it started almost at the same time as the visible Hα flare, which was around 18:10, and the peak was around 18:16 UT on the 9th of June.

Bracewell: With reference to the second-of-arc structure reported by Kundu to possess temperature of 50 000 K I wish to report observations made at Stanford University by John Grebenkemper with the fast interferometer which enables nine different baselines to be observed simultaneously at a wavelength of 2.8 cm. On a number of occasions he has found structure around 5″ in size, in active regions where no flare was in progress, with apparent temperatures around 7×10^6 K.

Kundu: One important thing that I'd like to point out is the following. For many years many of us thought that the gradual rise and fall type bursts at cm wavelengths were of thermal origin. Now, with interferometric determination of the sizes of such bursts which are 2″ or less, their brightness temperatures are found to be $\sim 10^9$ K, which implies they are of non-thermal origin. Further, the impulsive nature of the fringe amplitudes during the burst indicates their non-thermal origin.

Castelli: A year ago at AFCRL, we took a careful look at all gradual rise and fall type bursts recorded at Sagamore Hill for three or four years. By plotting the spectra of these bursts (recorded at numerous frequencies in the decimeter-centimeter range) at time of maximum emission, it was apparent that they did not have a thermal spectrum. Conservatively 80% of them had a decisive spectral maximum, as did the obviously impulsive bursts. Since the spectrum did not rise to a maximum and then simply flatten out in the high frequency direction, this would seem to rule out that the bursts were of thermal origin. Hence, Dr Kundu's recent finding tends to confirm what we already suspected.

Zirin: I just wanted to present some similar results to what Kundu has been showing, obtained by Ken Lang and myself, with the Owens Valley interferometer. This is made up of two 90-ft dishes and a 130-ft dish, with which we can get three interferometers, the longest base line is 4000 ft, with a resolution at 3.7 cm of 7″, and with a 1600-ft base line about 14″, give or take. When we first started observing last December we were extremely pleased to find that with a 1600-ft base line we found, as Kundu has remarked, that the Sun is covered with small elements which show fringes. Later on in the first run, we obtained a couple of flares which showed a strong enhancement of fringe intensity, again with a 1600-ft base line. In December we were able to use the 4000-ft base line, and found that, although there were fringes with the 1600-ft base line at about 13″, we could see no fringes with the 4000-ft base line, not even in active regions. However, during this period we began to see fringes with 7″ fringe spacing, and then a flare occurred which gave a large enhancement of these fringes as well as phase shift. In addition, there were rather sharp phase shifts and we noticed that they were associated with separate spikes in the burst. One might imagine that what we are seeing is separate lights twinkling on and off and we get phase shifts due to their different positions.

PART III

THE UPPER CHROMOSPHERE

FINE STRUCTURE OF THE UPPER CHROMOSPHERE

J. T. JEFFERIES

University of Hawaii, Honolulu, Hawaii, U.S.A.

Abstract. The paper contains a review of attempts to understand observational data on the inhomogeneous structure of the upper solar chromosphere, particularly the spicules. Observations over the electromagnetic spectrum from millimetric to the ultraviolet give equivocal results; and in spite of recent progress, models for these inhomogeneous structures still face difficulties in accounting for the observed radiation. New observations in the visible are needed in order to answer even the simplest questions of kinematics, while the interpretation of the line spectra still poses many difficulties. The acquisition of EUV data at high spatial resolution should lead to major progress in the field, just as it already has done in showing the EUV emission to be concentrated over the network boundaries.

1. Introduction

That the upper chromosphere is inhomogeneous is plainly evident from direct filter photographs taken at the limb or in the strong chromospheric lines on the disk. Spectra of the disk and chromosphere taken either during or outside eclipse show a corresponding structure – cf. Suemoto and Hiei (1962), Pierce (1965). The principal inhomogeneities are seen to lie on the network boundaries and to resolve into structures of the order of one arc second in size – as is most clearly shown in Dunn's Hα photographs taken with the Sacramento Peak tower telescope and illustrated, for example, in Figure 3 of Michard's review. These inhomogeneities are undoubtedly spicules. Other inhomogeneities, lying interior to the network, have been described by Liu (1972) and Sawyer (1972), but very little is known of these evanescent features, and they are generally agreed to form a less important aspect of the characterization of the chromospheric structure. This judgment may, however, reflect our ignorance rather more than a balanced assessment. While recognizing that the picture is certainly only a first approximation at best, we shall here follow what seems the normal practice and represent the upper chromosphere in terms of one region, lying along the boundary of the supergranular network, and another, more or less homogeneous in nature, lying interior to this. In these terms, we shall discuss recent efforts to elucidate the physical characteristics of such a model, and shall try to indicate some of the difficulties facing the various pictures which have been advanced.

Quantitative information on the inhomogeneous structure of the chromosphere has been derived from the disk observations in the EUV, and in the millimetric and centimetric radio, as well as from disk and limb observations in the visible. The latter data have the immense advantage of giving high spatial resolution, which has so far not been possible in either the radio or the EUV; nevertheless we shall see that some information on inhomogeneous structure can be obtained from analyses of spectral data in those wavelengths.

In Section 2 we discuss the inferences which have been drawn mainly from optical data, while in Sections 3 and 4, we concentrate on analyses of radio and EUV ob-

R. Grant Athay (ed.), Chromospheric Fine Structure, 71–88. All Rights Reserved.

servations respectively. This format has obvious limitations, but is at least partly justified in convenience as well as in reflecting the historical development of the subject. In Section 5, we shall endeavor to bring together the results obtained by the different observing techniques to show points of agreement as well as those of dispute.

2. Studies at Visible Wavelengths

2.1. OBSERVATIONS AT THE LIMB

2.1.1. *Line Intensities and the Spicule Model*

Direct, narrow-band photographs, such as those obtained by R. B. Dunn at Sacramento Peak, have provided the basic data for understanding such questions as the lifetime, size, and spatial distribution of the chromospheric inhomogeneities. However, spectroscopic information in lines of H, He, and Ca^+ has been the source data for most of our knowledge of the physical conditions – particularly the temperature, density, and state of motion. The spatial resolution needed for such studies is obtainable only under the best of conditions, while the faintness of the emission makes their observation above the sky background difficult in any but the strongest lines except for the rare occasions provided by total eclipses. Until observations from space vehicles allow us to acquire data with high angular resolution over a wider wavelength range, any direct inferences on the structure of spicules will necessarily be limited to regions where the temperature is less than 10000 to 20000 K; reflecting the conditions under which the strong lines of H, He, and Ca^+ are most emissive in the chromosphere. This is unfortunate, since the nature of the transition region around spicules is of profound importance to understanding the origin of the inhomogeneous structure of the chromosphere. A first indication of this transition region structure has recently been obtained by Brueckner and Nicolas (1973) on the basis of ultraviolet eclipse data; further observations in the ultraviolet will be eagerly awaited.

Giovanelli (1967) and Beckers (1968, 1972) have derived spicule models using intensity data gathered by different investigators (generally outside of eclipse) for different heights above the limb. Their approach has been essentially to compute the emitted intensity of a model spicule of given thickness as a function of electron density (n) and temperature (T), and then to determine the values of these parameters which are consistent with the observed intensity of the various spectral lines for which data are available. An example of Beckers' results is shown in Figure 1 for the height of 4000 km above the limb. The various curves should intersect at one point, which would determine a unique (n, T) if the spicule were representable in those terms. As can be seen from Figure 1, the data do determine n and T quite well and, bearing in mind the limited accuracy, all these intensity data can indeed be matched with a single temperature and density. This is consistent with the observations that the emission in these lines arises from the same feature in the solar atmosphere; see e.g., Suemoto and Hiei (1962), and Pasachoff *et al.* (1968).

Beckers' spicule model covers a height range of about 3000 to 11 000 km, over which the density varies from 2×10^{11} to 3×10^{10} cm^{-3}, while the temperature increases from 9000 to 16000 K with the largest gradient occuring in the lowest 2000 km. The conditions at low heights depend on a few difficult measurements of the H and He lines at 3889 Å, and cannot be regarded as very reliable – there is, therefore, no compulsive case in this analysis that a spicule's temperature does, in fact, vary with height.

An analysis of the observed K-line emission by Avery and House (1969) led to a

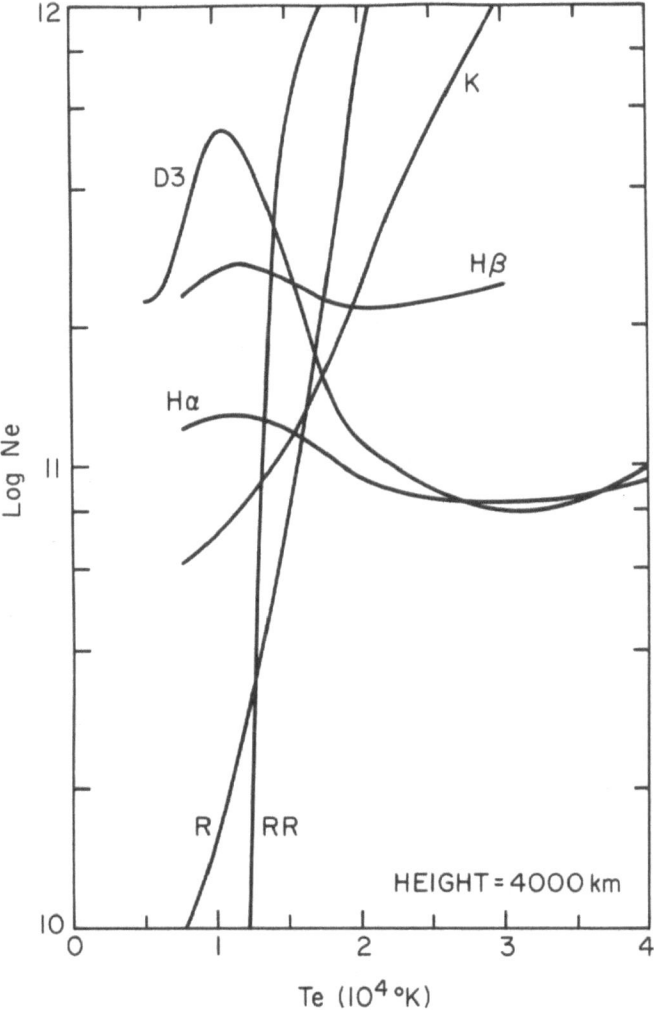

Fig. 1. The curves show the simultaneous values of electron density and temperature which are consistent with the observed intensity, in the indicated lines of the spicular emission at 4000 km above the solar limb. The curve labelled *R* refers to the ratio of the intensities of the helium and hydrogen line at 3889 Å, while *RR* refers to the ratio of the Ca$^+$ 8542 and the D$_3$ line – curves *R* and *RR* refer to eclipse data. After Beckers (1972).

rather different model with a generally lower temperature and higher density at a given height. The difference is probably due mainly to the different atomic data used for transition rates and to the fact that the analysis used information on only one spectral line, although the cylindrical geometry adopted would lead to systematic differences with Beckers' conclusions.

2.1.2. *Line Profiles*

In principle, one would also expect to obtain useful data from the shapes of the strong spicule emission lines at different heights. Although future work in the visible and EUV may achieve this hope, studies so far have been disappointing – little that is definitive to the spicule model has yet been obtained in this way, although several very puzzling questions have been posed.

The limb profiles in H, K, and Hα have central reversals at low heights and, according to Pasachoff *et al.* (1968), are frequently asymmetric. These authors maintain that the wavelength of the minimum is unshifted whether the profile is asymmetric or not. Since the central reversal would then not appear to share in the Doppler shift of the rest of the line, they conclude that the reversal has a non-spicular origin, arising in foreground attenuation – as Michard (1959) suggested – rather than in the spicule itself, which seems at first sight more natural. The case is far from clear; however this question is important since if the origin of the self-reversal were intrinsic to the spicule this would establish that its optical depth in the K and Hα lines were significant and so set some important constraints on the model. If it arises in foreground absorption on the other hand, then it places restraints on the characteristics of the (interspicular) matter.

The H and K lines in spicules are usually anomalously wide; cf. Zirker (1962), Pasachoff *et al.* (1968) and Athay and Bessy (1964). The origin of this broadening is as much of a mystery today as when it was first discovered over a decade ago; indeed, it is more so since it now seems clear that the width does not arise in overlapping Doppler-shifted features, as was first proposed, since it persists to great heights where the number of spicules is so small as to make overlap very unlikely. Attempts to explain the widths in terms of a non-thermal random velocity ('microturbulence') have not been convincing; we simply do not know what is doing the broadening, but it certainly does not seem that it arises in a Gaussian distribution of non-thermal motions. This question, also, needs added study.

2.1.3. *Line Tilts – Spicule Rotation*

A second strange feature is the 'tilt' of the line; i.e., the inclination of the line profile of the spicule to the direction of dispersion. These are invariably, and clearly, seen in H, K, and Hα spectra of spicules and again cannot be explained in terms of an 'overlap' of different features, since even under the finest seeing and the greatest heights they are present. The natural explanation of the tilt is that the spicule is rotating, and this is now widely accepted. The angular velocities are substantial, corresponding typically to a centripetal acceleration some 10 times that of solar gravity. This rotational motion must have some fundamental bearing on the origin of spicules though we have little idea just what. The existence of the rotation may, however, help to explain the anomalous line widths, since the rotational velocity of some 20 km s^{-1} is about the same as the K-line width. Again, further systematic study of this phenomenon is merited to ascertain whether the rotational velocity is the same in different spectral lines, to confirm whether it changes in sense, as Pasachoff *et al.* (1968) suggest, and to find how the rotation characteristics change with height.

2.2. Observations on the disk

Disk observations allow us to see the spatial distribution of the inhomogeneities on the solar surface in a way precluded by limb observations. Such data suggest very naturally the characterization – cf., e.g., Grossman-Doerth and von Uexküll (1973) – of the chromospheric model as a quasi-spherically symmetric background layer lying interior to the network (and for which spatial averages suffice to characterize the gas), together with a circumscribing boundary region representable only in terms of independent components – the bright and dark mottles.

For many years a debate has centered on just what disk features correspond to the limb spicules; there is still no clearcut resolution. In terms of their geometrical distribution, lifetimes, and such spectroscopic properties as Doppler width, source function, and optical depth, a rather clear case seems to exist for identification of the network 'mottles' with spicules. The line-of-sight velocity found for these disk features, however, is not as large as the limb observations would require, and this remains a matter of some concern. Perhaps Grossman-Doerth and von Uexküll (1971, 1973) are correct in believing this to be a consequence of our inability to resolve structures down to a few tenths of an arc second. Perhaps it is due to a substantial vertical velocity gradient being present in the spicule, which then places the fast moving material at small optical depths when viewed vertically downwards. Observations in the higher lying transition region lines might aid in resolving this question. Certainly, there is abundant evidence for vertical velocity gradients in the asymmetric profiles of the Ca II K and H lines. Until the limb and disk velocities are put into correspondence, however, this will remain a problem.

Disk spectra and spectroheliograms in the H and K line of singly ionized calcium have provided a rich field for discussion and dispute. The characteristic double-peaked profiles are widely held to originate in the chromospheric temperature rise, following the mechanism suggested by Jefferies and Thomas (1959). On this basis, we would interpret the observed K-line profile in terms of a basic background chromospheric layer, within which (at network boundaries) are located regions where the density and/or the temperature increases. A qualitative explanation is readily and naturally obtained with this theory for such characteristics of the K-line profile as the increase, towards the limb, of the K_2 separation, the variability of the K_2 emission from point to point, across the network and in the quiet regions, as well as the observed asymmetry.

Pasachoff (1970) has advanced an alternative suggestion for the K-line reversals

under which the line shapes are due to the superposition, on a basic underlying profile, of the emission from independent elements – some of which are moving up, some
moving down. The velocities must necessarily be within narrowly defined limits in
order to place the K_2 emission at so consistently the same position in the line as it
is observed to be. Pasachoff's picture was motivated by the observation that, when
viewed under high enough spatial resolution, the K-line profile often shows single
peaks either to the red, or, more frequently, to the violet. There is no doubt that
single-peaked profiles do occur, but we believe that these are simply the more extreme cases of asymmetry. There seems to be little doubt that double-peaked profiles
do exist. Indeed, according to Liu (1972), single-peaked profiles arise only in the inhomogeneities interior to the network (though double-peaked profiles occur there
also) while the network boundary elements (spicules) give double profiles in which
the K_2 violet peak is commonly the stronger. The quiet interior region has, at best,
a weak double reversal. Pasachoff's explanation faces several other difficulties (some
indeed anticipated by its author). The asymmetries which originally stimulated the
idea can be easily explained in terms of vertical velocity gradients – cf., Kulander
(1968), Athay (1970), Cram (1972). We believe that the chromospheric mechanism for
the formation of H and K is firmly enough established that it can be reliably used as
a basis for analysis both of the line intensity as well as its asymmetry. Even so, it is
not easy to visualize the geometrical structure of the chromospheric temperature rise
associated with the network elements. Is it, in fact, localized at the base of these
'spicules'; does the temperature transition occur along the whole vertical height of
the spicule; does it occur in a sheathing region? Neither the visible observations nor
indeed those in the XUV and EUV have yet yielded much information on the three-
dimensional temperature structure in the network element.

3. Inferences from Radio Observations

Several attempts have been made to obtain information on the structure of the solar
atmosphere using center-to-limb intensity variations (usually obtained at eclipse) in
the millimetric radio region. The interpretation of the results is not straightforward,
however, since invariably the limb brightening observed, if present at all, is smaller
than would be produced by a homogeneous model. Thus, as Coates et al. (1958) first
discovered, the millimetric radio data forces an inhomogeneous model for the upper
chromosphere.

As Simon and Zirin (1969) point out, this is, in fact, a problem common to most
observed wavelengths from the X-ray to the decametric radio. For the upper chromosphere, the conflict is particularly clearly shown from the simultaneous appearance of a brightness temperature increasing with wavelength in the millimetric to
centimetric range (i.e., temperature increasing with height where this radiation is
formed) without any significant limb brightening.

Millimetric observations obtained at the 1970 eclipse have been discussed by Hagen
et al. (1971) and by Simon (1971) with somewhat different conclusions. Hagen and

his collaborators obtained data at 3.2 and 8.3 mm, from a site in the band of totality, using an antenna with a beamwidth across the line of contacts of about 6' and some 2° along the contact line. To determine the specific intensity near the limb, they first averaged over the scans across the two limbs and then performed a deconvolution of the complex two-dimensional average over the uncovered solar surface and the antenna beam pattern, which yield the measured flux at each instant. The deconvolution along the line of contacts amounts to a differentiation of the flux-versus-time curve, however this yields only an average over a crescent shaped area contiguous to the Moon's limb and a second deconvolution is needed to obtain the surface brightness, or specific intensity. Hagen *et al.* completed this second step assuming an azimuthally symmetric intensity distribution of the Sun. In this way, they finally obtained a radial brightness distribution, having a double peak near the limb. They suggest this may arise from a peculiar temperature structure – perhaps an inversion – in the upper chromosphere.

Simon (1971) questions the reality of this double peak, particularly since the innermost maximum at 1 min from the limb does not appear on disk radioheliograms taken with the NRAO millimeter telescope at Kitt Peak where the eclipse was only partial. His data, supplemented by measures at longer wavelength, indicate simply a 'rough' chromosphere without limb brightening and with a scale size for inhomogeneities smaller than 20000 km, and with a small number of such inhomogeneities in an area of 20000 km². His data are in fact "...consistent with a multicomponent model of this region in which spicules protrude from the low chromosphere with the inbetween volume filled in by coronal material."

Whether or not a double-peaked radial intensity distribution is real (and we are bound to say that this seems unlikely) can only be found from observations with substantially higher spatial resolution.

However, a significant problem arises in attempting to account quantitatively for the center-to-limb intensity variation, as Vernazza and Noyes (1972) point out. The radiation temperature in the millimetric range is of order 6000 K and increases with wavelength (i.e., with height of origin). Thus, we would predict that the sun would be limb brightened at these wavelengths. The fact that the observations are not consistent with this expectation – cf. Kundu (1971), Lantos and Kundu (1972), Beckman and Clark (1973) – has been attributed to absorption near the limb by cold spicules which, of course, become more and more significant the closer one gets to the limb. A difficulty with this picture is simply that the spicule model derived by Beckers has nowhere a temperature as low as 6000 K and were that model correct we would actually expect to find an excess (rather than the observed diminution) of limb brightening.

Lantos and Kundu (1972) have shown, however, that it is possible to account simultaneously for the brightness temperature and limb darkening at three wavelengths (1.2, 3.5, and 9 mm) by accepting the lower temperature model of Avery and House (1969) and by modifying (below 3000 km) Beckers' estimate of the fraction of the solar surface at a given height which is covered by spicules. Since this fractional coverage can only be well estimated above about 5000 km, where individual spicules become visible, and since the temperature density model is, at best, only poorly known below 5000 km, Lantos and Kundu's explanation of the observed limb darkening seems very reasonable, but still requires independent confirmation, especially in regard to the physical structure of spicules at low heights.

4. Ultraviolet Studies

Direct information on the structure of the upper chromosphere is given by EUV spectroheliograms such as those obtained by Tousey (1971) and his collaborators. Such data are most important in showing that the network structure, seen so clearly in the visible, is also prominent in the EUV, extending in fact into the transition region – at least up to temperatures of 250000 K, where O v is formed. Whether or not a fine network can be discerned in the coronal line spectroheliograms is still under discussion. Most contemporary opinion holds that the coronal and chromo-spheric emission maps the magnetic field distribution. It is, therefore, very important to our understanding of the near and of the far corona to determine the emission pattern in a range of coronal lines, and so to follow, in particular, the change in the field configuration from the tight patterns seen in the chromosphere and the transi-tion region, into the broad, diffuse patterns which seem characteristic of the corona.

Just as for radio observations, center-to-limb intensity measurements from rockets and satellites in the EUV have shown the need for an inhomogeneous model in the upper chromosphere. Indeed, the reasoning leading to this conclusion is quite similar. Furthermore, the lack of spatial resolution, again limits the degree to which we can determine the precise details of the physical structure of the inhomogeneous layers.

As for the radio data, a purely homogeneous model does not allow us to account simultaneously for the observed intensity distribution among different spectral lines and for their center-to-limb variation. Thus as shown by Athay (1966), and Dupree and Goldberg (1967), the relative intensities of different lines lead in a straightforward way, and with few assumptions, to a specific temperature-versus-height distribution which a homogeneous atmosphere must have to account for the line intensities. From such a model it is a simple matter to compute the limb darkening or brightening in the various (optically thin) lines and to the extent that this disagrees with observa-tions, we must examine our assumptions. That most frequently questioned is, of course, that of homogeneity.

Examples of such studies are found in the work of Withbroe (1970a). In the first of these, he studied spectra of Li-like ions from N v to Si xii (lines probably all formed in the upper part of the transition region or the corona), and found no conflict be-tween the measured center-to-limb variation and that computed using the homo-geneous Dupree-Goldberg model. Presumably, therefore, inhomogeneities play only a slight role in the lower corona – at least insofar as the spectroscopic properties are concerned.

Systematic differences between prediction and observation do occur, however, for some lines formed wholly in the transition region as Withbroe (1970b) showed in a later study in which he essentially repeated the above analysis and found that all of the lines with wavelengths shortward of the Lyman limit (911 Å) had a predicted limb brightening greater than that observed. The natural explanation is that the radiation in these lines is absorbed by the neutral hydrogen in the inhomogeneities (spicules) which lie between the more or less homogeneous basic background layer. This argument is acceptable in terms of the known geometrical height of spicules (> 5000 km), in terms of the inferred height of the transition region – about 2000 km according to Athay (1966) – and also from the fact that spicules cannot significantly affect the intensity near the disk center, since they cover a very small area whereas, near the limb, projection effects can make them all-important.

Withbroe also used the difference between the degree of absorption found for the transition region lines and that for a coronal line (like Ne VIII), together with the known geometrical properties of spicules, to confirm the height of the transition region of 2000 km (with a substantial uncertainty) above the visible solar surface. Were the transition region to lie lower than this, the coronal lines with $\lambda < 911$ Å would show spicular absorption, contrary to observation; were it substantially higher, the degree of absorption in the transition region lines would be less than observed.

Since spicules extend characteristically to heights of 6000–10000 km, their surroundings over most of their height would be distinctly coronal in character on this model. Correspondingly, they should themselves be sheathed in a transition region between the more or less chromospheric spicular gas and the more or less coronal surroundings. However such a transition region will emit radiation – an effect neglected in Withbroe's work. If his picture is to be consistent, therefore, the transition zone must emit far less radiation per unit area than the interspicular transition region – cf. Noyes (1971). This, in turn, might arise from the zone being thin, or of low density.

Brueckner and Nicolas (1973), however, reach an opposite conclusion from their rocket spectra obtained at the 1970 eclipse. Indeed, they find that certain of the 'transition-region lines' extend to heights of at least 11000 km above the white-light limb and that the data can be well accounted for on the basis of a spicular model in which each spicule is sheathed in a transition region having essentially the same physical characteristics as that determined from EUV data with a homogeneous model.

The inhomogeneous structure of the upper chromosphere has also been studied in the Ly-c itself by Vernazza and Noyes (1972). They have adopted the same model of a basic homogeneous layer in which are embedded a random distribution of absorbers which are optically thick at the Ly-c head and are identified with spicules. Proceeding from the observed center-to-limb variation at various wavelengths within the Ly-c, they correct first for the shielding by spicules (assuming them again to be 'cold' absorbers) to obtain a center-to-limb variation representative of the homogeneous background. An empirical analysis of these data at different frequencies then allows them to determine the characteristics of the homogeneous background model. Since this agrees with the homogeneous model derived on theoretical grounds by

Noyes and Kalkofen (1970) using the central disk intensity distribution in the Ly-c, Vernazza and Noyes believe themselves justified in adopting a model of a homogeneous layer through which spicules protrude.

It turns out, however, not to be easy to reconcile this picture with the spicule models derived by Beckers (1968). Thus, both Giovanelli's (1967) calculations and those of Beckers show that the computed Ly-c source function in the spicules is very much larger than that in the quiet homogeneous chromosphere, which would imply that optically thick spicules would increase, rather than diminish, the Ly-c intensity towards the limb. For consistency, the inhomogeneous absorbing features introduced above must have temperatures well below those implied by the visible (Hα, H, K) emission. Vernazza and Noyes suggest that this clear conflict may be resolved if the spicule has a steep temperature gradient increasing from 5000 K near its base (~ 1500 km), which cannot be seen in the visible, to 15000 K at the heights ($\gtrsim 6000$ km), where the visible radiation has its origin.

One might hope to be able to test the credibility of such a model through a close comparative study of the relative variation of intensity toward the limb of a set of lines, or continua, whose heights (or temperatures) of formation varied in some known fashion. Indeed, if there is merit to this picture at all, one can readily envisage a scheme which would use the relative limb darkening of EUV lines with $\lambda < 911$ Å to diagnose not only the structure of spicules, but also the temperature/height relation in the interspicular region. This idea was, in fact, applied by Withbroe (1970b), but much more precision is needed to define the intensity variation near the limb before this can be applied with confidence.

5. Models

The model adopted to explain the EUV spectra consists simply of a homogeneous layer through which cold spicules protrude. The layer has the familiar photosphere-chromosphere structure and at about 2000 km a thin (~ 200 km) transition region occurs with an essentially coronal gas lying above. This model has the difficulties outlined above in reconciling the required properties of the spicules with those found from the limb spectra. Insofar as the background interspicular model goes, some observations favor it, some are in direct contrast.

. That the corona extends down to a few thousand kilometers from the limb can be inferred from the variation of intensity with height on slitless spectra taken at eclipses. Differentiation of such data yields a surface brightness, which in turn gives some indication of the variation of the emissivity with height. In this way, Athay and Roberts (1955), analyzing the observations of $\lambda 7892$ of Fe xi obtained at the 1952 eclipse, were able to show that the coronal temperature extended far below 10000 km. This conclusion has been strengthened by Weart (1968) and by Kanno et al. (1971) – from coronal red- and green-line data in both cases – who agree in placing the 'base of the corona' much below 10000 km. In fact, Kanno et al. conclude that the maximum surface brightness in these coronal lines occurs near the base of the interspicular region (~ 2000 km); they also find the red-line surface brightness to increase down-ward faster than that of the green line, indicating an outward temperature increase in the interspicular region.

In direct conflict with this however is the model derived by Beckers (1968) of the homogeneous interspicular region. He obtained a height scale from the electron den-sity distribution, $n(h)$, derived from K-corona eclipse measures, and adopting a tem-perature profile $T(h)$, computed the radio brightness temperature $T(v)$ for millimetric and centimetric wavelengths. By adjusting $T(h)$ for best agreement with the observed $T(v)$ Beckers obtained a temperature distribution having a far smaller gradient than that derived from the EUV analyses. Thus, in Beckers' model we find the temperature to be sub-coronal (250000 K) at 16000 km. The observations and their analyses are sufficiently straightforward that we must regard the conflict as real – the most im-mediate way out of it is to assume that the radio emission has its origin in a part of the gas which is entirely different from that where the EUV or forbidden coronal lines originate.

In fact, we know that the EUV lines are concentrated over the network boundaries so that the EUV analyses must presumably be referring to conditions in the general area where the spicules are formed. In that case, the applicability of the model under-lying the analyses by Withbroe and Vernazza and Noyes must be questioned. A more natural model indeed would be one in which the radio radiation arises from the regions interior to the network while the coronal-line radiation arises from a hotter, denser gas lying above a transition region which is part of the network boundary. The spicules lie within this region.

Filter photographs show that spicules move up (and sometimes down) in the sky plane at an apparent velocity of about 25 km s^{-1} – an individual feature rising typ-ically to 10000 km before fading or occasionally returning to the chromosphere. Spectrograms taken above the limb show these same features to have a Doppler motion of about 10 km s^{-1}. These two velocities are consistent with what we know about the inclinations of spicules to the radial direction, and so suggest that the movement in the sky plane is a real material motion along the axis of the spicule and not a passage of an excitation front. While Nikolskii and Platova (1971) main-tain that a large velocity transverse to the spicule axis is typical, most opinion is agreed that it is uncommon to find transverse velocities greater than a few kilometers per second. The question is fundamental to spicule origin and evolution, however, and needs to be clarified.

More significant perhaps than the simple state of motion, is the fact that spicules carry a large mass into the corona, if they do in fact move radially with high velocity. Thus, with a typical velocity of 25 km s^{-1}, and a proton density of 10^{11} cm^{-3}, spic-ules carry 2.5×10^{17} protons per cm^2 of their area each second into the coronal levels and if, following Beckers (1972), we suppose that they occupy 6×10^{-3} of the surface of the Sun, then 10^{38} protons per second find their way into the corona via spicules. Since the total coronal particle content is of order 10^{42}, the spicules could replenish the corona in only a few hours. This material does not escape however; its flux is two orders of magnitude more than that contained in the solar wind. It must therefore return to the chromosphere. However, the concensus seems to be that far less cold material is seen returning (in Hα for example) than is seen going up. If this is really so, then we must ask more closely the fate of the material ejected into the corona. The dynamic aspects of the circulatory pattern which must be set up to satisfy the mass balance could well be of fundamental importance in the overall energy balance of the corona-chromosphere. It is idle to speculate, however, before we have more detailed data on the state of the matter returning to the chromosphere after ejection as spicules into the corona.

Piddington (1972) has introduced a dynamic model including an upward hot-gas flux, which to some extent counteracts the downward conductive flux thus leading to a transition region which is thicker above the network boundary tending to concentrate the EUV emission there. Piddington's picture seems to require that the net movement of spicules, as observed in Hα, be downward in order to compensate for the upward mass flux which carries the thermal energy. This does not appear in keeping with observations.

6. Summary and Conclusions

Neither in terms of the observations nor of their analyses are we able to discern much agreement between the different workers. Even after many years of observations of chromospheric inhomogeneities we lack definitive data on the kinematics, an understanding of the line profiles, or even of the line intensities. In spite of recent progress, models still face difficulties in accounting simultaneously for the observed radiation in the EUV, radio, and visible, and about all that does seem accepted is that a homogeneous model is not adequate to account for the data in any one of these spectral ranges. Given the appearance of spectroheliograms in chromospheric lines this is scarcely an unexpected result.

The acquisition of EUV data at high spatial resolution will surely lead to major progress in this field. Such data have already allowed us to eliminate a class of models which places the EUV emission over the interior network regions; comparison between lines reflecting different heights of formation will be most instructive in clarifying the interaction between inhomogeneities and corona. Data in the EUV at high spectral and spatial resolution will help us to understand the evolution of the spicule gas after it is projected up toward the corona.

In the visible many (indeed, most) critical observations have been disputed. As a simple example, does cold spicular material return to the chromosphere, indeed, does the matter in a spicule really move upward? This, as indeed most aspects of spicule dynamics, is still an unsolved question.

References

Athay, R. G.: 1964, *Science* **143**, 1129.
Athay, R. G.: 1966, *Astrophys. J.* **145**, 784.

Athay, R. G.: 1970, *Solar Phys.* **11**, 347.

Athay, R. G.: 1971, *NATO Advanced Study Institute on the Physics of the Solar Corona.*

Athay, R. G. and Bessy, R. J.: 1964, *Astrophys. J.* **140**, 1174.

Athay, R. G. and Roberts, W. O.: 1955, *Astrophys. J.* **121**, 231.

Avery, L. and House, L. L.: 1969, *Solar Phys.* **10**, 88.

Beckers, J. M.: 1964, *Astrophys. J.* **140**, 1339.

Beckers, J. M.: 1968, *Solar Phys.* **3**, 367.

Beckers, J. M.: 1972, *Ann. Rev. Astron. Astrophys.* **10**, 73.

Beckman, J. E. and Clark, C. R.: 1973, *Solar Phys.* **29**, 25.

Brueckner, G. E. and Nicolas, K. R.: 1973, *Solar Phys.* **29**, 301.

Cram, L. E.: 1972, *Solar Phys.* **22**, 375.

Dubov, E. E.: 1968, in K. O. Kiepenheuer (ed.), 'Structure and Development of Solar Active Regions', *IAU Symp.* **35**, 255.

Dubov, E. E.: 1971, *Solar Phys.* **18**, 43.

Dupree, A. K. and Goldberg, L.: 1967, *Solar Phys.* **1**, 229.

Giovanelli, R. G.: 1967, in J. N. Xanthakis (ed.), *Solar Physics*, Interscience, London, p. 353.

Grossman-Doerth, W. and Von Uexkull, M.: 1971, *Solar Phys.* **20**, 31.

Grossman-Doerth, W. and Von Uexkull, M.: 1973, *Solar Phys.* **28**, 319.

Hagen, J. P., Swanson, P. N., Haas, R. W., Wefer, F. L., and Vogt, R. W.: 1971, *Solar Phys.* **21**, 286.

Jefferies, J. T. and Thomas, R. N.: 1959, *Astrophys. J.* **129**, 401.

Kanno, M., Tsubaki, T., and Kurakawa, H.: 1971, *Solar Phys.* **21**, 314.

Kopp, R. A. and Kuperus, M.: 1968, *Solar Phys.* **4**, 212.

Kulander, J.: 1968, *J. Quant. Spectr. Radiave Transfer* **8**, 273.

Kundu, M. R.: 1971, *Solar Phys.* **21**, 130.

Lantos, P. and Kundu, M. R.: 1972, *Astron. Astrophys.* **21**, 119.

Liu, S.-Y.: 1972, Thesis, University of Maryland.

Michard, R.: 1959, *Ann. Astrophys.* **22**, 547.

Nikolskii, G. M. and Platova, A. G.: 1971, *Solar Phys.* **18**, 403.

Noyes, R. W.: 1971, *Ann. Rev. Astron. Astrophys.* **9**, 209.

Noyes, R. W. and Kalkofen, W.: 1970, *Solar Phys.* **15**, 120.

Pasachoff, J. M.: 1970, *Solar Phys.* **12**, 202.

Pasachoff, J. M., Noyes, R. W., and Beckers, J. M.: 1968, *Solar Phys.* **5**, 131.

Pierce, A. K.: 1965, *Publ. Astron. Soc. Pacific* **77**, 137.

Sawyer, C.: 1972, *Solar Phys.* **24**, 79.

Simon, M.: 1971, *Solar Phys.* **21**, 297.

Simon, M. and Zirin, H.: 1969, *Solar Phys.* **9**, 317.

Suemoto, Z. and Hiei, E.: 1962, *Publ. Astron. Soc. Japan* **14**, 33.

Vernazza, J. E. and Noyes, R. W.: 1972, *Solar Phys.* **22**, 358.

Weart, S.: 1968, *Thesis*, University of Colorado (JILA Report No. 96).

Withbroe, G.: 1970a, *Solar Phys.* **11**, 42.

Withbroe, G.: 1970b, *Solar Phys.* **11**, 208.

Zirker, J. B.: 1962, *Astrophys. J.* **136**, 250.

DISCUSSION

(*Ed. Note*) The discussion of Dr Jefferies' review paper was essentially interwoven with his speech. The presentation of the discussion centers around the various parts of the review paper in which discussion arose and has been summarized by J.-C. Pecker.

1. The first point which led to some discussion was the *diagnostic of the observations in the visible* (Section 2.1.1.) and can be abstracted as follows:

(a) The first question concerned the accuracy of the $n_e T_e$ determinations:

Pecker: This analysis (Figure 1) relies on two types of data: (1) the measurements (i.e., D₃, Hα, K line, etc. as measured), (2) the assumptions used to compute those curves. The fact that they intersect in a rather vague area but not in a point means either that there is some dispersion in the measurements, or that there is some inaccuracy in the computation. In other terms, when you describe the 'spicules' (let us say 'features') you may have the tendency to say that there is some average model of the features, $T_e n_e$ being physically well-defined; or on the contrary, there is possibly no such thing as an 'average feature' (a feature being like a star in a cluster; no two stars in a cluster are identical; hence different features might be the same object, at different stages of its evolution). How do you see this dilemma?

Jefferies: The dispersion in the figure is certainly both a dispersion of the measurements (sometimes a considerable dispersion) as well as an uncertainty in the calculation of the emergent intensity.

Beckers: These curves are based on the average intensities of very diverse spicules. The Hα observation refers to a different spicule than the Hβ observations but this was all that was available at the time when I made the calculations shown in the slide. There is a study now on the way by Allissandrakis and myself to look at the spectrum of individual spicules and determine the intensity of a number of individual spicules and the changes with time and height.

Pecker: I am glad to learn that! From yesterday's discussion I got the impression that we are seeing trees of various species – some pine trees, some bushes, a lot of things – and if we are to see what the forest is we should not take an average tree!

Beckers: We carefully calibrate those spectra! Some of these curves cross close to one point, but some of the others do not; this is not only because of measurement inaccuracies but also, of course, because of theoretical inaccuracies.

(b) But can we determine *directly* n_e?

Athay: In order to predict the D_3 intensity, you have to know what the local ultraviolet radiation field is, which you don't know except in an average; so that's certainly one source of error! There is a way to get the electron density more directly – this is the way used by M. Makita (1972, *Solar Phys.* **24**, 59) where you deduce n_e from white-light images of the spicules. The result Makita obtained was $n_e \sim 1.8 \times 10^{11}$. We failed to improve on this data at the last eclipse (bad seeing) but the method is worth repeating.

(c) The second way of asking the same question is: "Is the same volume responsible for all emission lines?" This question, asked by *Brueckner*, was replied to as follows:

Jefferies: Yes, it is assumed in these computations. But that of course is one of the questions that one wishes to answer.

Beckers: I have assumed in my calculations *homogeneous* spicule temperatures and densities. If the different (N_e, T_e) curves do not agree with one another, the spread being more than could be tolerated in view of the observational and the theoretical inaccuracies, one has a basis for inferring an inhomogeneous spicule model. But if they do cross as in the figure (i.e. within the uncertainty of the observations and theory) one has no basis for an inhomogeneous spicule model. I do not claim that the spicule is homogeneous, but that we have insufficient data to derive an inhomogeneous spicule model.

Zirin: This method seems a very clever way of defining the physical regions where these lines are coming from; looking at the curves it is clear that one must expect a very high change in intensity, for some of these lines as a function of temperature and density; yet the spectra of these phenomena look monotonously the same. The changes with height do not seem to be well marked either. This seems to me to be pointing at the existence of 'plateaus' and stable physical conditions – the sort of stuff that Athay and Thomas talked about, years ago.

Jefferies: I don't think we should get carried away by this figure. I think that, as Beckers says, the way to do it is to look at individual spicules and see how a diagram of this kind works for those individual spicules. But Figure 1 is representing a broad range of different observations – averages, with a large variation from one observer to another; it is the best that could be done at that particular time.

(d) The numerical accuracy of the determination on n_e, T_e is also obviously a function of the choice of the parameters. This point was then discussed, as follows:

Jefferies: I think indeed that a sensible way of approaching the problem is to get diagnostic indicators that are more or less at right angles to one another (the old problem of surveying with á long baseline!). In our case, a line like Hα measures the electron density but doesn't give any information on the temperature. Some other lines, or line ratios, give, on the contrary, good information on temperatures but nothing on electron density.

Pecker: It would be very useful to assign an 'error-width' to each of those lines, derived both from experiment and from theory, but I gather it is pretty uncertain!

Jefferies: I think that's really pressing this curve beyond the point where it was intended – the method is the important thing.

(e) Whatever may be the basic correctness of the methodology, it relies upon theoretical computations. Are they foolproof? The question is raised by *Cannon*:

Cannon: The limitations to a model like this is that you are neglecting lateral transfer of radiation. I have very little faith or trust in what can be derived from such a diagram in those conditions!

Jefferies: I think you have the wrong idea of the model. At least above 5000 km, the spicule model is a simple slab. One could take a cylinder, but I doubt it would make any significant difference at the level of accuracy of interest here.

Cannon: I just don't think you can take a slab and put it in an atmosphere and say: "OK, let us bombard that slab with radiation, and compute its equilibrium." You have got to take into account how the slab interacts with the medium!

2. The second important point to be discussed was *the interpretation of self-reversal* (Section II, A, 2) in spectral lines. Are they the effect of interspicular medium, or are they basically part of the spicular spectrum, and reflect the trend of source functions within the spicule itself? Are there other interpretations, as suggested by *Wilson* – other spicules –? The discussion went essentially as follows:

Athay: It seems to me that it is incumbent upon those who propose the 'cool absorbing mechanism' to suggest some way that it can be done. I personally don't know how you can reduce the source function in Hα down to that point because it is determined essentially by the radiation coming from below, and I just don't see any mechanism for reducing either the radiation intensity or the source function.

Wilson: Isn't there a third possibility for the self-reversal? If you have a spicule forest and if you are observing them at the limb, you are also looking through some absorbing material of the same nature as the spicule on the way through to the particular one that you are seeing. This indeed seemed to come out in Michard's spectra of Hα that were shown yesterday by Rösch: above the limb, at heights of about 6000 km, there was no central reversal, but at 2000 km there was a clear central reversal and it appeared also that the width of the whole emission was increased*. That would indicate that, near the line center, you are not just looking through a whole forest of spicular material; this is forcing you to look further out into the wings of the line in order to see the intensities in that particular spicule.

Giovanelli: Another type of observation can help in the interpretation (though not make it unique): the central reversal does not show the structure that is present in the wings. In Michard's spicule spectra, the individual features can be traced right across the center of Hα at the greater heights, whereas low down the spicules are seen in the wings but not in the line center; the central absorption appears to be rather amorphous, any structures present being of much lower contrast. It appears from this particular type of observation that the absorption occurs outside the spicule – but there is no proof either that the absorption is distributed uniformly – or that it occurs in the tops of optically thin spicules (such as Peter Wilson has suggested).

Newkirk: Is this effect seen on eclipse spectra or exclusively on out-of-eclipse spectra? If this is not seen on eclipse spectra one would be inclined to believe that it is due to scattered light in the terrestrial atmosphere rather than the low layers of the solar atmosphere.

Giovanelli: I don't know the reply.

Athay: I don't see why an explanation based upon overlapping spicules in the line of sight doesn't hold. One could say that in the center of Hα the spicules have enough opacity so that you see a combination of spicules in the line of sight and you simply lose the structure. You pick it up again in the wings of the line where they become transparent. Then you don't need the interspicular material!

Zirin: It's clear experimentally in the pictures of Dick Dunn (scannings across the line). They show very clearly that you cannot see spicule structure in the limb band inside about 7/8 Å from the line center in the region between 0 and 3000 km, and therefore the spectrum that we are looking at, between + and − 7/8 Å is the 'general chromosphere' and not the spicular structure. We are seeing, when we go far off the center of the line, most of the spicules, and that is confirmed by looking higher up.

The discussion makes clear that the 'general chromosphere' is not a clear concept for everyone: as *Zirin* comments, if things are clear at about 4000 km, where we see a few spicules, what do we see at lower altitude: a 'general chromosphere', more or less homogeneous, or a 'forest of spicules'?

Jordan: If you take the available spicule statistics and calculate the areas that they add up to at the limb,

---

* *Rösch* made the following comment to the editor: "one slide showed Hα wider at 4000 km than at 6000 km, and without reversal; another one showed Hα and Hγ obtained simultaneously at about 2000 km: a strong reversal is observed in Hα not in Hγ, which demonstrates, as pointed out by Michard, that it cannot be due to scattered light. The reversal is fairly symmetrical and uniform along the slit, whereas the wings are not."

you find that they would merge completely at a height of about 3000 km. On the other hand, if you take EUV data and look at the height of the transition region, you come out to about 1700 km, so I think that what you are seeing from 1500 say to 3000 km is simply merged spicules because the EUV data show that there is no 'general chromosphere' there.

Giovanelli: I was just going to report that the central absorption in the strong lines of the Hα and H and K lines does not appear above 3000 km; these lines have flat tops at about 3000 km. It is only below this height that central absorption occurs, although it is strong at about 2000 km, and below.

Pierce: We are certainly seeing an interspicular homogeneous chromosphere by the observations of the lines of the rare earths: they have absolutely no spicular structure. Therefore unless you postulate separation of elements, the interspicular chromosphere must exist.

Thomas: I think you have to think very carefully about this geometry. I made a lot of calculations in about 1960 on this 'transition region' that you are talking about now. Only there we were getting it from the eclipse data. One of the things that was most important to the analysis was the optical thickness of Hα as a function of height. I tend to agree with Carole Jordan that in that region between 1000—2000 km and 1700–1800 km you have an enormous change in the opacity of any interspicular region in the Balmer lines, which is what you are looking at. So somehow it is needed to consider an effective bunch of spicules emerging. Now if that happens, then I think you ought to think very carefully about the diagrams that Cram put on the board yesterday; because if I interpret them in terms of velocity fields I have a choice – either we need rotation within a single spicule or we just need a radial velocity coming from the spicule inclination. And if I am looking at a bunch of spicules there is no reason that they are all inclined in the same direction. So along a tangential line-of-sight what I am looking at is several columns, whose line-of-sight component is sometimes up and sometimes down. This is not a single spicule, this is a super-position of many things. If I could persuade Cram to show some calculations that he made, I think you will see how difficult it is (even when you have these two emission peaks) to look at the velocities which you observe from these two peaks as compared to any 'central velocity' observed from the displacement of the central self-reversal – you get all kinds of things!

3. The third point which gave rise to a lively discussion was that of the *interpretation of the single-peaked features observed in the K line* (Section 2.1.).

This discussion being essentially already summarized in Jefferies' paper, we shall omit it here.

Another exchange followed later (Sivaraman, Jefferies, Athay) about the relation between double-peaked lines and supergranular cells: single peaks are generally found inside of a cell, not at the boundary.

4. The velocity fields in spicules as deduced from both disk and limb observations (2.2.) – can they be reconciled? *Grössman-Doerth* asks the following question:

Grössman-Doerth: Jefferies suggested that the fact that one does not see high velocities on the disk spectra may be attributed to a high gradient in the velocity. Maybe if you look at the disk you see further down, so you have perhaps lower velocities; if you observe at the limb, you see individual features and therefore higher velocities. Now, considering the fact that the spicules seen at the limb have an optical thickness of the order of unity across their diameter, should you not expect to see this gradient if you look along the axis from above, i.e. on the disk?

Jefferies: I don't really know that the optical thickness measured through the spicule at 5000–6000 km in the K line is in fact ∼1. I don't think we know enough about the physical conditions in these objects to have much idea what the optical thickness is either vertically (as on the disk) or horizontally (as on the limb).

Beckers: I would like to warn against assuming that the velocity of spicules outside the limb reaches 25 km s^{-1}. Those are the apparent motions, they are to some extent substantiated by Doppler shift measurements, but those Doppler shift measurements are generally made from strong lines. With weak line observations it seems that Doppler velocities could be much less, but I don't think that these have been measured yet.

An exchange between *Wilson, Jefferies* (essentially included in the paper, hence excluded here) points to the fact that when one observes a red peak, the absorbing region has not been necessarily shifted to the red, as 'conventional wisdom' would lead to (see *Cram*'s intervention, in Section II of the symposium).

5. The UV studies (Section 4 of the review) gave place to a discussion about the influence of optical depth effects upon the analysis of the UV lines, between *C. Jordan* and *Jefferies*. The meaning of the word

'spicule' in Vernazza-Noyes' analysis is also discussed: Jefferies' reply to this question essentially included an additional doubt put on the sentence "are identified with spicules" in Section 4 of this paper, in the paragraph starting by "the inhomogeneous structure..." etc. *Brueckner* added the following comments to the discussion:

Brueckner: I would like to make a remark here about this problem. In one of the recent issues of *Solar Physics* we published a paper about the 1970 eclipse. We can resolve the problem by assuming that the cool spicule center is surrounded by a transition zone similar to the one which all researchers have assumed since the transition zone became fashionable. And then when you integrate along the line-of-sight, using the optically thin lines like C IV, Si III, N V, and plot their intensities as the Moon covers the limb, you get a good agreement between the observed change and the prediction – if you assume Jacques Becker's spicule count as a function of height. The eclipse data can never be explained by a homogeneous transition shell around the Sun. I want however to warn about these observations: they are not very accurate.

6. Then some discussion about the final models took place (Section 5 of the review paper). *Athay* and *Beckers* do not seem to think that the discrepancies noted by *J. T. Jefferies* between two types of models (such as Kanno's and Beckers') are so difficult to remove. The Kopp-Kuperus model is criticized by *Gabriel*, as follows:

Gabriel: In the Kopp and Kuperus model, the thick transition region they obtain in the network center is the result of an error in their treatment. They choose as a 'typical' path a singular field line through a neutral point, which is not at all typical of most of the network center region. In addition to this comment, I have two criticisms to make of the model you put forward:

(1) The suggestion that the network center transition region is very thick and rarefied is not acceptable on the grounds of pressure balance. I would prefer to assume that it is normal density, but much thinner than in the network regions. This is possible since it is closely parallel to the magnetic field, which inhibits thermal conduction across it.

(2) If the upward flow in spicules of 20 km s^{-1} is balanced by a downward flow in the same region of coronal temperature gas, then the downward velocity has to be $100 \times$ higher or 2000 km s^{-1}. This is clearly not correct. Even if the area occupied by the downward flow is allowed to increase, this still leads to very high and measurable velocities.

Some attention was paid (*Giovanelli, Thomas, Jefferies*) to the exact location of the 'base of the transiton regions'. It is around 12–1500 km, where n_e is equal to about 10^{11} and above; T_e rises quickly from about 9000° to about 20000°, remains near 20000° for a few hundred kilometers then rises quickly to coronal values.

During the general discussion which followed Jefferies' presentation *Giovanelli* introduced a question as to the homogeneity of the region below 1500 km, as such:

Giovanelli: I would like to show a slide dealing with the homogeneous region – I think the top of the homogeneous region – we'll just see how homogeneous a homogeneous region is. This slide shows two photographs, one obtained at H$\alpha + \frac{1}{4}$Å, the other at H$\alpha - \frac{1}{4}$Å. There is a little spot group (just a day or two old) and shortly outside, the region is rather undisturbed. Over the center of supergranules, one can see a *granular* region. The grains also appear between the fibrils that stretch out either from the active region or from the supergranule boundary. These grains are clearly at a lower level than the overlying fibrils. They are not photospheric grains, but definitely chromospheric! An exchange (*Gabriel, Thomas*) deleted here because of duplication with what has already been expressed shows that the important problem of mass balance is not simple!

The problem of the 'inhomogeneities' of the 'homogeneous' region is the object of some interventions (*Pasachoff, Pecker, Athay, Schmidt, Kundu*) from which the following can be extracted:

Pasachoff: A very quick calculation I made is the following: Taking a single spicule as a cylinder 500 km across and 5000 km high, multiplying that by some mean number of spicules on the Sun (approximately half a million) we come out approximately the same surface area on the side of spicules as there is in between.

Kundu to Jefferies: Shouldn't your model show radio waves – if it is millimeter waves that you are referring to – as coming from below the transition zone?

Athay: I want to make a comment about the homogeneous model below 1500 km. Nearly everybody agrees that the structure is inhomogeneous – you can see all kinds of structures in all kinds of lines. Somehow

we have to reduce the problem to some kind of quantitative estimate of the importance of those inhomogeneities. One way of doing this is to look at different models of the chromosphere constructed from
different kinds of data. We are in the fortunate position now of being able to construct chromospheric
models from optically thick data, seen at the center of the disk, from optically thin data, and spectral
features observed at the limb during eclipse; we have radio data, continuum data, line data – a great variety
of data. I would argue that if the inhomogeneous structure is really an essential feature of the model we
ought to see different models in different spectral features. We ought to see a different model in the optically
thick features than we see from the thin features, and we ought to see a different model on the limb than we
see in the center of the disk. The point of this to those of us engaged in model-building is that, as time goes
on, the different models are all converging into a more-or-less single type of model or very close to it.

The new models, based primarily on optically thick data observed at the center of the disk, are in very
close agreement with models based on optically thin data (such as the one Dick Thomas and I derived
fifteen years ago, from limb data). Until we find some clear discrepancy between the models, from the
center of the disk out to the limb, we have no basis for saying that the inhomogeneous structure is affecting
them in a pronounced way.

FURTHER OBSERVATIONS OF THE STRUCTURE OF THE CHROMOSPHERE-CORONA TRANSITION REGION FROM LIMB AND DISK INTENSITIES

W. M. BURTON, C. JORDAN, A. RIDGELEY and R. WILSON*

SRC Astrophysics Research Division of the Radio and Space Research Station, Culham Laboratory, Abingdon, Berkshire, England

Abstract. Further observations of limb and disk intensity ratios of emission lines in the EUV solar spectrum were obtained on a Skylark rocket flight on 5 August 1971. Analysis of the data has shown that the observations support the existence of a steep rise in temperature in the transition region between $T_e \sim 6 \times 10^4$ K and 3×10^5 K. The average absolute height of the transition region above the visible limb has been measured with a greater accuracy than previously possible and is 1700 km $\pm$ 700 km. An independent method using arc-length measurements of spectrum lines gives an absolute height of 2100 km $\pm$ 850 km. Emission from lines which are optically thick in spicules is observed to extend to heights of 10000 km above the transition region. The observed decrease of Fe II emission with height is consistent with current spicule statistics.

The limb to disk intensity ratios of pairs of lines from a given ion, one of which lies at a wavelength shorter than the head of the Lyman continuum and the other at a wavelength longer than the head of the Lyman continuum, have been studied, and the observed absorption at the limb by the Lyman continuum can be explained by spicular material with $N_e = 1.4 \times 10^{11}$ cm^{-3} and $T_e = 1.1 \times 10^4$ K.

The abundance of iron has been derived from the chromosphere emission lines of Fe II. This value agrees, within the experimental accuracy of a factor of two, with the value of $N(\text{Fe})/N(\text{H}) = 4 \times 10^{-5}$ found from previous analyses of photospheric and coronal lines.

Reference

Burton, W. M., Jordan, C., Ridgeley, A., and Wilson, R.: 1973, *Astron. Astrophys.* **27**, 101.

DISCUSSION

After C. Jordan's paper, a lengthy discussion took place involving mostly the point of knowing whether one can assume the chromosphere below 1700 km to be homogeneous or not, and what forces are commanding the spicules to be in equilibrium – more or less – with the interspicular medium above 1700 km.

Sturrock raised the two following points:

Sturrock: I would like to make two comments:

(1) The thickness of most of the region around the spicule depends very much upon what you may assume on the conduction coefficient. If you have a model of the spicule's cool gas moving along the magnetic field the heat conduction is transverse to the magnetic field, and the thickness of the spicular transition region would be very much less than one would otherwise expect.

* Dept. of Physics and Astronomy, University College London.

(2) The models which you (C. Jordan and J. T. Jefferies) have drawn show spicules going down to the photosphere. Of course gas is being shot up, so some force is moving the gas up. This may mean that there is a bottom to the spicule, so that it is like a bullet of gas being pulled up. I think there are two possible forces which can move the gas upwards. One is the magnetic stress, and the other one is gas pressure. If the latter is true, it means that below the cool gas in the spicule there is some very hot gas pushing it up. This is possible if there is a mechanism for converting magnetic energy into heat. It might possibly raise the temperature, locally, to 10^6 K and the pressure increase then shoots up the mass of cool gas like a bullet. If this is the case then there will be a thin transition region on the underside of this mass of gas which one must incorporate into the models.

Jordan: We can't say anything about the spicules below 1700 km.

Thomas noted that the Jordan description fits the rather old model derived from the analysis of the H continuum. He commented that there must be heating near the bottom of the transition region, where T_e rises quickly from 9000 to 20000°. Such behavior must hold also across the spicules. What fixes the temperature variation across the spicule is the heating mechanism, and the behavior of the Lyman continuum opacity.

Schmidt: I maintain my statement that only magnetic stresses really can provide the necessary forces, not the pressure gradients.

Meyer: I just want to ask a question about the transition zone at the surrounding of the spicule. I have the impression that what Dr Thomas said was that you would have there the normal transition zone and you would also have the interspicular region, whereas in the remarks that Dr Schmidt made yesterday and which Dr Sturrock just brought up again, would indicate that the transition zone around the mantle of the spicule should be very small and it would be hard to believe that there could be any contribution of that to the interspicular medium. Would you agree with that statement?

Jordan: I don't mind how thin it is around the spicule, there is no inconsistency with having it any thinner.

Thomas: Be careful what you are talking about; you are not talking in terms of geometry, you are talking in terms of opacity! So the need here is to balance whatever energy input there is against the behavior of temperature and the ground-state population of hydrogen, i.e., against the rate of fall-off of Lyman opacity. But I have to couple the opacity with temperature and all I say is that it is not going to be uniform across the whole spicule.

Brueckner: In a paper (which was just published two months ago) we assumed a very similar model and tried to fit the eclipse observations into a transition zone like the one assumed by Jordan. The statistics of Beckers were used to derive the transition zone intensities as function of height as observed from the eclipse. Now we agree that a large part of the emission has to come from these spicules – the question is open whether the layer under the spicule is a very thin layer with very high temperature gradients, and we have here one point where we do not agree with Carole Jordan's results. At this altitude of approximately 1700 km, we would observe, during the eclipse, a jump in the integrated intensity when the Moon moves over the 1700 km altitude lines; we have tried everything that we could to find this jump and we cannot find it! It is a very smooth curve, the intensity of these transition zone lines goes very smoothly out – there is no jump at 1700 km. . . .

The discussion ended there – more or less – on a disagreement on the eclipse interpretation; *Thomas* noted that the second derivative of the eclipse curves is involved, which makes Brueckner's point apparently rather weak. *Brueckner* then stated that ATM observations might solve the apparent dilemma of the abrupt (or not) change of character of the photosphere above 1700 km.

NON-STATIC STRUCTURE OF THE CHROMOSPHERE-CORONA TRANSITION REGION

PHILIPPE DELACHE

Observatoire de Nice, Nice, France

Abstract. The emission measure Ne² dh, which is so useful in reducing XUV as well as radio data, is redefined as $f(T)\,dT$ where $f(T)$ is called 'thermal emission measure'. Theoretical predictions for $f(T)$ on the basis of a simple, one dimensional, steadily expanding atmosphere are presented. Depending upon the boundary conditions, and essentially upon the mass flux, two very different behaviours show up.

(1) With a mass flux compatible with an extrapolation of the solar wind flux, $f(T)$ would correspond to a transition region controlled by a constant conductive flux. It can be fitted, as was shown e.g. by Athay (1966), with the XUV observations of the integrated disk for lines formed at temperatures higher than about 2×10^5 K.

(2) But only with a mass flux enhanced by a factor of 50 to 100, can we interpret the radio spectrum and the XUV observations of lines formed at temperatures less than 2×10^5 K.

The suggestion is made that a single model can reconcile both behaviours: below 2×10^5 K the emitting region is squeezed so that it covers 1–2% of the surface that it occupies at higher temperatures where all structures begin to merge, filling the whole corona. This variation with height compares with usual models derived from spatially resolved observations; as usual the underlying magnetic field is expected to support this channelling.

DISCUSSION

Kundu: I'd like to make a comment about the radio model of Lantos and myself based upon center-to-limb brightness distributions at 1-, 3-, and 9-mm wavelengths. It will also answer the remark made by Dr Jefferies to the effect that the radio observations are not in accord with optical models. Indeed, Lantos and myself have shown that using the model of Avery and House on the spicular density and temperature and using a much higher relative surface area occupied by spicules (40–50%) below the transition region, the millimeter brightness distributions can be explained. It seems that the magic parameter is the relative surface occupied by spicules which increases rather sharply from about a few per cent to about 40–50% below about 2000 km – the transition layer.

Schmidt: I think what Dr Kundu has said is completely right but it has to be separated completely from solar wind fluxes, which concerns much more external layers.

Zirin: In response to both these remarks, I would like to point out that although one can perhaps match the lack of millimeter limb brightening, the point that Simon and I made some years ago is that we have the same problem through the entire radio range, i.e. the lack of limb brightness at centimetric and decimetric wavelengths. In that range also we have a lack of limb brightening and we do not have a model to explain it. It may be that the amount of hot material is considerably less than the existing models attribute, and that leads me to this business of the hundredfold discrepancy with the solar wind. I find it very difficult to understand the existence of such a discrepancy when we hardly know really what spicules are, and I don't see any point in getting upset about it.

R. Grant Athay (ed.), Chromospheric Fine Structure, 91. All Rights Reserved.
Copyright © 1974 by the IAU.

ON THE DESIGN OF CHROMOSPHERIC MODELS

U. GROSSMANN-DOERTH

Fraunhofer Institut Feiburg, Germany

Abstract. The basic features of the so-called 'cloud-model' (Grossmann-Doerth and von Uexküll, 1971) are discussed. The model permits us to derive approximate values of certain properties of the elements of the chromospheric fine structure by comparing the observed profiles of strong Fraunhofer lines with a simple theoretical formula involving four free parameters. The method is only applicable if two basic conditions are met: (1) The absorption coefficient varies with wavelength according to a gaussian with a broadening parameter that is independent of height, (2) the variation of the source function S_ν with height is sufficiently small so that

$$S(\lambda) = \int_0^\tau S_\nu(z)\, e^{-t}\, dt \left(\int_0^\tau e^{-t}\, dt \right)^{-1}$$

may be replaced by a constant. The validity of these assumptions, in particular the latter one, is discussed. It is shown that in the case of Hα there are reasons to believe that the assumption is valid provided the features are not very close to the limb ($\sin\theta = 0.9$ may constitute the border line). There is also some evidence to suggest that the cloud model will not work for the Ca II K-line.

In the second part of the paper the presently used methods of constructing multi-component models of the chromosphere are considered. It is proposed to apply the radiative transfer theory to selected parts of the chromosphere by comparing the model predictions with spatially resolved spectra and filtergrams instead of computing spatial averages and comparing those with integrated spectra.

Reference

Grossmann-Doerth, U. and von Uexküll, M.: 1971, *Solar Phys.* **20**, 31.

DISCUSSION

Thomas: I am sorry to be negative on this paper. Jack Zirker and I looked at this thing about twelve years ago and we found then that Grossmann-Doerth's method cannot work.

Giovanelli: In the photograph that I showed earlier this morning, there were grains which have similar appearances to, and are rather larger than photospheric granules. These are in the centers of supergranule cells while there are other structures, the fibrils, that extend over them. The fibrils fade out towards the centers of the supergranule cells. I would expect the fibrils to be well described by a cloud model, but not the grains.

Grossmann-Doerth: I think I have not made myself very clear. I did not mean to suggest that we could apply the cloud model to the interior of supergranule cells; on the contrary, I said the cloud model is not applicable there.

PART IV

MOTION AND EXCITATION IN THE CHROMOSPHERE

MOTIONS OF CHROMOSPHERIC FINE STRUCTURES

EDWARD N. FRAZIER

Aerospace Corporation, Los Angeles, Calif. 90009, U.S.A.

Abstract. This review concerns itself with the measurement of the effects of chromospheric motions and the diagnosis of those motions themselves over approximately the last ten years. The different types of observational techniques are described. The different size regimes of motions are reviewed and their possible effects on observable quantities are discussed. The different types of motion in the lower chromosphere are reviewed, including microturbulence, 'layered' motions, and observations and interpretation of asymmetries in the core of the Ca II K line. The observation of motions in Hα mottles on the disk and spicules on the limb are reviewed, from the standpoint of both line profile analysis and broad-band measurements. The interpretation of these motions and attempts to construct empirical models of chromospheric fine structures are discussed.

1. Introduction

A review of motions of chromospheric fine structures is a little difficult because motion is such a fundamental property of the fine structures. In fact, the most important single aspect of these fine structures is that they are dynamic, not static phenomena. Thus, motions should really be discussed only as an integral part of the general problem of fine structures, and indeed, we have seen discussions of motions being raised throughout this symposium in relation to every other aspect of the overall fine structure problem, as they should have been. However, for purely organizational reasons, this review will be limited to the observations of motions *per se*; how they are observed, what the results are, where the discrepancies and unresolved questions lie, and what should be done to further clarify the state of motion of the chromosphere.

I shall start with a brief summary of general observational techniques and a crude attempt to provide an organizational framework for considering the various phenomena according to their sizes. Consideration of the specific problems themselves are grouped according to that most unnatural separation; disk and limb. Finally I attempt to rectify that sin and provide as unified as possible a picture of chromospheric motions.

2. Techniques

2.1. OBSERVATIONAL TECHNIQUES

One should start by realizing that there are four main independent variables, or coordinates, each of which are important in the measurement of motions; λ, x, y and t. Historically, observations have fallen into three broad categories which can be distinguished by the coordinates that are well resolved. These can be seen in Table I. It is important to note that historically the analysis of the data in each category has been specialized and developed almost independently of the other categories. Only recently have there been observational and analytical techniques developed which tend to eliminate the artificial boundaries between these categories.

R. Grant Athay (ed.), Chromospheric Fine Structure, 97–135. All Rights Reserved.

EDWARD N. FRAZIER

TABLE I

Categories of instrumental techniques

Category	Resolved coordinate	Instrumental example
'Profile-blind' mapping	x, y, t	Filtergrams, spectroheliograms
Low resolution line profiles	λ	Photoelectric spectra
High resolution line profiles	λ, x	Photographic spectrograms
	$x, y, t, (\lambda)$	Tunable filters
Improvements or combinations	λ, x, t	Time-sequence spectrograms
	$\lambda, x, y, (t)$	'Stepped' spectrograms
	λ, x, y, t	Image slicer w/spectrograms, Multi slit spectrograph

The first category is what might be called 'profile blind' mapping. Filtergrams and spectroheliograms are the prime examples. Such observations are poorly suited for the deduction of local physical variables, because they contain no information about the line profile, but they are an excellent means of *discovering* dynamical phenomena and identifying their general nature. For example, such phenomena as the 5-min oscillations, the supergranulation and Moreton waves were discovered with this technique (Leighton *et al.*, 1962; Athay and Moreton, 1961), so we may expect profile blind mapping to remain at the forefront (and *only* at the forefront) of further study of dynamic phenomena.

Low (spatial) resolution profiles have been virtually forced upon us by those who recognized the need for first achieving a first order theory of the structure and radiation of the chromosphere. Since there is no real instrumental requirement that maintains this separate category, one should expect that when the first order chromosphere is understood this category will be allowed to pass into history. Indeed, since most of the recent articles on chromospheric radiation have dealt with inhomogeneities in one way or another, we might be able to say that this time has already arrived.

The category of high (spatial) resolution profiles has had a long history observationally but the analysis of the data has lagged behind, awaiting the completion of category two. An important observational improvement to this category is the use of reflecting entrance slits viewed through narrow band filters. This enables one to at least identify which type of spectral feature is under study.

There are a number of technical improvements which lead to combinations or extensions of these categories. They are listed in Table I without detailed comment. There are two general comments that should be made however. First, one can see that it is technically quite possible to obtain data that is a function of all the principal independent variables. This should free us from the need to tailor the analysis technique to the instrumental limitations. Secondly, this 'fully generalized analysis' has not yet really come to fruition. Why not? Simply stated, because no one seems to know what to do with all that data. The volume of data is massive and the data reduction problem is serious. This can be solved however with enough money and effort. What is really needed is a sound theoretical framework within which one can analyze all these fluctuations.

2.2. SIZE SCALES AND REGIMES

Continuing in the same spirit as in the previous section, I would now like to categorize the various possible motions according to their sizes, since their size largely determines the visible effect that they will produce. We probably are all familiar with the three classical size regimes; microturbulence, unresolved macroturbulence and macroturbulence. The dividing boundary between these regimes is the mean free path of a line photon, L, and the instrumental resolution element, R, respectively. (cf. de Jager, 1959, p. 93). Since the term macroturbulence connotes an additional property to mass motions which no longer seems appropriate, it would seem preferable to change the names of the last two categories to simply 'unresolved mass motion' and 'mass motion', respectively. Note that the boundary between these last two categories refers only to a horizontal plane. In the vertical plane we have a different situation (referring only to disk observations now). The intermediate regime becomes what might best be called 'layered motion', and the boundary between this regime and that of mass motion becomes an ill-defined quantity which is probably best called, 'the thickness of a line forming layer', Δz. So there are four regimes, depending on the horizontal and vertical size scale of the motion, and two of these regimes (layered motion and unresolved mass motion) overlap each other considerably. Only in the simplest regime – mass motion – is one justified in making a one-to-one correspondence between Doppler shift and line-of-sight motion. To infer velocities in the other regimes, one must postulate a complete model of the moving elements, calculate the emergent radiation (taking into account the radiative interaction between the elements) and adjust the model parameters to fit the observed profiles. Since this is a tedious precedure, and does not guarantee uniqueness, it is usually ignored.

One can make rough estimates of which regimes contain chromospheric motions. For strong chromospheric lines, $L \approx 10^2$ km, for most spectrographs, the best R is about 10^3 km. Δz is ill-defined, but from the results of Kulander and Jefferies (1966) and Athay (1970) we take $\Delta z \approx 5 \times 10^2$ km. These numbers should not be taken too seriously for several reasons. The boundaries are not all sharp; not only are they very broad, there is presently no theory to tell us how broad they are. Furthermore, these boundaries vary considerably as a function of the instrument used, the seeing, the line used and even the feature viewed. Despite these uncertainties, it is readily apparent that *all* four regimes will contain significant chromospheric motions.

First, let us consider microturbulence, and what its status is in the chromosphere. First, should we expect it to exist? Yes. Theories of acoustical generation in the convection zone predict the possibility of sound waves with wavelengths as short as a kilometer. Crude estimates based on viscous dissipation of turbulent eddies predict eddy wavelengths as short as a meter! We don't know for sure if it is there, but we really should expect it. Secondly, can we observe it in the chromosphere? Not in the classical sense. By original definition it is observable only by its effect on equivalent widths, but the curve of growth method and its descendant, the Goldberg-Unno method are inappropriate for chromospheric studies. However, the concept is preserved by the study of synthetic line profiles through the use of the usual equation in computing Doppler broadening:

$$\Delta\lambda_D = \frac{\lambda_0}{c}\left[\frac{2kT}{m} + \xi_t^2\right]^{1/2} \tag{1}$$

ξ_t is the microturbulent velocity and is the parameter specifying one component of a Maxwellian velocity distribution: $f(\xi) = (1/\pi^{1/2}\xi_t)\exp(-\xi^2/\xi_t^2)\,d\xi$. Thus, microturbulence is treated by assuming that there is no formal difference between turbulent motions and thermal motions, a situation which has not changed in nearly forty years.

The intermediate regimes, particularly unresolved mass motion, have frequently been lumped in with microturbulence under the heading of 'non-thermal motions'. This really should not be done because the use of Equation (1) assumes a gaussian distribution, which is clearly invalid for the larger scale motion. Work has proceeded on these regimes recently however, through the study of multi-component models. Here, specific models of differentially moving elements have been postulated (or inferred from observations of the size and amplitude of inhomogeneities) and the radiative transfer through such models has been solved explicitly. This technique has been applied to the analysis of both unresolved mass motions, through the use of two and three-column models, and to layered motions through the use of models with differentially moving lauers. The results are still very numerical and too specific to constitute a general analysis technique of these intermediate regimes, but the progress is encouraging. The specific models will be discussed in the following sections.

A word about asymmetries, which seem to be fairly ubiquitous companions to velocity studies. Velocities alone do not produce asymmetries. Asymmetries are produced when other local physical variables fluctuate in correlation with velocities. Since these variables obviously are a function of height, layered motions almost invariably produce asymmetries. Almost by definition, resolved mass motions should not produce asymmetries. The other two regimes, however, are not so clear. Asymmetries may or may not be present depending on the details of the temperature-pressure-velocity relationships. This calls for a much broader attack on the problem than the mere consideration of the size and amplitude of the motions.

3. Disk Phenomena

3.1. MICROTURBULENCE

We start the study of disk phenomena with a brief review of the first order chromosphere, and this means microturbulence. The most logical place to start is with low resolution profiles. This approach measures the total power contained in *all* the regimes and calls it microturbulence or 'non-thermal motions'.

3.1.1. *Low Resolution Studies*

The most appropriate starting point is probably the review by de Jager (1959) in which all the then available values of ξ_t are collected. There it was shown that all measurements were in reasonable agreement with a linear rise of ξ_t from about 3 km s^{-1} at $h=0$ to about 15 km s^{-1} at $h=3000$ km. This result includes a number of different observations, techniques, and resolutions. Can we believe it? I think that if you keep in mind that microturbulence is essentially a category for 'leftovers' you can see that this result is useful as an upper limit. Our job now is to see how far down we can pull this curve.

There is a fairly reliable measurement in the region of the temperature minimum. Canfield (1969, 1971) measured the half-widths of rare Earth lines formed in the wings of the H and K lines. This is a rather pretty measurement because (1) he showed that they are formed at the same height as the H and K wings, (2) the lines are weak, so the measured Doppler widths are the actual Doppler widths, (3) being so massive, these ions have a thermal speed much less than the turbulent speed. His results can be seen in Figure 1 (Canfield, 1971): ξ_t is about 2 km s^{-1} throughout the temperature minimum. This is still an upper limit, since this refers to power from all the regimes.

Higher in the chromosphere there is a result from analyses of the O I λ 1304.9 Å line. Athay and Canfield (1969), using a model with a microturbulence of 6–7 km s^{-1}, had predicted that this line should show a strong central reversal. However the observations show it to be nearly flat-topped. Jones and Rense (1970) concluded that, in addition to the microturbulence, a macroturbulence (unresolved mass motion) was necessary for the model to reproduce the flat top. The best fit was achieved with unresolved mass motion of 6.8 km s^{-1} (Figure 2).

Throughout the chromosphere one can infer ξ_t from synthetic models whose parameters are adjusted to fit various sets of observational data. Fitting the profiles of the Ca II H and K lines is the classic problem, and the homogeneous case seems to have reached the ultimate in the review paper by Linsky and Avrett (1970). To reproduce the wide Ca II profiles a turbulent velocity is incorporated into the model as shown in Figure 3. A much more ambitious model (Vernazza *et al.*, 1973) seeks to fit a number of hydrogen, carbon and silicon lines as well as EUV and microwave continuum data. In this model a similar turbulent velocity is derived (the Vernazza *et al.* turbulence is slightly higher in the upper chromosphere).

It should be noted that, for consistency, both these models incorporate a 'turbulent pressure' term:

$$P_{\text{total}} = P_{\text{gas}} + P_{\text{turb}}, \qquad P_{\text{turb}} = \tfrac{1}{2}\varrho\xi_t^2. \tag{2}$$

In the mid-chromosphere P_{turb} is an order of magnitude smaller than P_{gas} or less, and so has a negligible effect on the model. But at heights of about 2000 km, P_{turb} becomes almost as great as P_{gas} and thus increases the scale height in this region. The result of this is to raise the altitude of the transition zone. The magnitude of this altitude shift is 300 km in the Vernazza *et al.* model. So it appears that the inferred existence of high

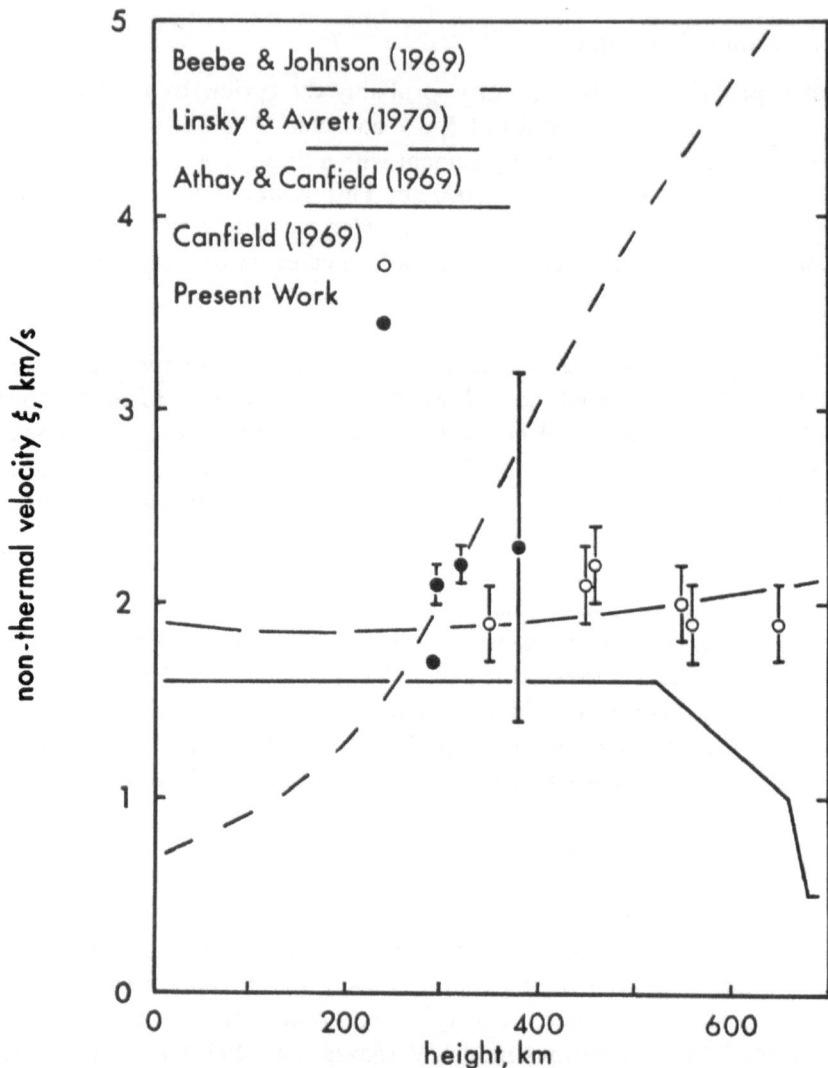

Fig. 1. The 'non-thermal velocity' as determined from the half-widths of rare Earth lines in the low chromosphere (from Canfield, 1971). The dashed lines refer to models used in synthetic K line profile calculations.

turbulent velocities (greater than about $\xi_t \approx 7$ km s^{-1}) in the upper chromosphere can be tested by observations of the altitude of the transition region, if sufficient height resolution can be achieved.

3.1.2. Inhomogeneous Models

There have been a number of two or three component models of the chromosphere, all of which employ microturbulence as a free parameter. All of these models suffer from a profusion of adjustable parameters, and therfore cannot be claimed to be

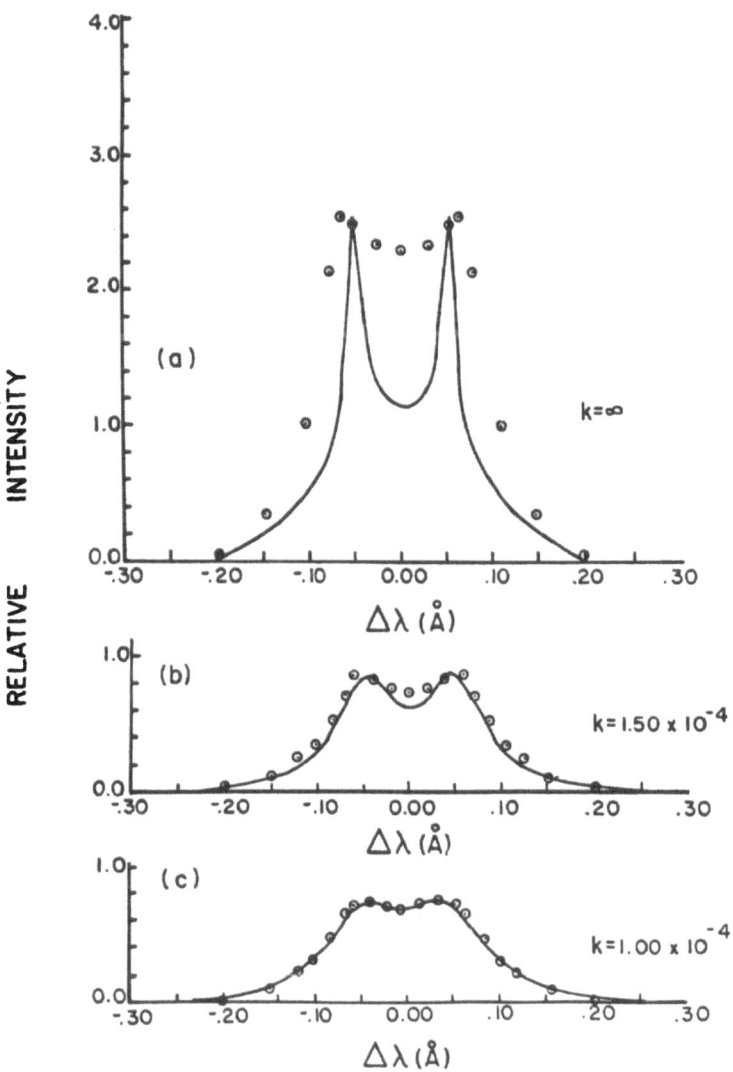

Fig. 2. The effect of unresolved vertical mass motions on the profile of the O I $\lambda 1304.9$ Å line (from Jones and Rense, 1970). The circles represent the observed relative solar profile; the solid lines are the calculated profiles.

$$k = \frac{1}{2\xi^2}. \text{ For } k = 10^{-4}, \; \xi \approx 7 \text{ km s}^{-1}.$$

unique. Beebe and Johnson (1969) and Beebe (1971) build a purely geometrical two component model for the emission of the K line and, after settling on a T_{min} of 4200 to 4300 K, infer reasonably small (depth and space dependent) values of ξ_t (Figure 4). Wilson (1970) builds a three component model with very large values of ξ_t. The need to consider explicitly the radiative interaction between the various components of such models has led to a series of papers in which a method of solving the two-dimensional equation of radiative transfer has been developed (Cannon, 1970, 1971a, 1971b; Cannon and Rees, 1971). This method was applied to the interpretation of

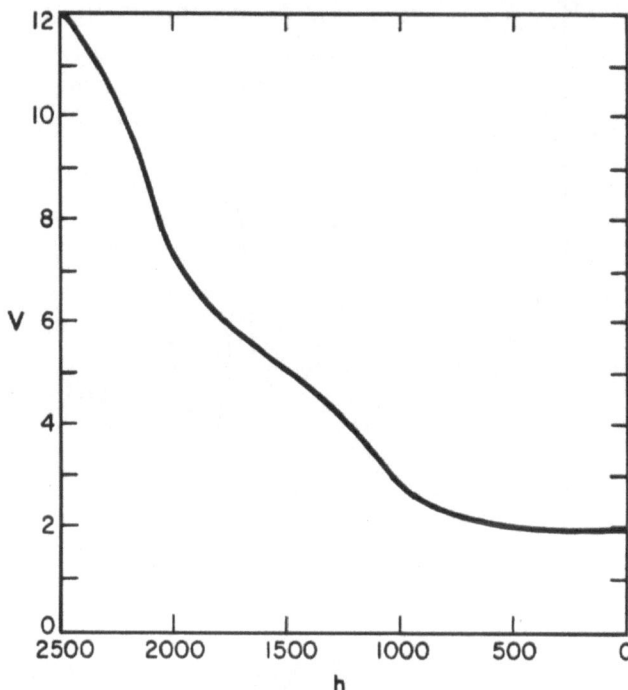

Fig. 3.   The microturbulence used in a homogeneous model to synthesize the profiles of the calcium lines
(from Linsky and Avrett, 1970).

observed fluctuations in the low chromosphere in terms of local density temperature
and velocity fluctuations.

Cram (1972) has computed a self-consistent two component model for the K line
in the cell interiors only. The two components correspond to (1) a background or cell
component with a monotonically decreasing source function and (2) a K_2 bright-point

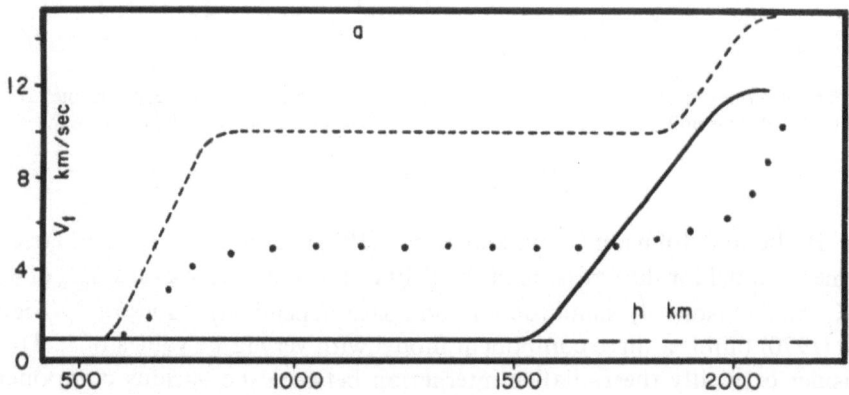

Fig. 4.   The turbulent velocities used in a two-component model of the K-line profile (from Beebe, 1971).
The dotted line represents the turbulence in the wall or network component, the solid line is the turbulence
in the cell component. The dashed lines represent two other test models.

component. With the explicit inclusion of layered motions, reasonable agreement with observations is achieved with $\xi_t = 5 \, \text{km s}^{-1}$ in the 'bright' component and $\xi_t = 7 \, \text{km s}^{-1}$ in the 'cell' component.

All these models have the common advantage of easily reproducing the limbward behavior of the K line (both darkening and broadening) in spatially averaged profiles. However at disk center, the individual components of these models should be compared directly with high resolution spectra. It makes a little sense to average several synthetic profiles, average spectra over a large area, then compare the two.

3.1.3. *Summary*

In summarizing the actual results on microturbulence the major difficulty lies in separating it from the other regimes, principally unresolved mass motion on the one side, and simple temperature effects on the other side. In separating microturbulence from unresolved mass motions, I have been able to find only one possibility; the analysis of central reversals (or absence thereof) in the chromospheric lines other than H and K. The example of the O I λ 1304.9 line quoted above is the prototype. It appears that the further addition of microturbulence cannot remove a predicted central reversal and that unresolved mass motion is necessary to reproduce the flat topped profile. Separation of ξ_t and T_e still appears to be done best by fitting a model to as wide as possible a set of observational constraints (e.g. H and K profiles, EUV and microwave emission). This is a rather arbitrary process though, and it is an open question whether or not it can survive the inclusion of inhomogeneities into the problem.

There are conflicting indications of spatial variations of ξ_t. Beebe (1971) finds a higher ξ_t in the network while Wilson (1970) requires a lower ξ_t there. Mein (1971), in his analysis of the infrared Ca II lines finds no change in ξ_t from cell to network. This question will certainly become a very important one as the fundamental difference between cell and network regions becomes more widely appreciated.

It is tempting to collect all the inferred values of ξ_t on one figure and draw a smooth curve through all the points, but that is premature now. Instead, Figure 5 just illustrates schematically the broad path where the various values are generally found. The sound speed is included for comparison. About all that can be said about the numerical value of microturbulence at this time is that is is very roughly Mach 0.2 to 0.3 throughout the chromosphere. Aside from the well-known defects in the theoretical foundation of microturbulence, there seem to be two main reasons for our present lack of accurate knowledge of microturbulence: (1) The spatial resolution of the various techniques varies greatly (e.g. what appears as microturbulence in one measurement is actually resolved mass motion in another measurement). (2) The inaccuracy introduced with the assumption of a homogeneous chromosphere is poorly known, and the possibility of spatial variations of ξ_t is not yet settled.

The idea of anitsotropic turbulence has been around for a long time because nearly all Fraunhofer lines broaden toward the limb. The interpretation of this broadening is entangled to a certain extent with the height variation of ξ_t, but primarily with the

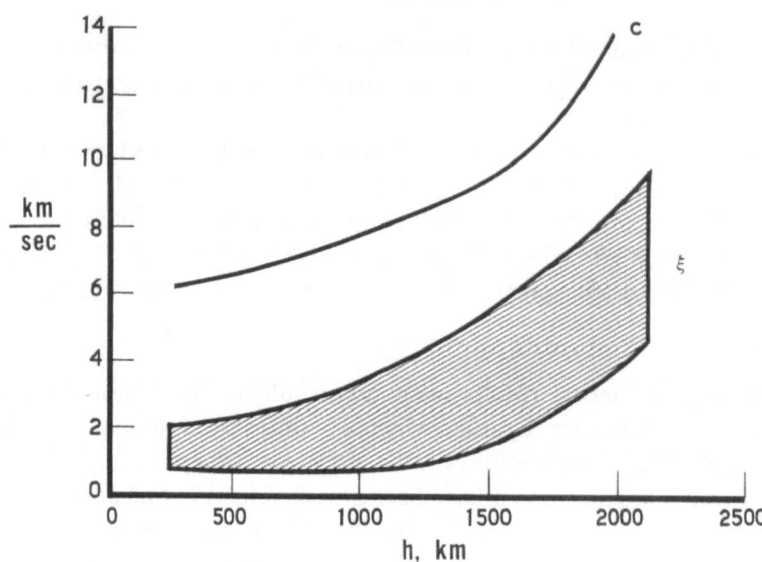

Fig. 5.   Schematic summary of the various calculations of ξ_t in the chromosphere. All reasonable values
fall roughly within the band denoted. c is the local sound speed.

effect of inhomogeneities. Both for the H and K lines and for the Na D lines (Wilson,
1973), it has been shown that the inclusion of inhomogeneities can completely reproduce the limb broadening. For this reason one can probably dismiss anisotropic
turbulence.

3.2. LAYERED MODELS

There have been a number of studies concerning the effects of differentially moving
layers. These have been motivated by an attempt to explain the asymmetries observed
in H and K, Hα and other strong lines. These models are still very exploratory and
numerical, but encouraging progress has already been made. We mention the work
of Kulander and Jefferies (1966), Kulander (1976, 1968), Beckers (1968), Athay (1970a,
b), Cannon and Rees (1971) and Cram (1972).

Differentially moving layers have two general effects on the line profile. First, the
source function itself may be changed, and secondly, the emergent intensity is changed
due to the shift of the absorption profile. Several of these studies have determined that
if the layered motions have a magnitude less than the Doppler width of the line, and
if the gradients in the velocity and the source function are small, then the source
function is not significantly affected by the motion. In this case the source function
from static models can be used; otherwise it must be computed explicitly. Of course
the shift of the absoprtion profile always has a strong effect on the emergent intensity.

From the different numerical cases that have been studied, several general conclusions can be reached. First, only motions in the uppermost layers produce line shifts
which could be accurately interpreted by means of, say, the line bisector technique.
And this will be the case only if the moving layer is thick enough. For example, if a

layer comprising the entire region where the Hα core is formed is moving, then the entire Hα core will show an accurate Doppler shift; however as this layer becomes thinner, the inferred Doppler shift falls below the true one. In addition to this result, there is the more general result that height discrimination (by observing different parts of the line profile) is lost. More explicitly, "Motion far above $\tau_v = 1$ may produce an apparent evidence of motion near $\tau_v = 1$ and motion near $\tau_v = 1$ may shift the profile at v by an amount corresponding to a velocity that is only a fraction of the true velocity." (Athay, 1970b).

If the moving layer occurs deeper, then the inferred velocity could be either smaller than or larger than the true velocity depending on whether the layer is located at $\tau_v = 1$ or not, or whether the velocity gradient is positive or negative. Furthermore, if the source function and optical depth vary independently of the velocity, then the direct inference of velocities becomes so ambiguous that even velocities of the wrong sign can be inferred. In this case, it seems essential to postulate a model in which all of these quantities are free parameters, and then fit this model to the entire line profile.

Two authors, Athay and Cram, have had the courage to publish actual asymmetric K profiles which could be compared with real data. Although Cram's model is much more sopisticated than Athay's, they do agree that a K_{2v} feature can be reproduced by a rising intermediate layer or a falling top layer, probably the latter. In either case, the velocities needed are of the order of the sonic velocity. It also appears that velocity effects alone are insufficient to reproduce the observed profile accurately.

3.3. K LINE ASYMMETRIES AND FEATURES

The observational study of K line asymmetries has seen a considerable renaissance recently. The number of interesting new results, coupled with the theoretical studies of layered motion, make this one of the more fascinating fields of solar dynamics that exist today. We start with a summary of the observed statistics of asymmetries.

3.3.1. *Asymmetry Statistics*

Three papers have investigated this topic recently (Bappu and Sivaraman, 1971; Liu and Smith, 1972; Pasachoff, 1970). We summarize their findings on Table II, which tabulates the percentage abundance of 5 rough classes of asymmetry. A separate sixth class includes the total absence of K_2. Since this sixth class is not mutually exclusive of the classes, 'no K_{2r}' or 'no K_{2v}', and since the authors use different terminology to report their results, there remains some ambiguity. However, some general trends can be noted. Viewing only the first two rows of Table II for the moment, we see that about one-half to three-quarters of all K profiles have double peaked profiles of some sort and that these profiles are found both in the network regions and in the cells. (The distribution between network and cell of these double peaked profiles may be very uneven however. This point has not yet been studied.) Of these profiles, the vast majority are asymmetric with the violet peak asymmetries outnumbering the red peak asymmetries roughly two to one. The remaining one-quarter to one-half of the profiles are missing one or both peaks *and are found preferentially within the cells.* Any

TABLE II

The percentage abundance of the 5 qualitative classes of K_2 asymmetries

	no K_{2R}	$I_{K_{2V}} > I_{K_{2R}}$	$I_{K_{2V}} = I_{K_{2R}}$	$I_{K_{2V}} < I_{K_{2R}}$	no K_{2V}	no K_2 AT ALL	SAMPLE SIZE
BAPPU AND SIVARAMAN (1971)	22.3 (dark K_3) (cell) 3	45.3	4.7	25	0.7	0.7	1 FRAME 148 PROFILES
LIU AND SMITH (1972) (preferred location)	20 (cell)		50 (network, cell)		10 (cell)	20 DARK SPOT IN CELL	200 FRAMES
PASACHOFF (1970)	80		10		20	?	1 FRAME

dynamical model of the chromosphere should be capable of reproducing these facts.

The cause of the asymmetries seems to be indicated from further results of Bappu and Sivaraman. They found a significant correlation of these classes with the observed Doppler shift of K_3 (Figure 6). This relation can be summarized with the interpretation that the asymmetry in K_2 is caused by a Doppler shifted K_3 component partially veiling one of the K_2 components. The same effect was noted by Pasachoff although he did not establish an absolute wavelength scale. This is precisely the behaviour predicted by Athay (1970a). Although both Athay and Cram (1972) have shown that a K_2 asymmetry can be produced either by a rising K_2 layer or a falling K_3 layer, the fact that K_3 is Doppler shifted away from the bright K_2 component indicates that the asymmetries are caused by motion in the K_3 layer. That motion is then of rather large amplitude (10 to 20 km s^{-1}) and predominantly *downward*.

There seems to be yet another feature in the K emission; the 'dark K_3 regions'. Both Bappu and Sivaraman and Liu and Smith find that the class of 'no K_{2r}' is correlated with a K_3 component that is not only red shifted, but also unusually dark. In fact Liu and Smith find that in this feature K_{2v} is also absent, in contradiction to Bappu and Sivaraman. This point should be cleared up (although one should give greater weight to the results of Liu and Smith due to their much larger sample) before one tries to understand the physical nature of this type of region. If there is no K_2 at all, then this could be the background component postulated by Wilson (1970) and Cram (1972), in which the source function actually decreases monotonically with height.

3.3.2. *The Pasachoff-Zirin Model*

The main discrepancy in the asymmetry statistics is that Pasachoff observed almost no double peaked profiles at all, and on the basis of this developed a radically different

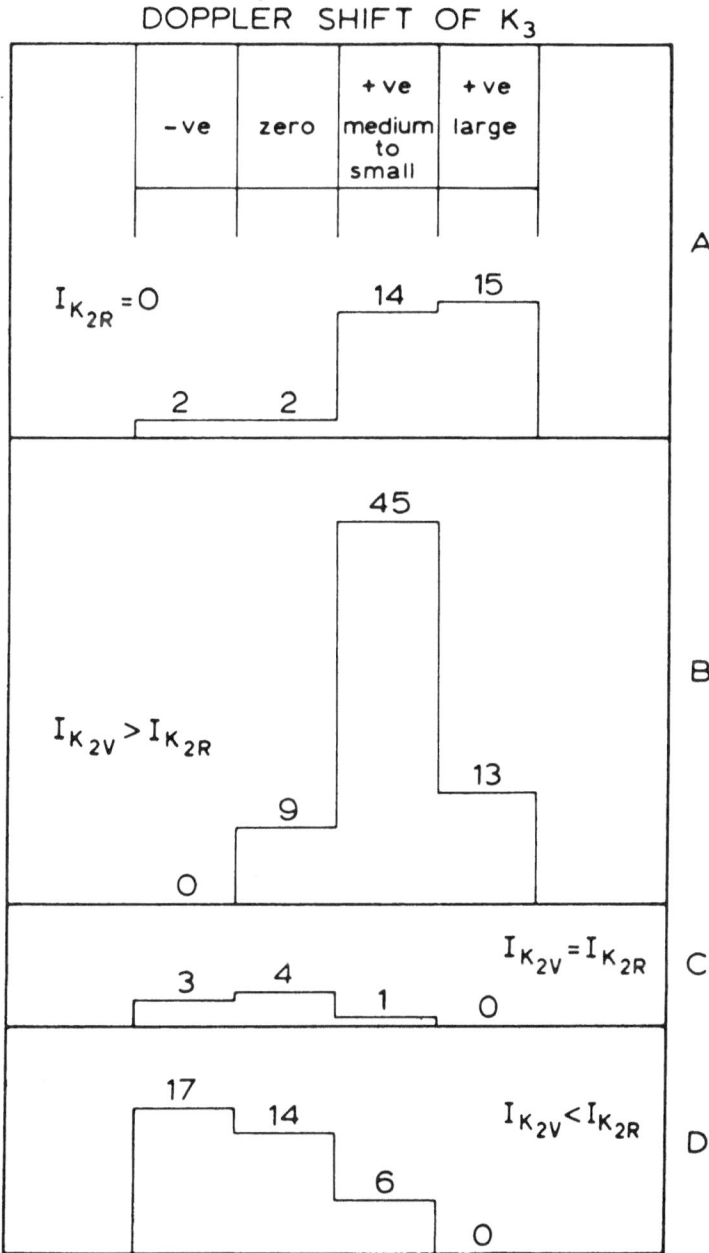

Fig. 6.   Histograms of the Doppler shift of K_3 grouped according to four of the K_2-asymmetry classes (from Bappu and Sivaraman, 1971). The numbers over the histograms represent the number of instances in each category. There is a definite correlation of the K_3 Doppler shift with the sense of the asymmetry.

model of calcium emission originally proposed by Zirin (1966). This model consists of a basic profile which has little or no K_2 emission, overlaid by small features which emit a single K_2 peak. These features must possess a binary velocity distribution of ± 10 to 20 km s^{-1}. This model has been investigated by many people; the most recent and detailed study is that of Cram (1972). He shows that the model cannot reproduce the observed limbward separation of the peaks. The data is also contradicted by the other two observations and by Wilson and Evans (1971). Thus it seems fairly clear that, while occasional features may radiate by this model, it is inadequate as a general explanation for calcium emission.

The question arises as to how Pasachoff obtained such an atypical distribution of asymmetries. Skumanich *et al.* (1972), by comparing the distribution of broad-band brightnessses of Pasachoff's sample, showed that it was very deficient in bright network elements (i.e. the slit fell by chance almost exclusively across cell regions). Liu and Smith showed that the single peaked profiles strongly prefer cell regions, so this case is an excellent example of strong accidental bias of a small sample.

3.3.3. *Time Behavior*

While the theoreticians have just started to grapple with steady-state asymmetries, the observers have opened up a fascinating and fertile new chapter in the seemingly endless saga of the K line. The first indications that the asymmetries have complicated histories came from Wilson and Evans (1971) and Wilson *et al.* (1972), who reported instances of single peaks evolving into double peaks and vice versa. The time scales were as short as 15 s. Also, Liu *et al.* (1972) outlined the time history of K_{2v} bright points from a time sequence of K_{2v} spectroheliograms (Figure 7). The K_{2v} points rise in brightness in 20 to 30 s and seem to fade out even faster! The study of such features with time sequence spectrograms should be very fruitful. In fact Liu (1973) has just observed a very systematic time development of the K profile through these events. Samples of his spectra are shown in Figure 8. The subtracted profiles (profile at time t minus profile at time zero) of event B are shown in Figure 9. The brightening starts in the K_1 wing, progresses in toward the center, produces a K_{2v} point, then fades out. The whole process repeats roughly every 180 s. Since K_3 was not observed to brighten significantly, Liu interprets this type of event as a direct observation of deposition of wave energy at the level of K_2 formation, i.e. the mid chromosphere. Further, since the 180-s period makes it fairly clear that this dissipation is the fate of the high frequency tail of the resonant oscillations, we can expect this behavior to be very common and to contribute significantly to the energy budget of the chromosphere.

3.4. Hα MOTTLES

The discussion up to now has concerned itself with altitudes less than about 2000 km. It seems that around this height even a quasi-homogeneous chromosphere vanishes and one is left with only isolated features which are imbedded in the corona. This model has of course been developed from a much broader context, but it is equally valid and useful when considering Hα mottles.

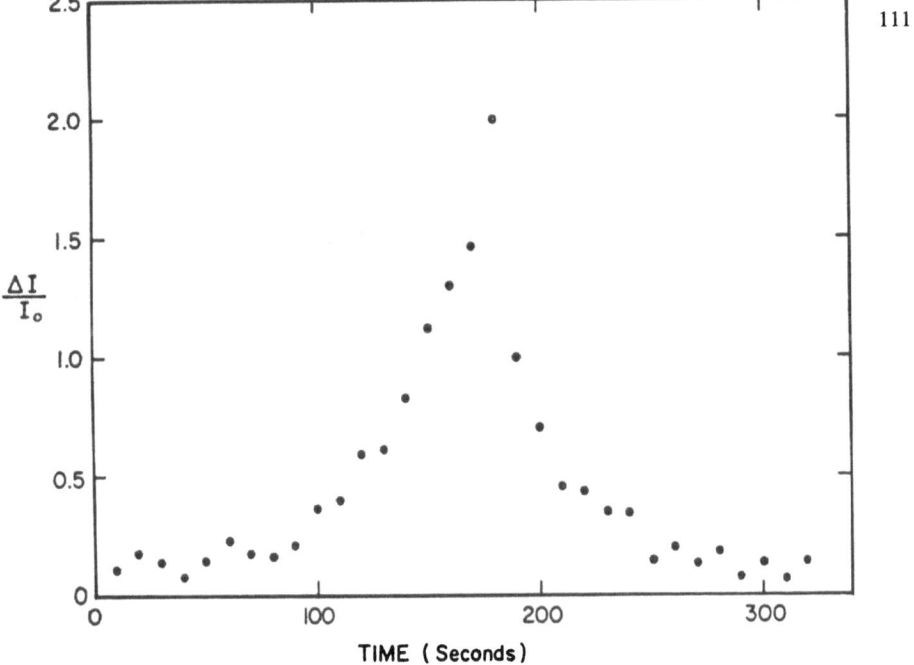

Fig. 7. The time history of the contrast $\Delta I/I_0$ of a typical cell point on a K_{2v} spectroheliogram (from Liu *et al.*, 1972). I_0 is the average intensity of the 'quiet' background.

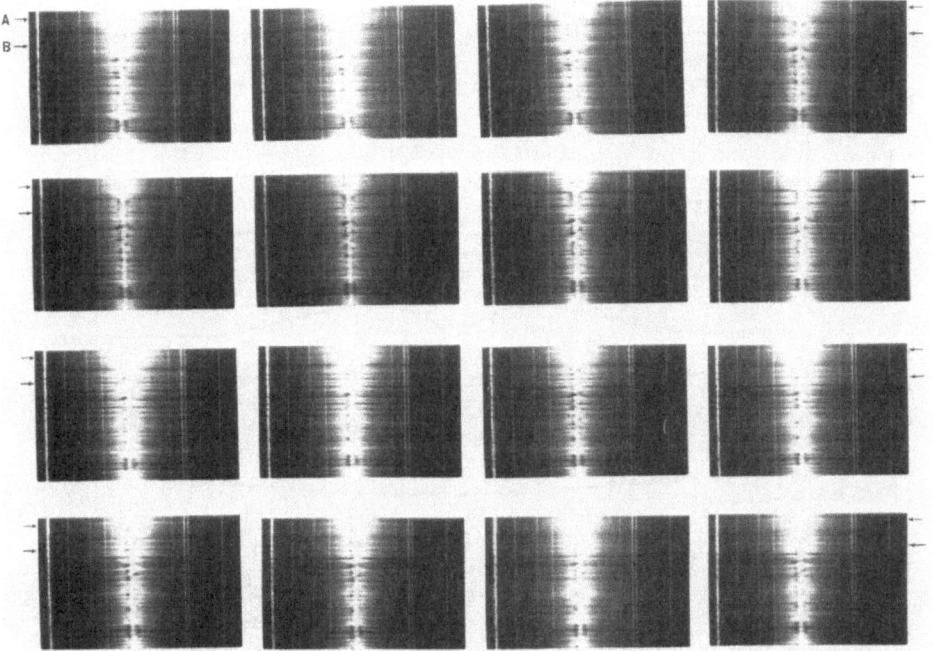

Fig. 8. A time sequence of K-line spectra of a quiet region at the center of the disk (from Liu, 1973). In this example Δt is 8 s between frames. The disturbance marked 'B' begins in the far K_1 wing (first row), proceeds to the inner K_1 wing (second row), produces an intense K_{2v} emission (third row) then fades out (fourth row).

EDWARD N. FRAZIER

3.4.1. *Phenomenological Studies*

Studies resulting from filtergrams, spectroheliograms, or direct attempts to interpret the Hα profile have had a long history. The benchmarks are perhaps the works of Bhavilai (1965) and Title (1966). Bhavilai inferred from Hα filtergrams that the mottles were moving downward close to the centers of the rosettes and were moving upward at greater distances from the rosettes. In many cases a connection between the two features was observed and loop-type motion was inferred.

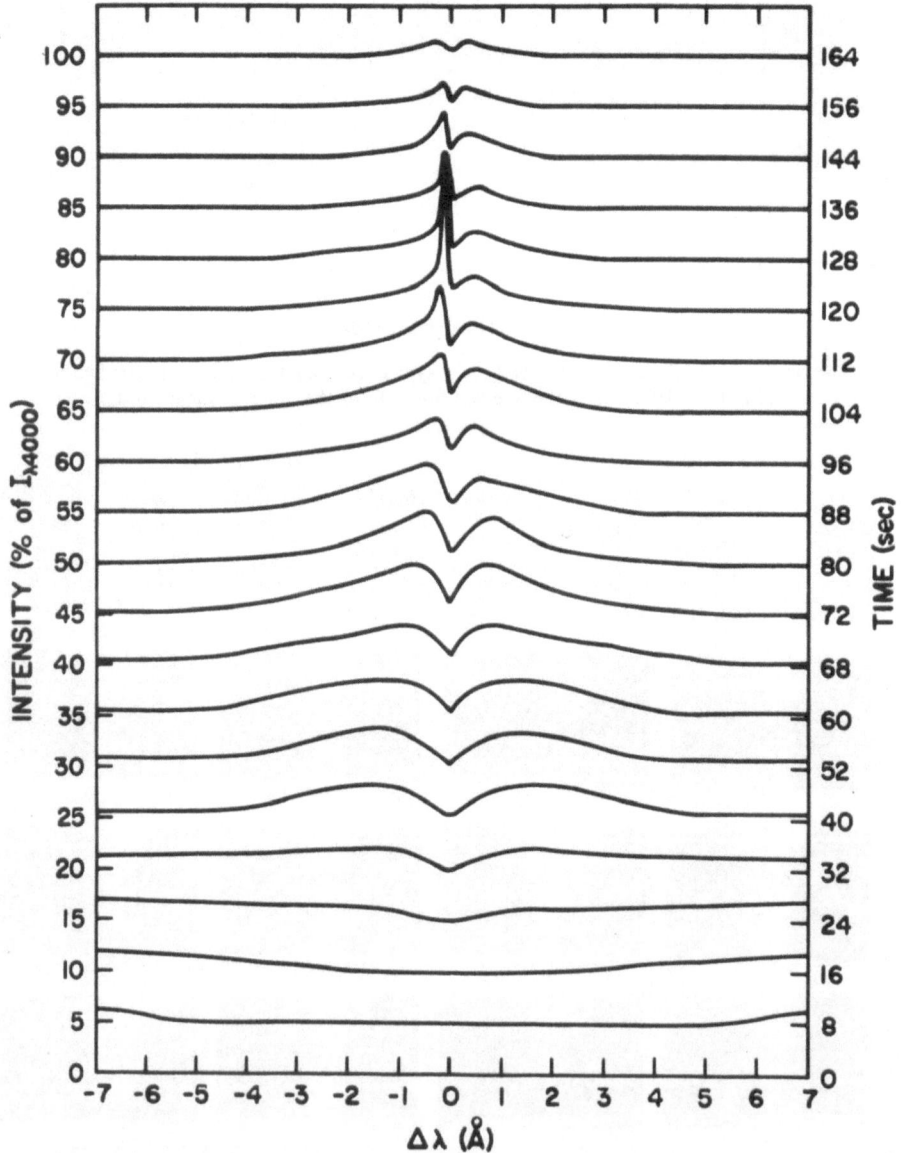

Fig. 9. Subtracted profiles (profile at time *t* minus profile at time 0) of event *B* of Figure 8 (from Liu, 1973).

Title confirmed the rough distribution of upflow and downflow elements, but found them to be completely different phenomena. Downflow elements were much longer-lived (6 to 9 min) than the upflow elements (2 min), and there was no correlation between the two features anaway. So this type of investigation has been left appropriately enough on a rather uncertain note.

3.4.2. $H\alpha$ Profile Analyses

It was shown by Beckers (1962, 1964, 1968) and Athay (1970) that direct inference of velocities from $H\alpha$ line shifts or asymmetries can lead to very inaccurate results. These ideas were tested and confirmed observationally in a pair of very important papers by Grossman-Doerth and von Uexküll (1971, 1973). Using data from simultaneous spectrograms and filtergrams, they tested both the 'cloud' model of Beckers and the 'velocity layer' model of Athay. The velocity layer model did not fit the mottle spectra, which is not surprising because it was so highly simplified (no variations in any of the other physical variables were allowed). However the cloud model fit the spectra very well, as shown in Figure 10. The cloud model assumes that a mottle is a separate cloud overlying the chromosphere. The physical variables are assumed to be constant throughout the cloud and the emitted profile is then fully determined by four quantities; S, τ_0, $\Delta\lambda_0$, v, the source function, the optical depth at line center of the cloud, the line width, and the bulk velocity, respectively.

The results of fitting these spectra to the four parameters of the cloud model $(S/I_c, \tau_0, \Delta\lambda_0, v)$ were quite surprising: (1) There was no qualitative distinction between dark mottles and bright mottles or any other kind of mottles. (2) There was no correlation between any of the four parameters. (3) The velocities inferred (rms $v = 3.6$ km s^{-1}) were considerably smaller than what would be expected if $H\alpha$ mottles were indeed spicules viewed on the disk. (It should be noted that this velocity which the cloud model infers is much smaller than has been frequently inferred directly from filtergrams. It is also larger than would be inferred by the line bisector method).

Recognizing the importance of the disparity between their rather low measured velocities and those that would be expected from the vertical component of spicule motion (≈ 30 km s^{-1}), they went on to show that the smearing due to finite seeing is so effective in reducing apparent velocities that it can account for the entire disparity; "A circular mottle of 1″ diameter (and a seeing parameter of 0.7–1.5″), for example, with a true line of sight velocity of 30 km s^{-1} would in most circumstances appear to move with only about 5 km s^{-1}". Krat (1972), measrued $H\alpha$ spectrograms with a resolution of 0.4″ (taken with the Soviet Stratospheric Solar Station) and concluded that the optimum size of $H\alpha$ mottles is indded 0.8″ to 1.1″. So this explanation of the low velocities would seem to be a very realistic one.

Bray (1973) and Loughhead (1973) have performed the same analysis of the cloud model, using instead of spectrograms the interesting technique of tuned filtergrams. This technique was first employed by Bhavilai (1966). Bray seeks to preserve the dark mottle-bright mottle terminology; Loughhead seeks to show that the cloud model breaks down near the limb. The point here is that, where velocities are concerned, Bray and Grossman-Doerth and von Uexküll are in agreement.

4. Limb Phenomena

We now finally discuss spicules *per se*. At the very start, I must mention the excellent reviews by Beckers (1968, 1972), which I shall be using rather heavily. As an introduction a few words should be said as to how the size regimes mentioned in Section 2 are

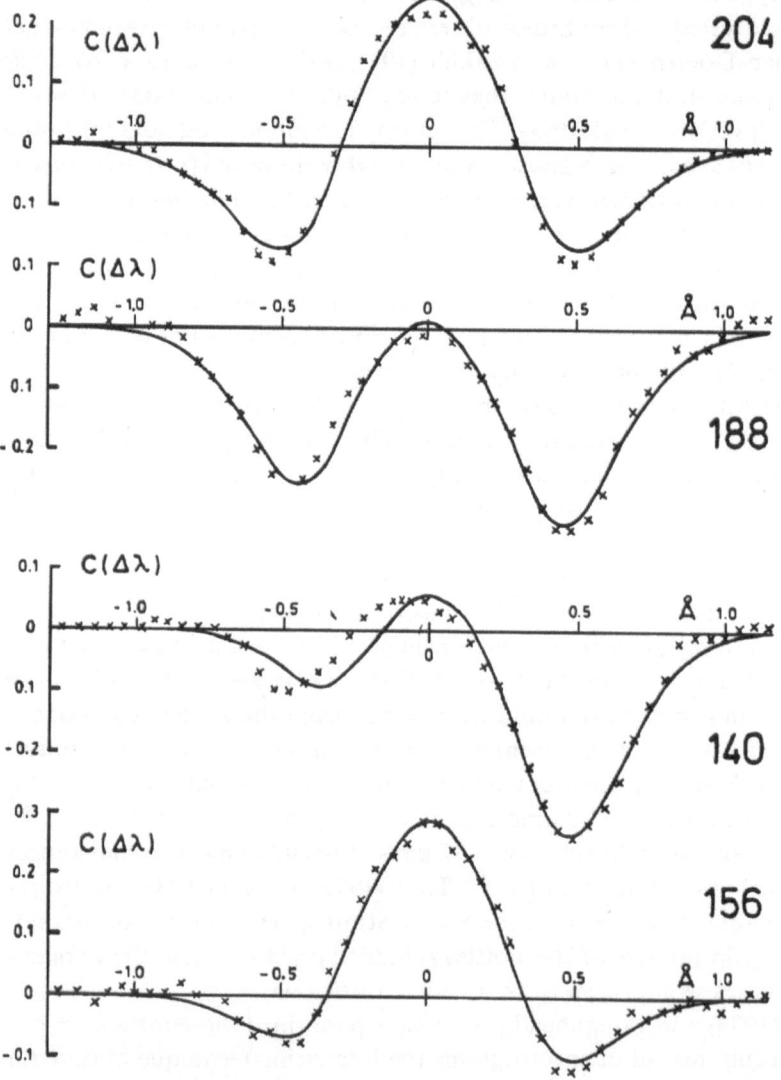

Fig. 10.   Examples of contrast profiles $(C(\lambda)=(I-I_0)/I_0)$ derived from Hα spectra (from Grossman-Doerth and von Uexküll, 1971). The crosses are the observed profiles, the solid lines are calculated profiles obtained by fitting Beckers' cloud model to the observations.

changed when applied to limb observations. Now all three spatial coordinates have separate effects, and any attempt to schematicize the regimes becomes three-dimensional and very complicated. Even more important, all the characteristic lengths are roughly the same as each other, *and roughly the same as the spicule diameters*. Furthermore, the optical depths of the spicules vary considerably with altitude and with the line observed. As a result, the interpretation of line shifts, widths, asymmetries, etc. in terms of detailed motions is on an even less sound basis than for disk features. Such terms as microturbulence or non-thermal motions should be used only in the broadest sense; they are little more than convenient terms to describe observable effects. It should also be noted that microturbulence occupies a unique position in that, as our understanding of the diagnostic problem worsens and as the observational data becomes more crude, a larger and larger fraction of the total motion is ascribed to microturbulence. Thus we can expect limb observations to imply larger values of microturbulence than disk observations. We should probably also expect large variations in all types of motions because the observable effects depend very strongly on instrumental parameters such as spatial resolution, height resolution, and even the spectral line observed.

4.1. RESOLVED MASS MOTIONS

4.1.1. *Proper Motions*

Almost all measurements of *vertical* proper motions were made in the 1950's and have been reviewed by de Jager (1959) and Beckers (1968). Most find a rise of about 30 km s^{-1} with occasional fast spicules. The time or height variation of spicule rises cannot be observed accurately enough to identify a particular type of trajectory, such as a rise at uniform velocity or a parabolic trajectory (Athay and Thomas, 1957), but there is a linear relation between maximum height attained and the velocity of ascent (Rush and Roberts, 1954). The fate of spicules after they reach maximum height however, is still uncertain. Lippincott (1957) observed roughly half of them to fall again; the other half simply fade away. In relation to this. Mouradian (1967) found spicules to diffuse at a rapid rate (10 to 20 km s^{-1}) while fading. Later measurements (Lynch *et al.*, 1973) contradict this. Since the later measurements were derived from the same observer using a better telescope, one must give greater weight to them and say that the diffusion of spicules is very much in doubt. Of course it is quite possible that spicules diffuse into the corona after they fade out and become invisible.

There are several recent reports of observed *horizontal* proper motions (Weart, 1970; Nikolsky, 1970; Pasachoff *et al.*, 1968; Nikolsky and Platova, 1971). One can easily imagine instrumental effects and particular spicule geometries that can create the appearance of horizontal proper motion, however all authors agree that the effect is definitely real. The observed speeds range from 5 km s^{-1} (the observational limit) up to 60 km s^{-1}. Nikolsky and Platova measure a distribution with a mean of roughly 20 km s^{-1} (Figure 11). They also observe the spicules to oscillate with a most probable period of ≈ 1 min. The observational constraints are such that this could also be interpreted as random motion. It isn't clear *how many* spicules exhibit horizontal proper motion. We only know the number is more than 5% (Weart, 1970). It would be most valuable to get better statistics of this important phenomenon.

4.1.2. *Doppler Shifts*

Most measurements of Doppler shifts have been reviewed by Beckers (1968). The rms velocities are about 10 km s^{-1}, while the maximum observed velocites are about

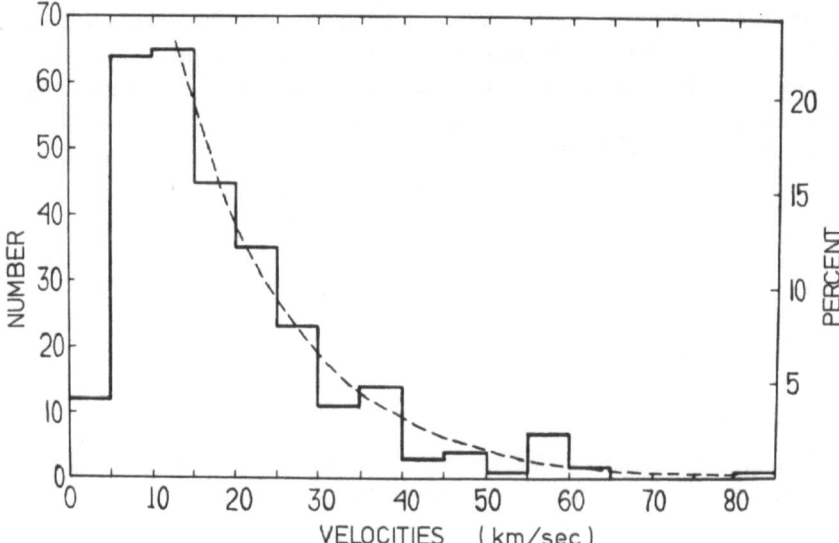

Fig. 11.   The observed velocity distribution of transverse proper motions (from Nikolsky and Platova, 1971). The peak at about 10 km s^{-1} is probably caused by an observational cutoff at low velocities.

30 km s^{-1}. More recently, Krat and Krat (1971) observe significantly higher Doppler shifts. (They make a further startling observation; that Doppler shifts of spicules measured simultaneously in Hα, H & K, and D$_3$ are not always the same. Table III lists the measured radial velocities of 15 spicules for which v_r was observed simultaneously in the Hα, K and D$_3$ lines. Spicules Nos. 1 and 2 ($h = 7000$ km) are the same spicules as Nos. 6 and 7 ($h = 8000$ km) respectively. One can see many cases of large differences in v_r between the three lines. We shall return to this point later.) The rms Doppler shifts, if interpreted as being caused by motion along the axis of the spicules, results in a mean axial velocity of the spicules of 25 to 30 km s^{-1} (Beckers, 1968).

There is contradictory evidence of the height variation of Doppler shifts. Mouradian (1965) finds a decrease with height, several others (Michard, 1959; Beckers, 1966) find an increase with height, and two groups Pasachoff *et al.*, 1968; and Nisolsky and Sazanov, 1966) find no significant change. It seems that the only tenable summary is that *on the average* there is very little height dependence, but that large fluctuations are possible. This is illustrated by the very wide scatter in Figure 12 (Pasachoff *et al.*, 1968). Here Doppler shifts at two different heights (simultaneous) are correlated.

TABLE III

Radial velocity of 15 spicules measured simultaneously in three different lines
(from Krat and Krat, 1971)

h (10³ km)	Spicule No. [a]	v_r (km s⁻¹)		
		Hα	Ca II K	D₃
	1	0.0	0.0	0.0
	2	+5.0	+8.0	+5.0
7	4	0.0	0.0	0.0
	5	+2.3	0.0	−13.3
	6	+7.0	+5.0	+8.0
	7	+3.0	+9.0	+14.0
	8	−28.6	−17.2	−29.4
8	11	−27.7	−19.7	−
	12	+37.5	+21.9	−
	13	−28.0	−21.2	−29.4
6	15	+11.5	+6.9	0.0
	18	+10.7	+10.0	+2.5
	20	−31.8	−31.8	−31.5
7	22	+11.0	+8.0	+9.1
	23	0.0	0.0	0.0

[a] Spicules whose numbers do not appear did not have their radial velocities measured.

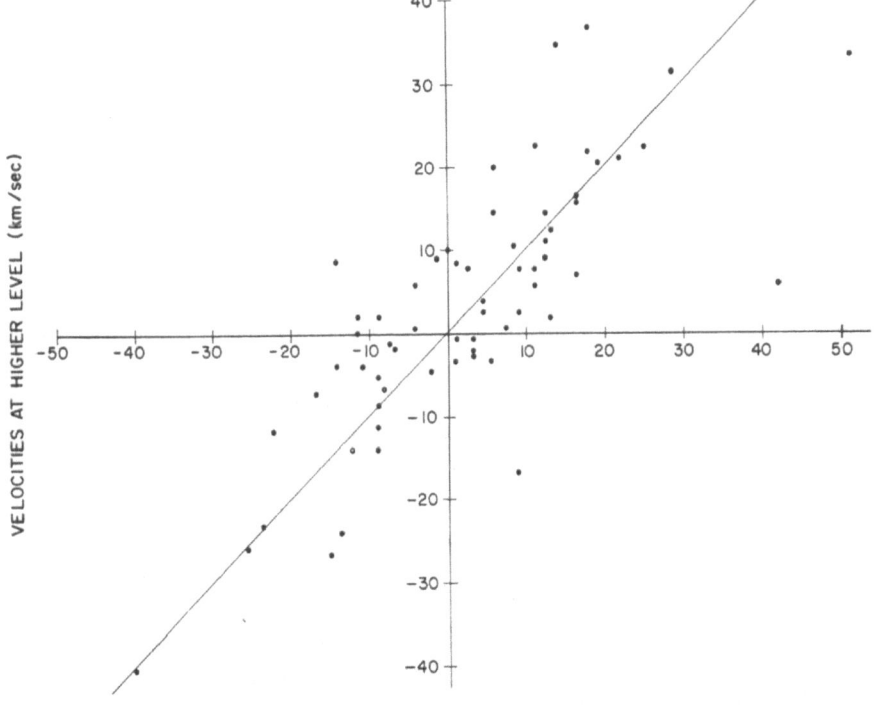

Fig. 12.   Comparison of radial velocities measured simultaneously at two heights in Hα (from Pasachoff *et al.*, 1968). The heights are separeted by 1800 km. Positive signs indicate velocities of recession. *On the average* velocities are the same at the two heights, but the variations are large.

The time variation of Doppler shifts is also an open topic. Most authors seem to agree that the *majority* of features (perhaps 80%) exhibit a simple Doppler history of uniform rise or fall. Several observers (Nikolsky and Sazanov, 1966; Zirker, 1967; Pasachoff *et al.*, 1968) find features which reverse sign, oscillate, or otherwise show complicated behavior. Figure 13 (Pasachoff *et al.*, 1968) displays a good sample of

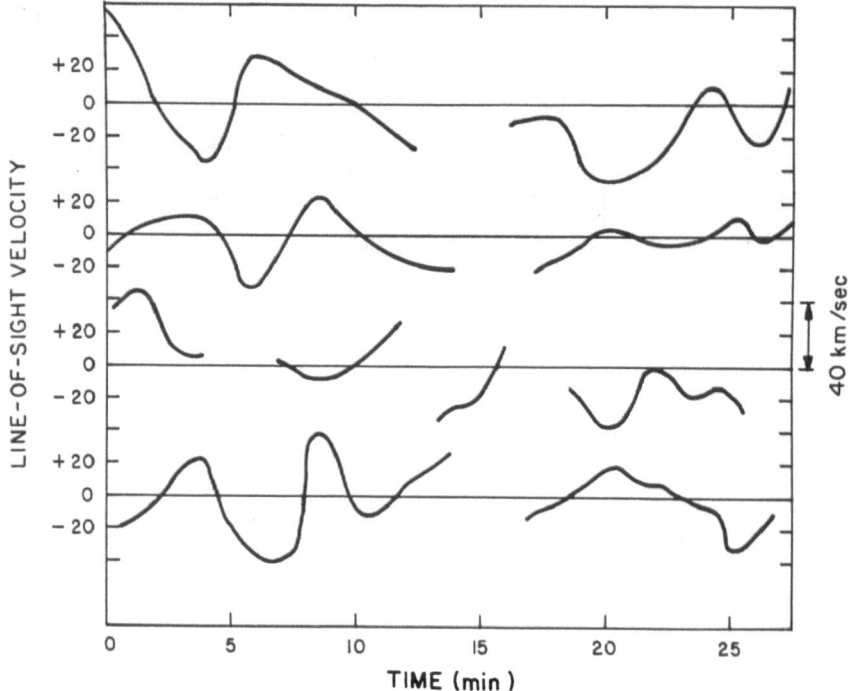

Fig. 13.   Examples of spicules whose radial velocities vary significantly with time
(from Pasachoff *et al.*, 1968).

such things. It is all very reminiscent of the observed horizontal proper motions mentioned earlier. In fact, Weart (1970) stated that the two are very closely coupled. In a number of cases, individual spicules showed complicated Doppler histories that were closely matched by simultaneous horizontal proper motions, indicating a transverse back and forth motion of the spicules along a direction inclined to the line of sight. Thus it seems that some unknown but probably small fraction of observed Doppler shifts must be interpreted as horizontal, not axial motion.

4.1.3. *Rotational Motions*

Emission lines which are tilted slightly with respect to the direction of dispersion have been observed by many people (Michard, 1956; Livshitz, 1966; Pasachoff *et al.*, 1968; Weart, 1970). Such a tilted line can be seen in Figure 14 (Noyes, 1965). This is a very atypical example; most tilted line profiles are far less obvious. In fact in many cases,

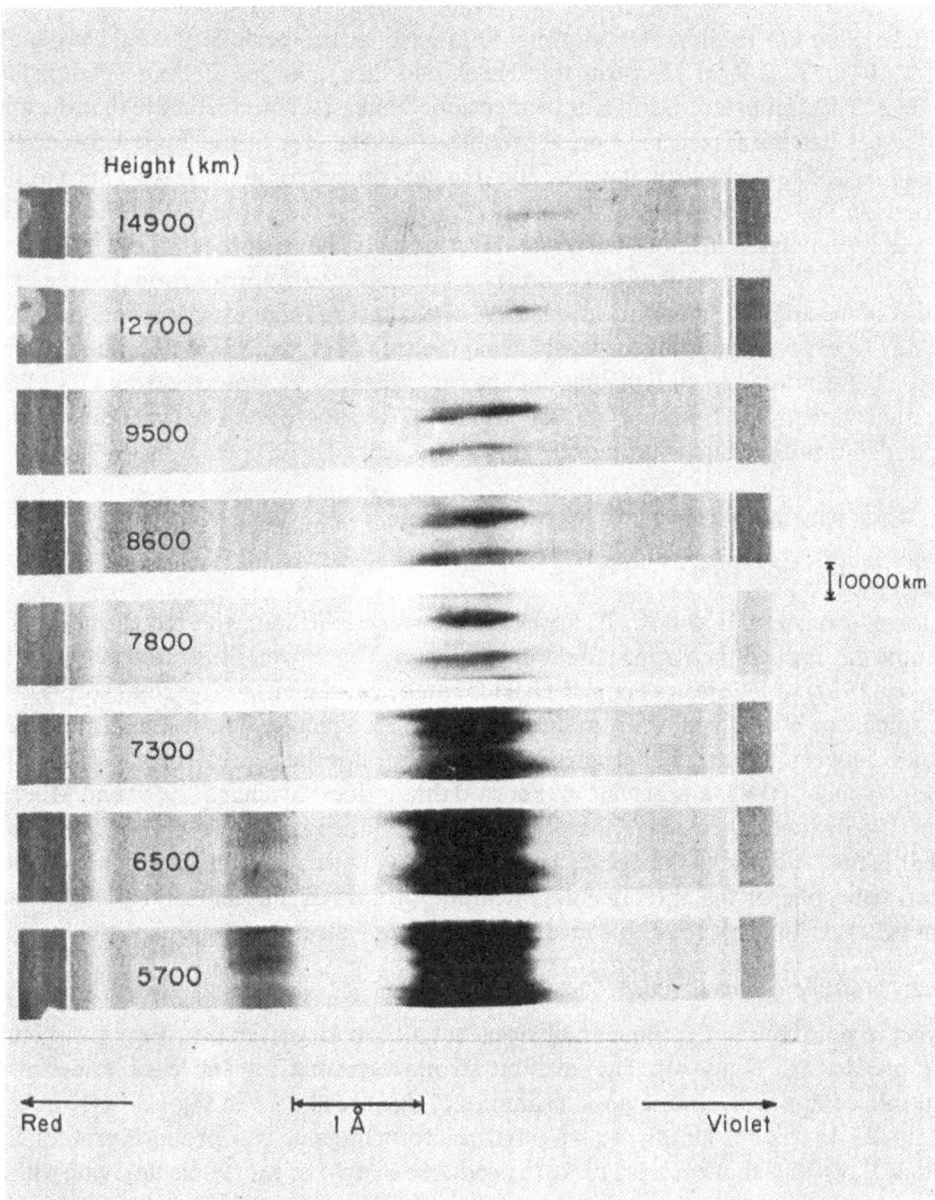

Fig. 14. The 'most outstanding example' of a tilted Ca II H line (from Noyes, 1965). The same feature is observed over a sequence of different heights. Note the apparent rapid increase of rational speed with height.

the apparent tilt could be the result of two close, unresolved spicules, each with greatly different Doppler shifts. However, there are many cases where it seems clear that this tilt is real and results from spicule rotation (Beckers, 1966, 1968). But nobody says what *fraction* of spicules show tilted lines. The size of the tilts is small (e.g. $\sim 2°$ for K, $\sim 1°$ for Hα, Pasachoff *et al.*) but quite important. Pasachoff *et al.* calculate that a column 1000 km in diameter rotating 30 km s^{-1} at the periphery would produe a tilt of 1.9° for K, 2.6° for Hα. Note that Hα should show a larger tilt than K, but in fact it is less. This important point will be mentioned later. (It is unfortunate that the unit of degrees has been used to express this tilt, since the size of the angle between the tilted line and the direction of dispersion depends on instrumental factors such as the dispersion and the plate scale as well as on the rest wavelength of the line. The tilt should be expressed in more universal units; namely the reciprocal of the transverse gradient of a Doppler shift, (km s^{-1}/km)$^{-1}$. This is of course proportional to the reciprocal of the angular rate of rotation, ω^{-1}, of a rotating spicule. So it would be much simpler to express these tilts in terms of ω.) Getting back to the specific example, the authors also note that, with this model, the centripetal force would be about 6 g! This is rather violent rotation, and in fact Weart (1970) saw four cases wherein spicules expanded rapidly while rotating.

4.2. Line profile shapes

4.2.1. *Central Reversals*

At lower levels Hα, H and K all show central reversals which, with increasing height, become flat-topped then approximately Gaussian. This is well illustrated in Figure 15 (Zirker, 1962) which shows Hα and H. This can be caused either by self-absorption in the spicule or by absorption by matter in front of the spicule. One test that can differentiate between these two alternatives is whether or not the central reversal shares the Doppler shift of the main profile. Zirker said that it does. Michard (1959) and Mouradian (1965) said that it doesn't. The latest and as yet uncontested vote (Pasachoff *et al.*, 1968) is that it doesn't. This observation leaves us with a curious component in the solar atmosphere: the interspicular medium. Although Pasachoff *et al.* speculate somewhat on the nature of this medium, it is largely an unknown quantity.

4.2.2. *Multiple Components*

I wish to point out at this time an obvious but all too forgotten fact. Very few spicule line profiles are Gaussian. They exhibit strong asymmetries and clear evidence of multiple components. Some good examples of this are shown in Figure 16 (Krat and Krat, 1971). There is simply no way that microturbulence can produce profiles like that. It is obvious that such profiles are produced by two or more spicules lying within the resolution element of the telescope, each with its own Doppler shift and line width. It is clear that the measurement of Doppler shifts and line widths of such profiles requires some care and judgment. One can measure the Doppler shift of the 'center of mass' of the entire profile and some sort of overall width. In this case the results will be abnormally low Doppler shifts and large widths. Alternatively, one could try to identify the individual components and measure them separately. This procedure is more realistic, but is also tedious and requires some subjective judgment. Both approaches can be found in the literature, which makes comparison of published values difficult. Some authors don't even confide which approach they used.

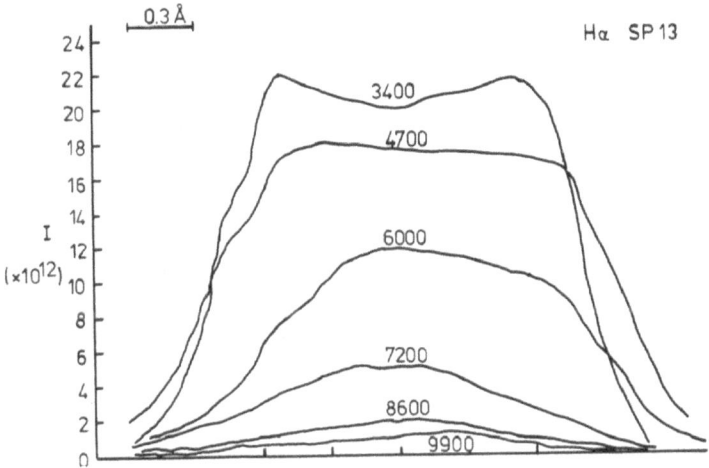

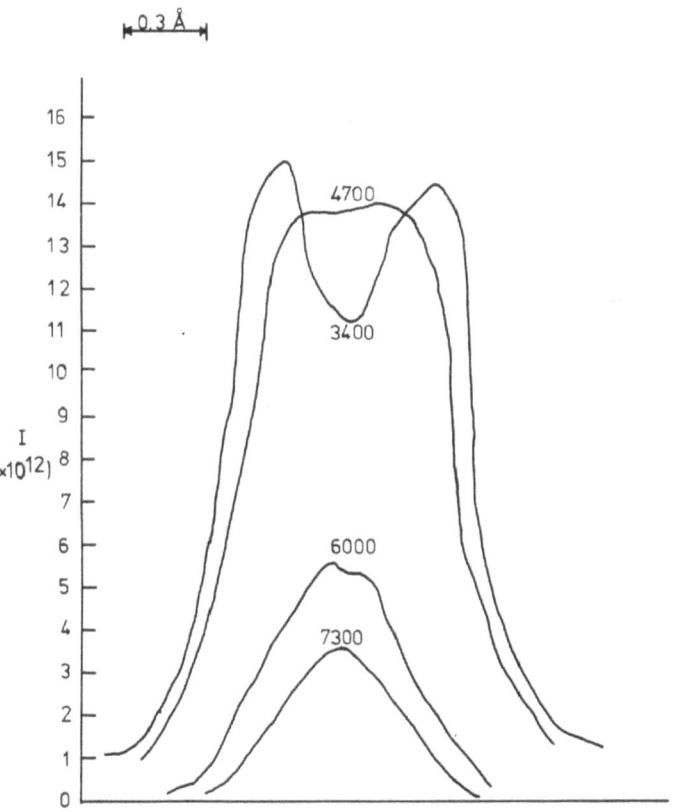

Fig. 15.   Typical profiles of Hα and Ca II H at different heights (from Zirker, 1962).

EDWARD N. FRAZIER

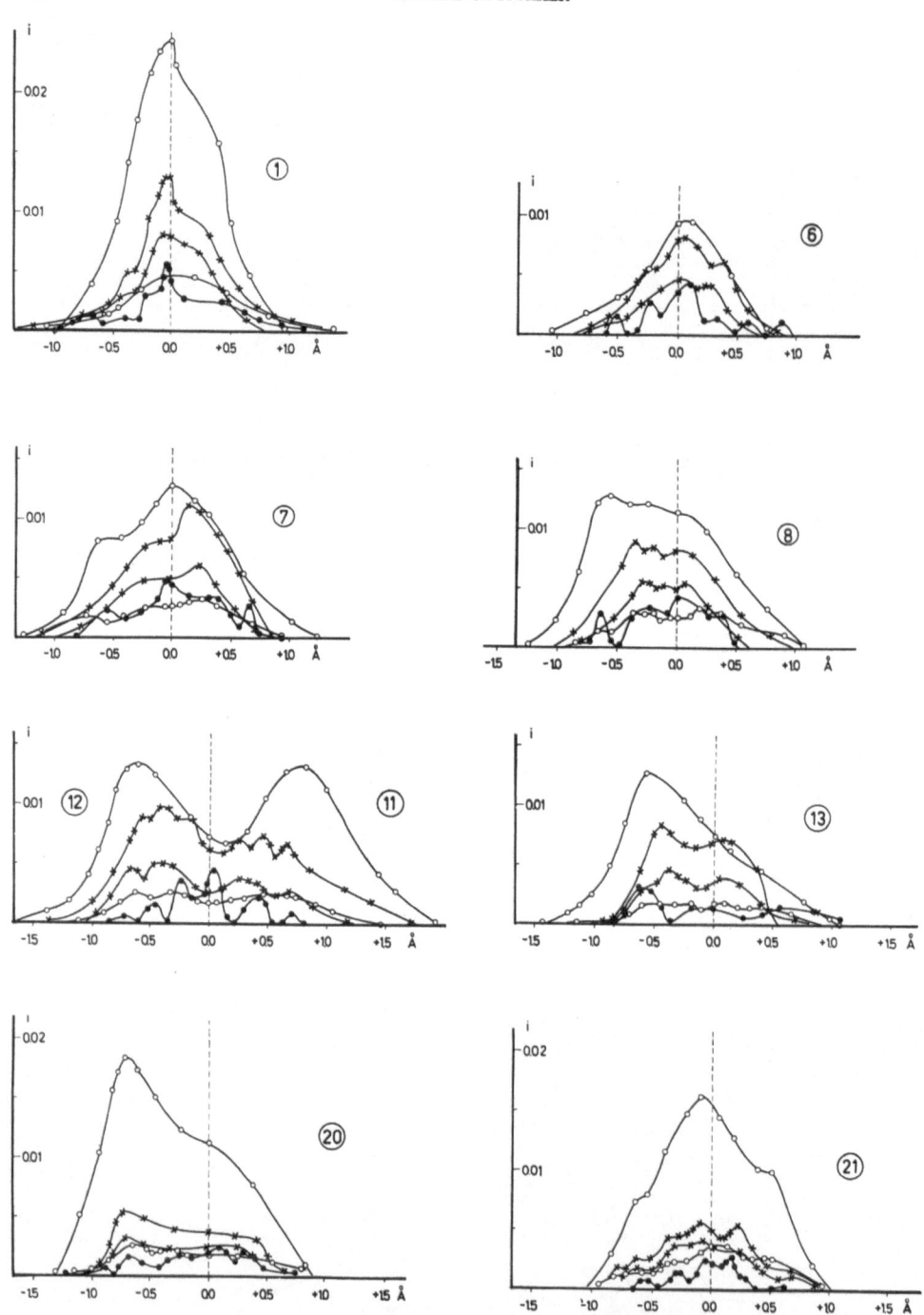

Fig. 16. Outstanding examples of spicule profiles with multiple components (from Krat and Krat, 1971). These were selected as being 'narrow faint line profiles'. Open curcles are Hα and Hβ, ×'s are H and K, filled circles are D_3.

4.3. LINE WIDTHS

In this section we will consider the variation of line widths first as a function of μ, (the mean atomic weight), then as a function of height. There are two ways to consider this variation; statistically (e.g. the average Hα width versus the average K width), and individually (i.e. simultaneous measurements of both lines in an individual spicule). In both cases we will consider the statistical method first and the individual method second. We will also adopt Beckers' notation for line widhts; $W = \Delta\lambda_2/\lambda$. where $\Delta\lambda_2$ is the full width at half maximum of an emission line.

There has been essentially no work done on this topic since the reviews by Beckers. This is in itself an important and surprising point. Where have all the observers gone, these last two years?

4.3.1. *The Variations of W with μ*

Following Beckers, we expect W to be composed of two parts; the thermal part and the non-thermal part. The thermal part varies as $\mu^{-1/2}$; the non-thermal part is a constant. So they should sum together as follows:

$$W^2 = W_{nt}^2 + W_t^2/\mu. \tag{3}$$

Figure 17 is a plot of $W^2(\mu^{-1})$ from Beckers (1968). One should be able to fit a straight

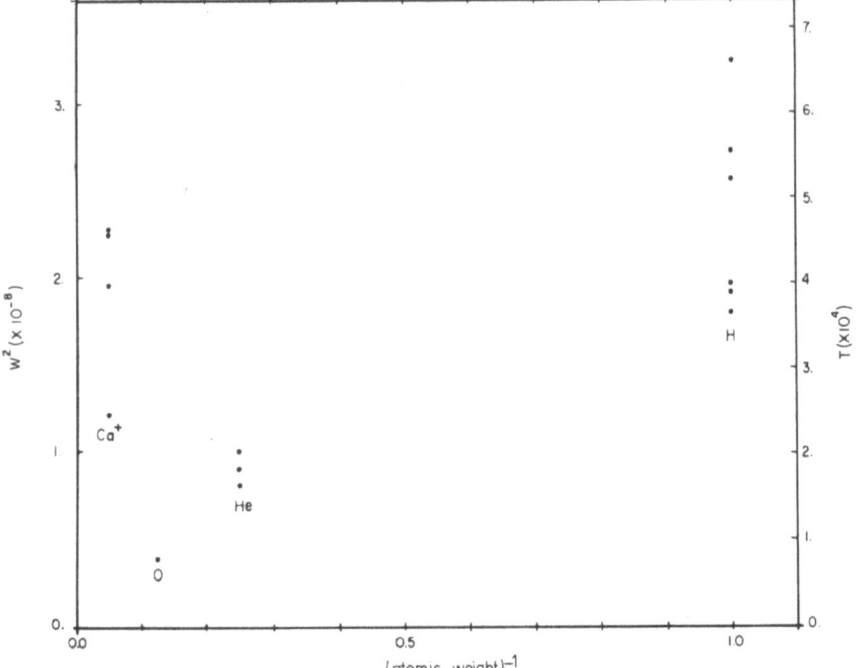

Fig. 17. Dependence of the half-widths of hydrogen, helium, oxygen, and ionized calcium emission lines in spicules on the atomic weight (from Beckers, 1968). The temperature scale on the right is for hydrogen atoms assuming the line width to be entirely thermal. The points represent the most likely value of W^2 for each observed line.

line to these data, the intercept yielding W_{nt}^2. One can see the problem. Actually, the situation is worse than this, because Beckers had already averaged many points so that each point on Figure 17 represents an average value of all measurements of a single line. (The widths are measured at about 3000 to 6000 km above the limb.) All the published values of W yield Figure 18. It is obvious that H and Ca display enor-

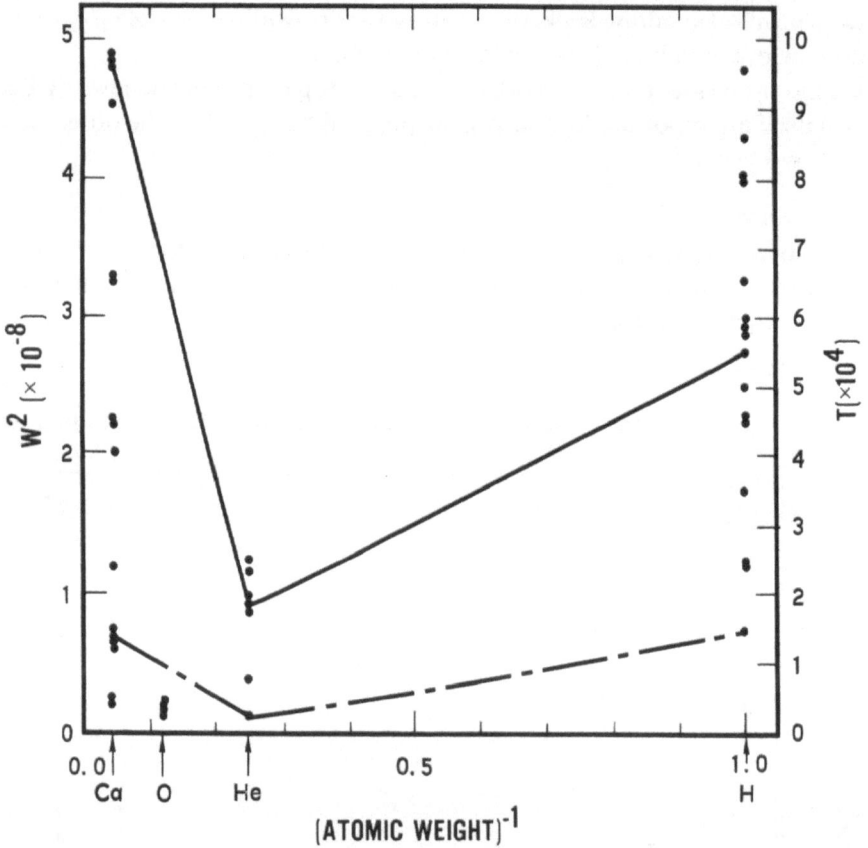

Fig. 18.   Identical to Figure 17, but with all published values of W^2 plotted. Note the wide variations of W for hydrogen and calcium. Solid line, average W^2 (for Hα, D$_3$ and K) of wide line spicules; Dashed line, W^2 of narrow line spicules (data from Krat and Krat, 1971).

mous variations, which led Athay and Bessey (1964), Beckers (1968, 1972) and Krat and Krat (1971) to speak of 'wide line spicules' and 'narrow line spicules' as if they were two different types. There is however no evidence that there is not a continuous distribution of W.

Now, one can conclude either that the calcium lines are anomalously broad (and T_e is high, ξ_t low) or that the helium and oxygen lines are anomalously narrow (and T_e is low, ξ_t high), or that the different lines are formed in different regions with different T_e's and ξ_t's.

4.3.2. *Observational Sources of Variations*

Before proceeding, I would like to stop for just a moment and list very briefly possible observational sources of these wide variations:

(1) Resolution (both seeing and instrumental). We know from Section 4.2.2 that finite resolution has a strong effect on W, but it is very difficult to evaluate.

(2) Are we even looking at the same spicules in the different lines? Faint lines may be formed preferentially lower; strong lines may be formed higher, or even only in foreground spicules. Pasachoff *et al.* (1968) claim to rule out this effect, but their data is open to debate.

(3) The faint lines are very sensitive to scattered light and even to effects of underexposure. I note that the oxygen lines are too faint even to allow the measurement of individual spicules.

(4) Since the range of wavelengths is great, there is a strong possibility of chromatic effects (differential seeing, chromatic aberrations, etc.)

(5) The methods of measuring W vary from observer to observer as mentioned earlier. For example, Athay (1961) seemed to measure a composite width, while Krat and Krat tried to separate various components of a profile.

4.3.3. *Simultaneous Observations of Different Lines*

The advantages of simultaneous observations are obvious. Krat and Krat (1961) compared $H\alpha$ and D_3, Gulyaev (1965) and Nikolskaya (1967) compared H_8 and $He I \lambda 3889$: Pasachoff *et al.* (1968) compared various combinations of H, K, $H\varepsilon$, D_3 and the Ca II IR triplet, but reported no width measurements. The most comprehensive work has been that of Krat and Krat (1971) who observed $H\alpha$, $H\beta$, H, K, and D_3 simultaneously. All these observations essentially confirm our impressions from Figure 18; compared to $H\alpha$, D_3 is so narrow as to imply almost no non-thermal velocities, and Ca is so wide as to imply huge velocities. A warning should be noted however when comparing lines of different strengths, as mentioned earlier in Section 4.3.2, item No. 2. In this case, D_3 is much fainter than $H\alpha$, $H\beta$, H or K, so one should be cautious when interpreting differences between D_3 and these other lines.

4.3.4. *Height Variations of W*

It has been mentioned by nearly everyone who has analyzed spicule spectra that the broad profiles may be caused by unresolved clumps of spicules, each with their own Doppler shift. It has been stated often that this effect can be tested by seeing if a broad profile of a particular spicule breaks up into several resolved narrower profiles at higher altitude, where the spicule count is lower and the spicules are more likely to be resolved. This is not really a valid test, since most components of a composite profile would simply fade out with altitude, not separate spatially. One can only measure $W(h)$.

Measured statistically, $W(h)$ does decrease for hydrogen and calcium and increase for helium. All observers agree at least on the sign but disagree on the magnitude. Some of the largest height changes were measured by Zirker (1963), Figure 19.

Measured *individually*, the situation is much different. Athay and Bessey, for example, state that profiles of individual spicules do not change with height; that the statistically observed decrease of W with h is caused by the fact that the relative population of narrow line profile is greater at high altitudes (i.e. narrow line spicules are *taller* than broad line spicules). This is in fact the origin of the two 'types' of

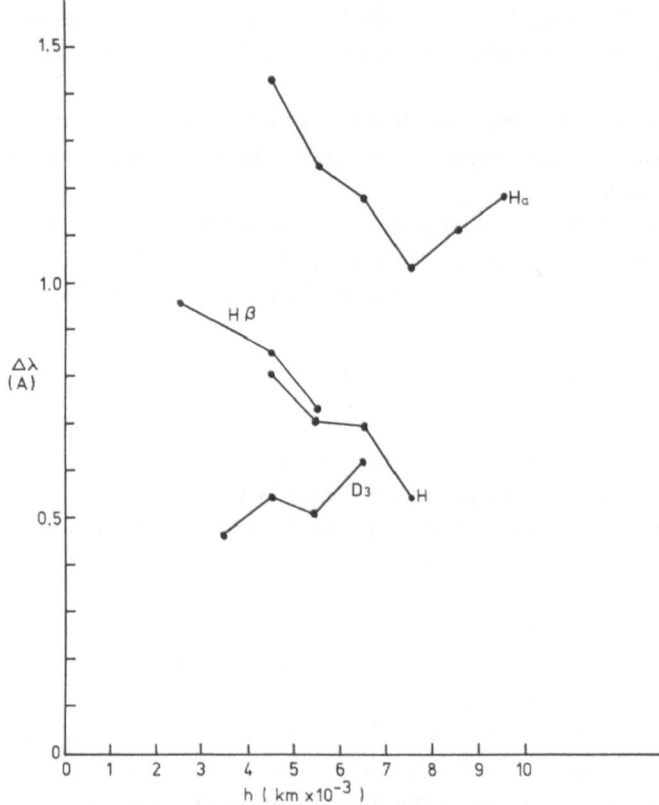

Fig. 19.   The height variation of the *average* half-widths (in ångströms) of spicule line profiles (from Zirker, 1962). Other average measurements show less dependence on h.

spicules. This point seems to be quite a source of argument. Pasachoff *et al.* (1968) have joined Athay and Bessey, while Zirker (1962), Gulyaev and Livshitz (1966) and Giovanelli *et al.* (1965) argue the opposite.

Actually the question of whether or not the broad line profiles come from unresolved clumps of spicules becomes somewhat academic if it is used as an explanation for the anomalously broad calcium lines. The reason for this is that, although one can find individual spicules with rather narrow calcium lines, those same spicules also have narrow hydrogen lines. This point was mentioned by Athay and Bessey, and shown rather clearly by Krat and Krat (1971). They measured the widths of 'wide line' spicules and 'narrow line' spicules for Hα, Hβ, K, H and D_3 (Table IV).

Although it is still not clear whether the wide line spicules are unresolved clumps of spicules or have their lines intrinsically broadened, it is clear that in the narrow line spicules, *all* the lines are narrow (in fact D_3 is so narrow as to put severe upper limits on T_e and/or ξ_t.) Thus, no matter which type of spicule one chooses as being 'the

TABLE IV

Mean widths of line profiles of spicules (after Krat and Krat, 1971)

Profile Characteristic	Line	Hα	Hβ	K	H	D₃
Narrow	No. of spicules	6	4	5	6	20
	$W (\times 10^{-4})$	0.87	1.11	0.84	0.82	0.32
Wide	No. of spicules	3	3	3	3	3
	$W (\times 10^{-4})$	1.68	2.08	2.19	2.14	0.97

real spicule', the $W^2(\mu^{-1})$ function remains non-linear. This can be seen on Figure 18, where Krat and Krat's data for the two types of spicules are shown.

4.4. SUMMARY AND INTERPRETATIONS

It is time to collect all the data mentioned in this section and attempt to combine it into a cohesive empirical model. First a few editorial comments are probably in order. The one thing that is clear from all this data is that spicules possess huge variations. It is also clear that observers possess an amazing proclivity for observational selection. These two points together spell danger and demand caution in any interpretation. Of course, the problem of observational selection is an unavoidable result of the spectrographic technique. Spicule spectra preferentially show spicules which are unusually bright, tall, more isolated or have larger Doppler shifts than the 'typical' spicule. But this only means that observers should be doubly careful to obtain a statistically meaningful sample of spicule spectra. And anyone who interprets such spectra should understand that he might be describing or explaining a very atypical spicule. A separate point is that, in contrast to the preceding section on disk phenomena, here there seems to be a clear need for more and better observations. Such observations are well within present capabilities; they will be mentioned in detail below.

4.4.1. *Mass Motions*

The picture which seemed rather clear a few years ago has now become if anything, a little more confused. The correspondence between proper motions and Doppler shifts in terms of axial motion of inclined spicules should be generalized somewhat to include the increased values of the Doppler shift and the observed horizontal motions. However, an improved empirical model of these motions is not possible at the present time due to the lack of detailed data. Furthermore, the observed diffusion of spicules into the corona is now in doubt. So the fate of the half of the spicules which are not observed to descend again is still an open question. The classical description of a spicule as rising along its axis at about 30 km s^{-1}, then either descending back down the same path or diffusing and descending somewhere else seems to remain reasonably valid as an average scenario. But there are significant uncertainties as to variations from this mean behavior which can and should be answered observationally. The tool to use is a time sequence of spectrograms combined with simultaneous entrance slit filtergrams.

4.4.2. *Internal Motions*

Interpretation in terms of internal motions centers on the explanation of the large widths of the spectral lines. There are two general possibilities: (1) the spicules are unresolved, and each spicule displays a different radial velocity, v_r. (2) the spicules possess structure, that is different lines are formed in different regions of the spicule, with different values of T_e, ξ_t and perhaps v_r. The most important point to make concerning these two possibilities is that they are not mutually exclusive. In fact, there is quite valid evidence which virtually demands that each possibility at least partially accounts for the line widths.

The first possibility has already been discussed in Section 4.3.4. To that discussion should be added two results from broad-band measurements. The measured diameters of spicules are about 900 km (Lynch, *et al.*, 1973; Dunn, 1960, 1965). This diameter is slightly less than the resolution of most spectrographs. Also, at least in Hα below 5000 km, there must be overlapping of spicules because spicule counts actually decrease in that region (Athay, 1959). So we must assume that all strong lines show effects of unresolved clumps of spicules below at least 5000 km. The composite profiles discussed in Section 4.2.2 confirm this. It has been argued (Athay and Bessey, 1964) that this effect cannot be the principal explanation for the large widths because then a decrease in widths would be accompanied by an increase in radial velocities of up to 30 km s^{-1}, and this was not observed. However these authors did detect at least a tendency for narrow widths to be correlated with large radial velocities, so this effect must be of secondary importance. The variation of line width with height furnishes evidence that is only inconclusive. The simultaneous measurements of Krat and Krat (Section 4.3.4) however furnishes conclusive evidence that lack of resolution will never explain the anomalous variation of line widths with atomic weight. So we can make two conclusions concerning this possibility of 'smearing broadening'; (1) Smearing does definitely artificially increase the widths of all lines by some as yet unknown amount. (2) Smearing broadening cannot account for the observed strange $W^2(\mu^{-1})$ behavior.

There is evidence for the second possibility of structured spicules in addition to a simple lack of an explanation of the $W^2(\mu^{-1})$ plot. First, the observation by Krat and Krat that some spicules show different radial velocities in different lines (Section 4.2.2). Further, the same authors note that D_3 spicules are more diffuse than Hα spicules and that "there appears considerable interspicular emission (in D_3) some-times nearly of the same intensity as the emission in spicules". There is supporting evidence, although weaker, from Pasachoff *et al.* (1968). The cross correlation be-tween the brightness in D_3 and in a wing of H or K is only 0.3. Even the correlation between Hε and H or K is only 0.6. Of course some part of the decrease in correlation is caused by differences in line strength, but still the correlations are rather low. The authors' own conclusions from this data should be noted; "A given spicule emits simultaneously in lines of hydrogen, helium and calcium. This does not mean that the emission comes from the same volume within that spicule." In summary, several different pieces of evidence seem to demand the conclusion that spicules are usually structured somehow, but lack of more detailed data prevents a specific model.

One of the simplest possible models is that of a spinning spicule, which has been investigated by Rodionov (1968) and Avery (1970). The hydrogen and helium lines fit a rotational velocity of 19–23 km s^{-1} (Rodionov); the calcium lines require 22–36 km s^{-1} (Avery). These calculations should be checked however not by line widths, but by line tilts (which measure ω^{-1}) of different lines *simultaneously*. The really im-portant aspect of the structured model is that different lines must be formed in dif-ferent regions, and there is not yet any direct evidence that demands this. Line tilts could be that evidence. Remember that, in the one specific observed example dis-cussed in print, the tilts of the Hα and the K line did not agree with each other if one were to assume a homogeneous spicule.

If we accept the fact that both 'rotation broadening' and 'smearing broadening' are at work simultaneously, we can to some extent understand the disparate observations of line widths and their height variation. Both broadening mechanisms will have distribution functions independent of each other, and both will be different functions of height. As a result of all the different possible combinations, we can expect to see a wide variety of phenomena. Broad lines could come from either rapidly rotating spicules *or* unresolved spicules and thus should be common. Narrow line spicules must be slowly rotating *and* well-resolved and thus rare. Lines which are narrow down low will stay narrow at greater heights. Lines which are broad down low could either remain broad at greater heights (due to a constant rotation) or become narrower (due to decreased smearing). All of this has been observed. The introduction of more com-plicated inhomogeneities or differential velocities, such as very narrow cores, sheaths, knots, etc. is of course a possibility and represents yet another level of complexity. At the present, this subject is a total unknown, both theoretically and observationally.

4.4.3. *Approximate Numerical Values*

Three basic different types of motion have been observed, all of which may or may not be lumped together as 'non-thermal' motion: radial velocity, v_r, rotational ve-locity, v_{rot}, and microturbulence, ξ_t. Each of these types of motion can vary from spicule to spicule independently. So it is not surprising to find widely differing values quoted in the literature. Any attempt to summarize actual values of these three types of motions necessarily requires a large degree of judgment (i.e. guesswork) and should include an indication of the range of variation that is possible from spicule to spicule. These are the current 'best guesses'.

$$v_r \approx 15 \pm 10 \text{ km s}^{-1}$$
$$v_{\text{rot}} \approx 20 \pm 10 \text{ km s}^{-1}$$
$$\xi_t \approx 7 \pm 7 \text{ km s}^{-1}$$

They all have large intrinsic dispersions. Note that there is no compelling evidence for supersonic microturbulence, but it is probable that mass motions are supersonic.

4.4.4. *Further Observations*

Due to the oft-mentioned wide variations, statistical analyses do not seem to be as powerful as analyses of individual spicules. But the 'individual' method *must* be made more comprehensive. More lines, particularly oxygen, must be observed simultaneously. For example, if someone were to observe Hα, D$_3$, the Ca IR triplet, the O triplet, and perhaps 10830 simultaneously, he could probably rewrite the book. Also, different heights should be observed simultaneously. This can be done with image slicers. Lastly, line tilts of the different lines should be observed simultaneously. This is very difficult, but there are certainly some secrets there, waiting to be discovered.

5. Comparison of Limb and Disk Phenomena

At this point it would be nice if one could combine limb and disk observations of fine structure motions into a single model. Unfortunately, this is not possible. In fact, I am troubled by several very large discrepancies between limb and disk observations. One can only hope that these discrepancies are not fundamental, but are rather a result of insufficient data. Thus, my purpose in this section will be to point out several of the important pieces of the puzzle that are conspicuous by their absence, and to suggest research which could possibly supply those missing pieces.

5.1. FLOW PATTERNS

One would hope that, through the study of resolved motions as a function of space and time, one could deduce flow patterns that would at least statistically result in a single model for limb and disk features. This is a classic puzzle problem, and there are too many pieces missing to proceed. On the disk, the profile-blind technique of Bhavilai and Title established the existence and spatial relationship of upflows and downflows, but left the connection between the two literally up in the air. The high resolution profile work of Grossman-Doerth and von Uexküll proved much better at measuring velocities, but did not address itself to the spatial relationships. Such work is sorely needed, and the 'tuned filtergram' technique is a perfect tool for this purpose.

On the limb, the question of the reality of apparent proper motions must really be considered open. The same can perhaps be said of the question as to what fraction of the observed Doppler shifts is due to axial motion and what fraction is due to horizontal motion. These questions can be addressed by some good solid analysis of tuned filtergram movies, for example, or time sequences of spectrograms combined with entrance slit filtergram movies. Furthermore, I note the fact that no one has really analyzed the horizontal motions of disk features. This is an obvious piece of work that should be done.

5.2. LINE PROFILES

5.2.1. *The Question of Velocities and Widths*

Both the velocities and line widths of Hα mottles on the disk, as measured by Bray and by Grossman-Doerth and von Uexküll, are considerably smaller than those seen on limb spectra. In fact, the measured widths on the disk, $W \approx 0.8 \times 10^{-4}$, agree with Beckers' spicule model, which has no non-thermal motions *at all*! Perhaps we should be thankful that we finally have a case where observed widths agree with computed widths without resorting to non-thermal motions, but I am troubled by it because neither case agrees with limb observations.

The disk velocities are almost an order of magnitude lower than the limb velocities. Grossman-Doerth and von Uexküll showed that this could be caused by seeing effects, i.e. loss of resolution. But, if this were the case, then the line widths should be correspondingly large, since an increased line width is the classic symptom of unresolved mass motion. But, as we have mentioned, the line widths are too small, not too large. So if the Hα mottles are deconvolved to achieve higher velocities, the widths would become smaller still and probably would even set severe upper limits on T_e. Unfortunately, this point was not investigated. It should be.

There exists an interesting possibility that the small disk widths and large limb widths imply that the unresolved mass motions in spicules are largely horizontal, as in rotational motion, or helical motion with a small pitch angle for example. This idea is however pure speculation at this time.

5.2.2. *Profiles near the Limb*

When confronted with a discrepancy, as in the line widths, between the disk and the limb, the logical place to look for an answer is in between; near the limb. Loughhead (1973) did exactly that and found that the Hα profiles showed characteristics very similar to spicule profiles; large widths, central reversals and asymmetries. Figure 20 shows contrast profiles of dark mottles at $\theta = 79°$, and they are starting to look very much like spicules. The widths are so broad that Loughhead couldn't even measure them. The central reversals violate the cloud model, as do the spicules. It is clear that a more sophisticated model than the cloud model is needed to provide the framework within which to analyze these mottle profiles. (And of course observations over a wide range of θ are also needed). However, if this could be accomplished it is obvious that this topic will be an extremely fruitful field of work.

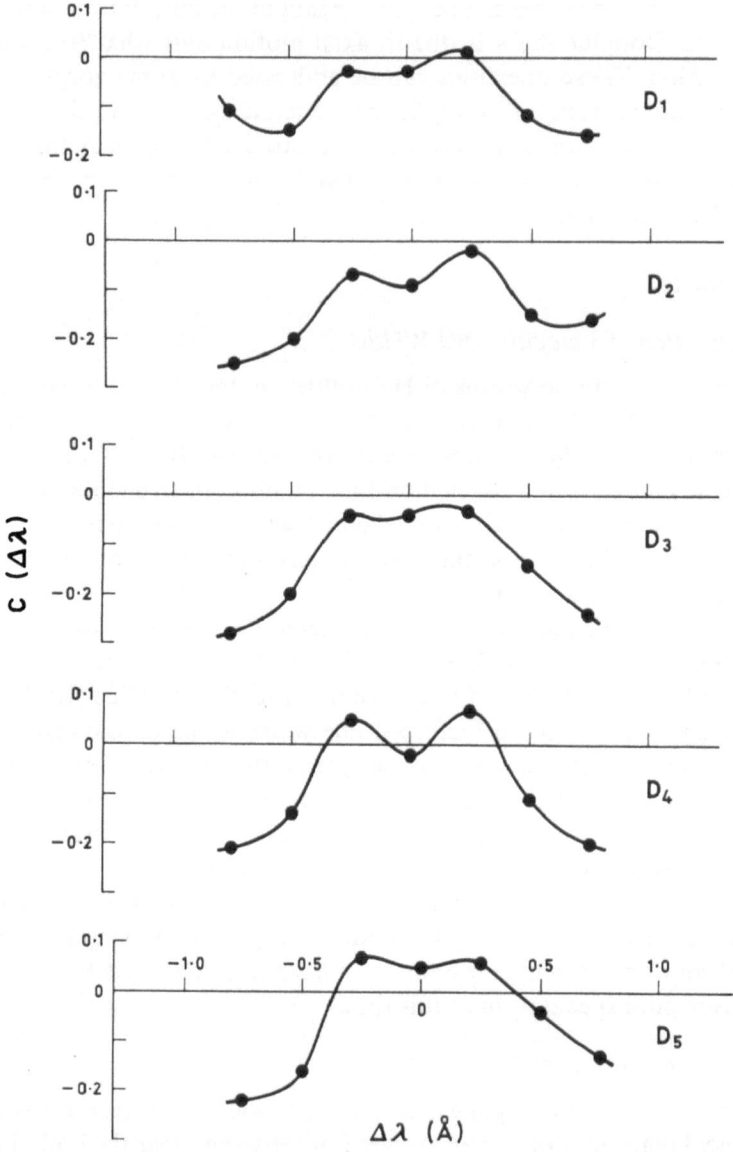

Fig. 20. Contrast profiles of dark mottles measured at $\theta = 79°$ (from Loughhead, 1973). The profiles exhibit central reversals, suggestive of spicules. The maximum contrast occurs beyond $\frac{3}{4}$ Å, the tuning limit of the filter.

5.3. Summary

There remains one last point of substance to be made. Lurking beneath the entire discussion of this last section is the question that when we compare disk and limb observations, are we really looking at the same features at all? It is perhaps appropriate to end this review on a negative note by pointing out the following possibility: On the disk we are looking at features which lie below about 4000 km and which are relatively highly inclined, whereas the features we see on limb spectra are primarily above 4000 km, are more vertical, move faster, and have larger internal motions. We don't see the high features on the disk because they are too thin optically; we don't see the low features on the limb because they are too crowded to be resolved.

Certainly such a situation would explain why we have such difficulty identifying spicules on the disk. There is actually only one strong argument against this idea: In spicule models (cf. Beckers, 1972) spicules are indeed optically thick in the strong lines, and that is perpendicular to their long axis. On the disk, one would be looking at an acute angle to the spicule axis, hence through a longer path length through the spicule. Thus the spicule would be optically even thicker and should be visible on the disk. Playing the devil's advocate though, one should point out that this argument is model dependent, and one must wonder how sensitive the predicted optical depth of a spicule is to reasonable changes in the model. Further, there is a possible geometrical rebuttal; perhaps spicules are actually much narrower than observed. Then their contribution to a spectrographic resolution element on the disk could be very small and easily overwhelmed by the underlying chromosphere.

This question is of course an open one and probably will remain so until we fully understand the profiles of Hα mottles as a function of θ. It is at least comforting to know that there is plenty of research yet to be done.

References

Athay, R. G.: 1959, *Astrophys. J.* **129**, 164.
Athay, R. G.: 1961, *Astrophys. J.* **134**, 756.
Athay, R. G.: 1970a, *Solar Phys.* **11**, 347.
Athay, R. G.: 1970b, *Solar Phys.* **12**, 175.
Athay, R. G. and Bessey, R. J.: 1964, *Astrophys. J.* **140**, 1174.
Athay, R. G. and Canfield, R. C.: 1969, in Groth and Wellman (eds.), 'Spectrum Formation in Stars with Steady State Extended Atmospheres', *U.S. NBS Special Report No. 332* (IAU Colloquium No. 2), p. 65.
Athay, R. G. and Moreton, G. E.: 1961, *Astrophys. J.* **131**, 935.
Athay, R. G. and Thomas, R. N.: 1957, *Astrophys. J.* **125**, 806.
Avery, L. W.: 1970, *Solar Phys.* **13**, 301.
Bappu, M. K. V. and Sivaraman, K. R.: 1971, *Solar Phys.* **17**, 316.
Beckers, J. M.: 1962, *Australian J. Phys.* **15**, 327.
Beckers, J. M.: 1964, Thesis, Utrecht.
Beckers, J. M.: 1966, in K. O. Kiepenheuer (ed.), *The Fine Structure of the Solar Atmosphere*, Franz Steiner Verlag, Wiesbaden.
Beckers, J. M.: 1968, *Solar Phys.* **3**, 367.
Beckers, J. M.: 1972, *Ann. Rev. Astron. Astrophys.* **10**, 73.
Beebe, H. A.: 1971, *Solar Phys.* **17**, 304.
Beebe, H. A. and Johnson, H. R.: 1969, *Solar Phys.* **10**, 79.
Bhavilai, R.: 1965, *Monthly Notices Royal Astron. Soc.* **130**, 411.
Bhavilai, R.: 1966, in K. O. Kiepenheuer (ed.), *The Fine Structure of the Solar Atmosphere*, Franz Steiner Verlag, Wiesbaden.

Bray, R. J.: 1973, *Solar Phys.* **29**, 317.
Canfield, R. C.: 1969, *Astrophys. J.* **157**, 425.
Canfield, R. C.: 1971, *Solar Phys.* **20**, 275.
Cannon, C. J.: 1971, *Solar Phys.* **16**, 314.
Cannon, C. J.: 1971, *Solar Phys.* **21**, 82.
Cannon, C. J. and Rees, D. E.: 1971, *Astrophys. J.* **169**, 157.
Cram, L. E.: 1972, *Solar Phys.* **22**, 375.
de Jager, C.: 1959, *Handbuch der Physik* **LII**, 80.
Dunn, R. B.: 1960, Thesis, Harvard.
Dunn, R. B.: 1965, *AFCRL Env. Res. Paper*, No. 109.
Giovanelli, R. G., Michard, R., and Mouradian, Z.: 1965, *Ann. Astrophys.* **28**, 871.
Grossman-Doerth, U. and von Uexküll, M.: 1971, *Solar Phys.* **20**, 31.
Grossman-Doerth, U. and von Uexküll, M.: 1973, *Solar Phys.* **28**, 319.
Gulyaev, R. A.: 1965, *Soln. Dann.* **6**, 51.
Gulyaev, R. A. and Livshitz, M. A.: 1965, *Astron. Zh.* **42**, 584.
Jones, R. A. and Rense, W. A.: 1970, *Solar Phys.* **15**, 317.
Krat, V. A.: 1972, *Solar Phys.* **27**, 39.
Krat, V. A. and Krat, T. V.: 1961, *Izv. Pulkova* **22**, No. 167, 6.
Krat, V. A. and Krat, T. V.: 1971, *Solar Phys.* **17**, 355.
Kulander, J. L.: 1967, *Astrophys. J.* **147**, 1063.
Kulander, J. L.: 1968, *J. Quant. Spectr. Radiative Transfer* **8**, 273.
Kulander, J. L. and Jeffries, J. T.: 1966, *Astrophys. J.* **146**, 194.
Leighton, R. B., Noyes, R. W., and Simon, G. W.: 1962, *Astrophys. J.* **135**, 474.
Linsky, J. L. and Avrett, E. H.: 1970, *Publ. Astron. Soc. Pacific* **82**, 169.
Lippincott, S. L.: 1957, *Smithsonian Contrib. Astrophys.* **2**, 15.
Liu, S. Y.: 1973, Preprint.
Liu, S. Y., Sheeley, N. R., Jr., and Smith, E. v. P.: 1972, *Solar Phys.* **23**, 289.
Liu, S. Y. and Smith, E. v. P.: 1972, *Solar Phys.* **24**, 301.
Livshitz, M. A.: 1966, *Astron. Zh.* **43**, 718.
Loughhead, R. E.: 1973, *Solar Phys.* **29**, 327.
Lynch, D. K., Beckers, J. M., and Dunn, R. B.: 1973, *Solar Phys.* **30**, 63.
Mein, P.: 1971, *Solar Phys.* **20**, 3.
Michard, R.: 1956, *Ann. Astrophys.* **19**, 1.
Michard, R.: 1959, *Ann. Astrophys.* **22**, 547.
Mouradian, Z.: 1965, *Ann. Astrophys.* **28**, 805.
Mouradian, Z.: 1967, *Solar Phys.* **2**, 258.
Nikolskaya, K. I.: 1967, *Astron. Zh.* **44**, 1043.
Nikolsky, G. M.: 1970, *Solar Phys.* **12**, 379.
Nikolsky, G. M. and Platova, A. G.: 1971, *Solar Phys.* **18**, 403.
Nikolsky, G. M. and Sazanov, A. A.: 1966, *Astron. Zh.* **43**, 928.
Noyes, R. W.: 1965, in R. N. Thomas (ed.) 'Aerodynamic Phenomena in Stellar Atmospheres', *IAU Symp.* **28**, 468.
Pasachoff, J. M.: 1970, *Solar Phys.* **12**, 202.
Pasachoff, J. M., Noyes, R. W., and Beckers, J. M.: 1968, *Solar Phys.* **5**, 131.
Rodionov, V. V.: 1968, *Vestn. Mosk. Univ.* **4**, 33.
Rush, J. H. and Roberts, W. D.: 1954, *Australian J. Phys.* **7**, 230.
Skumanich, A., Smythe, C., and Frazier, E. N.: 1972, *Bull. Amer. Astron. Soc.* **4**, 391.
Title, A. M.: 1966, Thesis, Calif. Inst. of Tech.
Vernazza, J. G., Avrett, E. H., and Loeser, R.: 1973, *Astrophys. J.* **184**, 605.
Weart, S. R.: 1970, *Solar Phys.* **14**, 310.
Wilson, A. M.: 1973, Preprint.
Wilson, P. R.: 1970, *Solar Phys.* **15**, 139.
Wilson, P. R. and Evans, C. D.: 1971, *Solar Phys.* **18**, 29.
Wilson, P. R., Rees, D. E., Beckers, J. M., and Brown, D. R.: 1972, *Solar Phys.* **25**, 86.
Zirin, H.: 1966, *The Solar Atmosphere*, Blaisdell Publ. Co., Waltham.
Zirker, J. B.: 1962, *Astrophys. J.* **136**, 250.
Zirker, J. B.: 1967, *Solar Phys.* **1**, 204.

DISCUSSION

(i) *Properties of the Chromospheric Structure as Seen in the Ca$^+$ Lines (H and K)*

There was some discussion on the relative abundance of regions with different types of H and K profiles (with single K2 peaks, double peaks, no peaks at all etc.) in spectra of the disk chromosphere (*Sivaraman, Wilson, Jensen*). *Pasachoff* pointed out that the spectrum he had analyzed was selected to show singly-peaked profiles. It was meant to demonstrate the reality of such profiles rather than to provide statistics valid for the whole Sun. *Newkirk* asked what the physical interpretation of the different kind of profiles is. The only reply to this was by *Cram* who described his view of the profiles of the supergranule center structures. He interprets the emission features called grains (greatly enhanced K$_{2v}$ peak) to be the results of propagating waves and shocks. The regions of completely quiet chromosphere between the grains have probably no K reversals at all (which could mean that the chromosphere there is transparent). *Cram* stresses that it is necessary to use time series of line profiles in the derivation of physical conditions. The grains, and the area in which they are embedded, show a strong 3-min oscillation in intensity (*Frazier, Zirin*), which again shows the need for a time dependence study. Stix points out that the 3-min period is precisely the type of period which falls just above the region in the (k, w) diagram where propagation is forbidden. Some inconclusive discussion followed by whether these waves therefore propagate or not (*Meyer, Frazier, Sturrock*).

(ii) *Line Profiles of Spicules*

Giovanelli described spectroscopic observations by Harvey, Hall and himself of the limb in the three optically thin $\lambda 10830$ (He), P_{β}(H) and $\lambda 8542$ (Ca$^+$) lines with 23″ resolution, thus including many spicules (see also Session A discussions). They find an increase in Doppler width with height, large internal motions (~ 20 km s^{-1}) and very low temperatures. *Athay* discussed the nature of the broad spectral profiles for some spicules. These, he believes, are not the result of clusters of unresolved spicules with narrow profiles, since one would expect in that case a systematically larger Doppler shift for the narrow profiled features. Often one sees the broad H and K line profiles break up in irregular, bumpy profiles (*Beckers*). The same is occasionally seen in Hα although it is not as pronounced there because of the larger thermal width. This is even the case when spicules are observed at great heights where the chances of unresolved spicules are very small. It suggests that the emission arises from inhomogeneities within the spicule. *Schmidt* sees no theoretical difficulties in producing such an inhomogeneous-multi-component-spicule.

(iii) *Transverse Motions of Spicules*

Schmidt and *Meyer* raised the point of the sideways (or transverse) motion of spicules. If the spicule consists of a 10000 km long column with 10^{11} cm^{-3} electron density in a field of a few Gauss one expects a sideways oscillation of a few minutes period (*Meyer*). *Nikolski* and *Platova* indeed claim to see oscillatory sideway motions of a few minute period (*Frazier*). *Pasachoff* pointed out that Doppler shifts seen in spicules could be due to these transverse motions (although he does not exclude the possibility of them being due to up and down motion or to a unidirectional motion along a magnetic field loop). *Zirin* and *Beckers* question the reality of the reported transverse motions of spicules. The same goes for the transverse motions (or the so called flagellant motions) reported to be associated with dark Hα mottles on the disk (*Zirin*).

WAVES AND OSCILLATIONS IN THE CHROMOSPHERE
IN ACTIVE AND QUIET REGIONS

R. G. GIOVANELLI

CSIRO Division of Physics, National Standards Laboratory, Sydney, Australia 2008

Abstract. Oscillations and waves have been observed in Hα in sunspot umbras and penumbras. They appear to be basically of the Alfvén type, but do not carry away sufficient energy to account for the sunspot energy deficit.

There is a pronounced difference in the velocity structures of supergranule centres and boundaries. In the former it is oscillatory, of shorter period but bigger amplitude than in the photosphere. In the network there are oscillations of the fibrils but, on the whole, little evidence of wave motion along their lengths except perhaps near their diffuse ends.

In plages, there are velocity oscillations which at times may be phase coherent over extensive regions up to 10^5 km apart; there are suggestions of a connection between these times and the occurrence of flares. There is also a report of a velocity pulse travelling along an active-region fibril with a velocity of propagation of 150 km s^{-1}.

1. Introduction

For several years following Leighton *et al.* (1962) discovery of photospheric oscillations, their behaviour was studied almost exclusively via time variations in spectra. But in the past four years there has been a growing tendency to use spectroheliographs and filters, as a result of which we are getting more-detailed two-dimensional information. The most spectacular result has been the discovery of waves propagating outwards over sunspot penumbras. We trace below the discovery of these waves and the associated umbral oscillations, and review present knowledge of oscillations and waves in quiet and active parts of the chromosphere.

2. Sunspots

Sunspot umbras exhibit vertical velocity oscillations at low levels. Beckers and Schultz (1972), using Fe 5434 which has zero Zeeman splitting, found a peak-to-peak amplitude of 1 km s^{-1} and a period of 178 s in one umbral centre, and a longer period of 255 s at the umbral-penumbral boundary; in four other umbras there were no apparent motions. Bhatnagar *et al.* (1972), using unidentified lines formed only in sunspots and also with zero Zeeman splittings, found peak-to-peak amplitudes of 0.5 km s^{-1} and periods of 448 s (6021 Å) and around 310 s (6252, 6910 Å) in one sunspot. In Fe II 4924, Sheeley and Bhatnagar (1971) had earlier found no umbral or penumbral oscillations, perhaps because of inadequate sensitivity, while Beckers and Schultz (1972) found no intensity oscillations in the continuum. There are therefore clear differences in the strengths of oscillations observed in different umbras and at different heights though, as Bhatnagar *et al.* (1972) indicate, the amplitudes are less than in the normal photosphere. We should reserve judgment on whether the reported differences in period in

R. Grant Athay (ed.), Chromospheric Fine Structure, 137–151. All Rights Reserved.

different lines are significant until observations have been made simultaneously at the one position in these lines.

The first evidence of sunspot oscillations in a chromospheric line was presented by Beckers and Tallant (1969), who described umbral 'flashes' in the K line, of diameter up to 2000 km, tending to repeat at intervals up to 145 s. Individual flashes lasted about 50 s, with a faster rise than decay. There were associated transverse velocities of some 40 km s^{-1} towards the penumbra, though these were thought unlikely to be due to material motions. Doppler shifts showed the vertical velocities to be mainly upwards at 6–7 km s^{-1}, though there was a brief, rapid change to a downflow just before the next flash. Flashes in the IR Ca$^+$ line occurred virtually simultaneously (Beckers and Schultz 1972), but while some umbral flashes occurred in Hα, these were much less frequent. No connection was found between flashes and lower-level umbral velocities in Fe 5434.

While the nature of umbral flashes remains vague, interest has swung recently to waves and oscillations observable in Hα. At Big Bear, Stein and Zirin (1972, and private communication) have observed sunspots through an Hα filter, finding waves propagating outwards over the penumbra at velocities typically 8–12 km s^{-1}. They reported these as usually seen best at line centre, but occasionally in the wing out to 0.7 Å. They occur in almost every sizable spot with a stable regular penumbra, but rarely in active spots with complex structure. In almost all cases the period lay between 4 and 5 min, the shortest observed being 150 s in a spot of particularly strong magnetic field. They inclined to the view that these represented sound waves.

Independently, at Culgoora, Giovanelli (1972) studied sunspots through an Hα filter, using a polarizing beam-splitter to secure photographs at effectively 0.25 Å from line centre simultaneously in opposite wings. Subtraction of one from the other yields the line-of-sight velocity distribution. The resulting films of sunspots near disk centre also showed waves propagating outwards over the penumbra (Figure 1). The first sunspot measured at Culgoora had periods of the order of 210 s and velocities of 15–25 km s^{-1}. Measurements on this spot over a longer time interval and on other spots favour 15 km s^{-1} as more typical (Table I), though wavelength (which is not easy to measure) and period vary appreciably. The peak line-of-sight velocity, v, measurable only with difficulty, is of the order of 1 km s^{-1}. The waves observed at Big Bear and Culgoora are undoubtedly the one phenomenon though the measured properties differ somewhat; some at least of the differences are due to real differences between one spot and another.

The wave mode can be studied using sunspots at different positions on the disk. Longitudinal modes, as in acoustic waves, would give Doppler shifts for spots near the limb when the waves propagate along the disk radius vector. Transverse waves with horizontal displacements can be studied in spots near the limb when the waves propagate at right angles to the line of sight. We have looked for modes of these types but failed to find any trace; at most, they are below our resolution limits.

Searches have also been made for waves with an 0.7 Å filter centred on Hα; with this pass band, pure Doppler shifts without an associated change in depth of the line

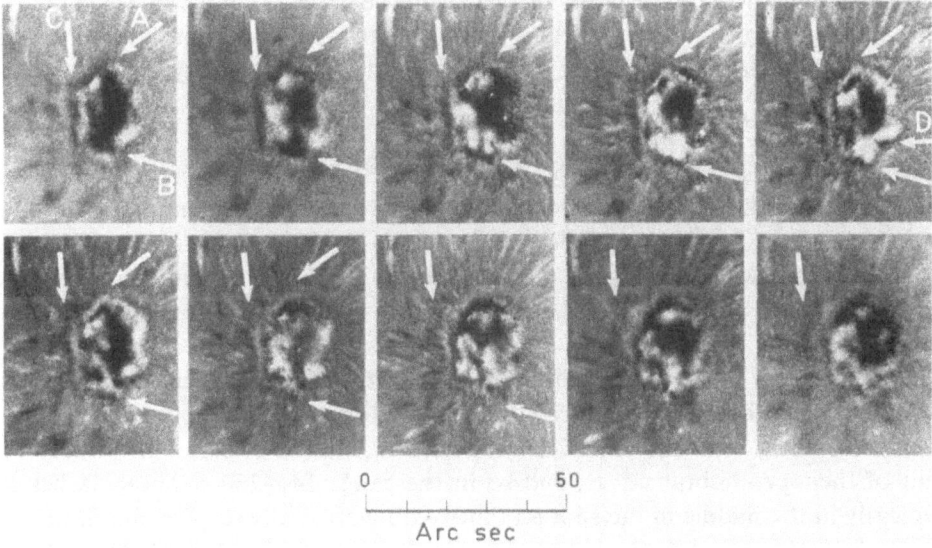

Fig. 1. Hα velocity distributions across a sunspot and surrounding regions at 30-s intervals, obtained by subtracting simultaneous photographs obtained in opposite wings at ±0.25 Å from line centre. Dark indicates downward velocities, bright upward. *A*, *B* and *C*, and the corresponding arrows on the various frames, point to and lie in the same line as progressive positions of wave fronts moving outwards from the umbra across the penumbra.

should produce no change in the penumbral intensity. No intensity waves have been detected, in conflict with Stein and Zirin's result. Nor have such waves been detected in the continuum or in velocities in Fe 6569. We have searched for the latter using the 1/8 Å filter and a polarizing beamsplitter giving images at ±1/16 Å from line centre simultaneously, but whether velocity waves are absent in this line or the observing technique has been inadequate is just not known. Further observations are needed.

The nature of the wave follows from these results. The velocity observed by Giovanelli exceeds the sound velocity considerably, though Stein and Zirin's velocity is close to that of sound. Preferring my own results, I infer that they cannot be basically acoustic waves – and in any case the observed displacement is transverse, not longitudinal as in acoustic waves. Gravity waves propagate much more slowly. There remain only

TABLE I

Properties of penumbral waves (author's observations)

Sunspot date	Period (s)	Wavelength (10^3 km)	V (km s^{-1})
71–VII–4	240	3.35	14
71–XI–3	210	3.1–5.2	15–25
	200	2.35	12
72–III–10	230	2.75	12
72–IV–21	180	3.8	21

Alfvén waves, which can certainly propagate in the observed mode. We may note that in the penumbra the magnetic field is radially outwards, and inclined at only a small angle to the horizontal; typically, $H \approx 1200$ G. The Alfvén velocity is

$$V_A = H/(4\pi\varrho)^{1/2},$$

and with the observed velocity of $V_A = 15$ km s^{-1}, the density ϱ becomes 5×10^{-8} gm cm^{-3}. This fits about as closely as we could guess to the penumbral density at $\tau \approx 1$ for H$\alpha \pm 0.25$ Å. We conclude that the waves have been identified as Alfvén. The total flux in the wave, $\frac{1}{2}\varrho v^2 V_A$ per unit area normal to the direction of propagation, is 4×10^8 erg cm^{-2} s^{-1}, much less than the photospheric radiation flux of 6.4×10^{10} erg cm^{-2} s^{-1}.

Is the velocity uniform or accelerated? One cannot give a definite answer as yet, though at times one has the impression of a positive acceleration outwards.

In the umbra, Stein and Zirin have reported Hα flashes of period invariably $\frac{1}{2}$ to $\frac{2}{3}$ that of their penumbral waves, and so in the range 140–200 s. These flashes start typically in the middle of the spot and move outwards; as every second flash moves to the edge of the umbra, the inner edge of the penumbra brightens and a new wave begins. It is by no means established whether these are related to the umbral flashes in K. However it is almost certain that they are related to umbral oscillations observed in Hα by Giovanelli (1972). Using the beamsplitter technique described above, he has found two interesting properties of all umbras studied:

(i) Near disk centre there is a quasi-oscillatory velocity distribution (Figure 2) whose structural elements have a size of the order of 3″, a period typically 165 s, and velocity amplitude of the order of 3 km s^{-1}. In a 30-min set of observations, oscillations of individual points were followed for almost the whole sequence, but the pattern as a whole broke down after 1 or 2 periods since the individual points have differing periods. In one umbra, periods of 137 and 185 s were recorded at different points.

(ii) There is a fascinating behaviour of the locus of points of the same phase, i.e. the wavefront. This undergoes continual change. New wavefronts often develop near the centre of the umbra, though they may be eccentric or appear at the edge of the umbra and move across to the opposite side. They sometimes swirl around the umbra, and the sense of swirling can change rapidly from clockwise to anticlockwise or *vice versa*, demonstrating that at a given time the period varies across the umbra, and changes with time at a given point. On the whole, the wavefronts progress outwards towards the penumbra, but their behaviour is far more complicated than a mere symmetrical expansion from the centre outwards.

Giovanelli, also, finds that there is a close relation between umbral oscillations and penumbral waves; the latter run smoothly out of the umbral waves. Both form part of the one wave system (Giovanelli's measured differences in period are almost certainly due to sampling, and of no physical significance apart from forming additional examples of the variability of period from point to point and in time). Specifically, this demonstrates that there is not merely energy storage in the oscillations, but a flux of energy out of the umbra and across the penumbra. Simultaneous observations are

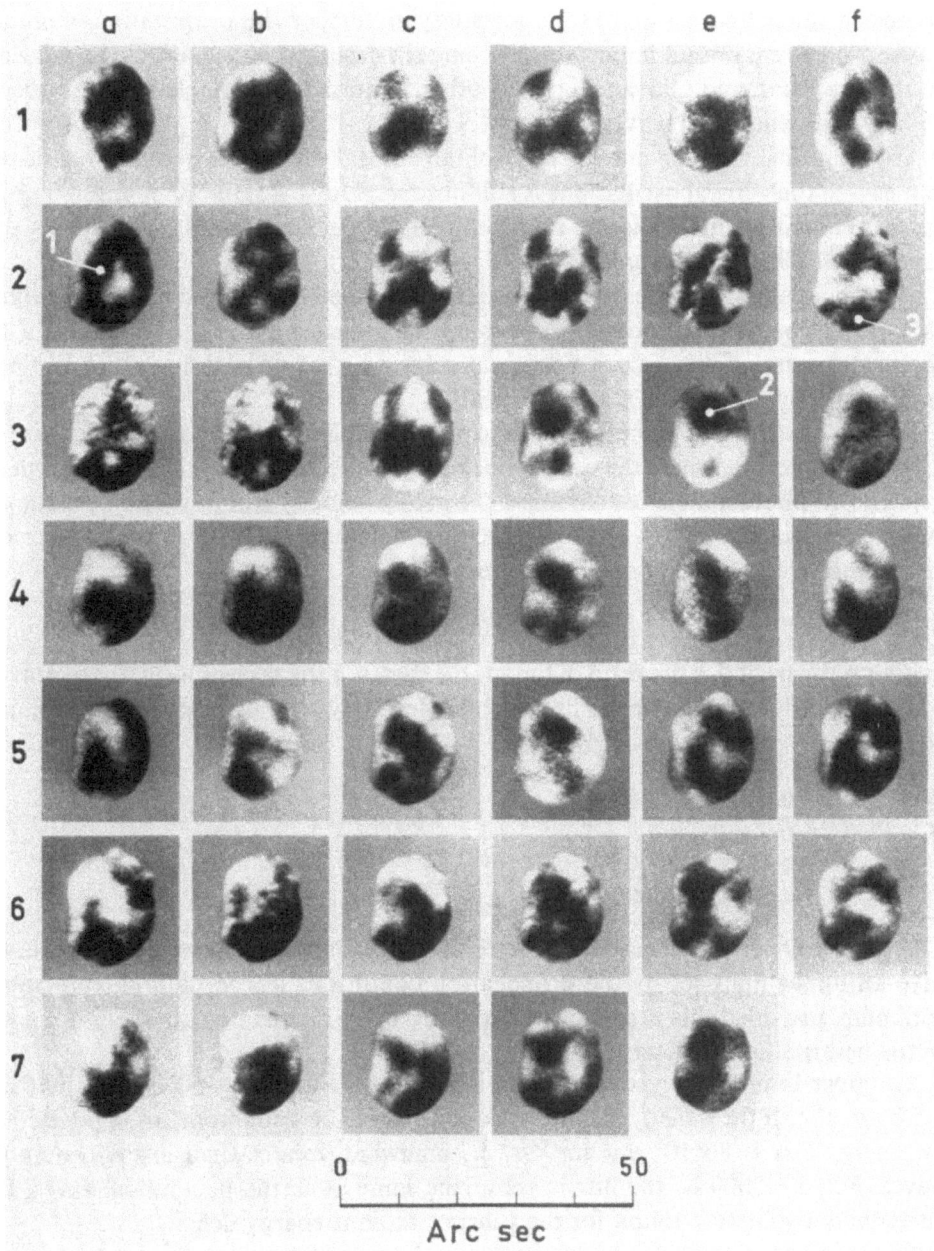

Fig. 2.   Array of Hα velocities in an umbra at 30-s intervals. The sharp boundary is due to instrumental saturation. Point No. 1 is oscillatory with maximum downward velocities on frames 1b, 2a, 2e, 3e, 4b, 5a, 5f, 6d. Two other oscillatory points are 2 (frames 1a, 2a, 2f, 3e, 4d, 5d, 6e) and 3 (frames 1a, 2a, 2f, 3f, 4f, 6a, 7b).

needed in other lines as well as Hα for studying further the propagation of umbral waves. There remains an important difference in description between Stein and Zirin on the one hand, and Giovanelli on the other, as to the precise relationship between umbral and penumbral waves.

Any relationship of the fine structure of the umbral velocity pattern – 1″ on the best frames – to the general oscillations is still obscure. However, the Hα intensity in both the umbra and penumbra is not related directly to the pattern of oscillations – they have quite differing structures.

The modes of umbral oscillation have been examined in large umbras near the limb. No evidence of any oscillations has been found, indicating that the amplitude of any horizontal component is again comparatively small. Thus both in umbra and penumbra the displacement is mainly vertical.

The oscillations are therefore more-or-less parallel to the magnetic field in the centre of the umbra, where the wave mode must be effectively identical with that in the well-known photospheric oscillations. The group velocity can be found by assuming them to be adiabatic. For a region where the undisturbed temperature is isothermal, the vertical group velocity may be written

$$U = C_s (1 - \tfrac{1}{16}\pi^{-2t2}\gamma^2 g^2 C_s^{-2})^{1/2},$$

where t is the period, γ the ratio of the specific heats, g the acceleration due to gravity, and C_s the velocity of sound. The latter is effectively that in neutral hydrogen, and $\gamma = 5/3$. Propagation occurs only when

$$t\gamma g < 4\pi C_s,$$

or

$$t < 228 \text{ s at } 5000 \text{ K},$$
$$< 177 \text{ s at } 3000 \text{ K}.$$

Umbral temperatures may be close to 3000 K, and the observed periods should have an upper limit set by the propagation conditions at the umbral temperature minimum, provided this lies below the region from which Hα escapes. This may well be the best method of determining T_{min}.

An upper limit to the group velocity is given by taking T as 5000 K, so that U is $\leqslant 5.8$ km s^{-1}. If the density is about the same as in the penumbra, an upper limit to the energy flux is 8×10^8 erg cm^{-2} s^{-1}, somewhat greater than in the penumbral waves. For $T = 3000$ K, the flux is about the same as in the penumbral waves. It is inadequate by far to account for the sunspot radiant energy deficit.

The penumbral waves are obviously due to cross-coupling between the effectively vertical gas motions in the umbra and the sunspot magnetic field, which becomes more and more inclined from the vertical on going towards and into the penumbra. The force tubes are waved up and down, generating Alfvén waves with a predominantly vertical displacement.

It is not yet known why the waves disappear at the outer penumbral boundary. The decreasing density outwards should result in an outward acceleration of the penumbral waves. While this has not been observed, it is a difficult measurement to make because of the diffuseness of the waves. Perhaps this question will be resolved when we can observe a very large sunspot under favourable seeing.

3. Quiet Chromosphere

There are obvious differences between the oscillatory behaviours of supergranule centres and boundaries. Most investigators have used spectrograms, the test of location on supergranule structures being the intensity in the K or Hα line.

Over supergranule centres, the period of velocity fluctuations decreases with height; there are analogous intensity fluctuations throughout the K line and in Na 5896 (Orrall, 1966; Tanenbaum et $al.$, 1969; Liu and Sheeley, 1971). At the height of formation of K_1 and K_2, the intensity period is about 250 s (Jensen and Orrall 1963). At the level of K_1, intensity maxima lead upwards velocity maxima in photospheric lines by about 90° in phase, indicating adiabatic conditions and standing, rather than running, sound waves. Tanenbaum et $al.$ (1969) found similar conditions for Na 5896; they also estimated the coherence length (the size of the region of uniform phase) to be 3000 km. The period is significantly lower in the cores of Hα and K_3, in the range 150–210 s (Jensen and Orrall, 1963; Orrall, 1966; Elliott, 1969; Bhatnagar and Tanaka, 1972), while the velocity amplitude in K_3 has risen to about 1.6 km s^{-1} as against 0.25–0.5 km s^{-1} in the photosphere (Orrall, 1966). Deubner (1971) has reported a more-or-less uniform distribution of amplitudes up to 3 km s^{-1} in Hα, so that Hα and K_3 give concordant mean velocity amplitudes of about 1.5 km s^{-1}.

A wider spread of periods has been found at supergranule boundaries, for which the results appear to be somewhat more discordant. Liu and Sheeley (1971) found the period of intensity fluctuations in K_{2v} to be 5 min, substantially longer than in supergranule centres, and they reported bigger amplitudes also. In K_3, Orrall (1966) found many periods of velocity fluctuations grouping around 180 s, and there was also quite a wide spread from 5 min upwards. In Hα, Elliott (1969) found velocity fluctuations having periods around 150–210, 287 and about 900 s, while Bhatnagar and Tanaka (1972) reported 312 ± 56 s in the bright centres of rosettes. However, the agreements in respect of supergranule centres and mild disagreements in respect of the boundaries hide an array of puzzles. For example, in Hα, supergranule centres contain two quite distinct structures, the fibrils (or threads) and the grains. The latter are at a substantially lower level than the fibrils, and may be seen clearly when the fibrils become faint or disappear because of inadequate opacity. The grains also appear between the ends of fibrils under favourable conditions. Do the reported velocity oscillations apply to both these structures, or to one only? Supergranule boundaries too contain quite a pronounced structure, and one wonders what differences there may be from point to point.

Velocity films in Hα hold out the hope of clarifying such issues. Films of this type obtained at Culgoora – similar to those of Ramsay at Lockheed – show a vigorous state of oscillation, with quite distinct patterns in different regions.

(a) *The Grains.* In the central portions of supergranule centres – i.e. in those regions where we can see the grains clearly – the regions of oscillation have no particular shape, and appear to be distributed more or less at random. I have sought evidence of any propagation of these oscillations across the surface of the sun but have found none. It would be interesting to compare the patterns of oscillation in Hα and K_3, and in Hα and lines from underlying regions, using simultaneous filtergrams or spectroheliograms obtained at intervals of at most 15 s and preferably much less.

(b) *The Fibrils.* Where the fibrils are well developed, the system of motion is quite different and is still inadequately studied. However the regions of uniform phase appear to be elongated and aligned parallel to the pattern of the fibrils, and this is particularly noticeable where the fibril alignment varies rapidly from place to place – here the individual elongated oscillating regions copy faithfully the varying fibril directions.

It is not eacy to locate the place where pattern (a) gives way to pattern (b). The former seems to extend somewhat into the fibril zone, but whether this means that the diffuse ends of the fibrils partake of the oscillations of the grainy region or that we are able to see the underlying region through or between fibrils is uncertain.

On the whole, the oscillations in the regions of the fibrils do not exhibit obvious wave motions along their lengths. However, we occasionally find places where there seem to be motions along the fibrils and these are generally located near the diffuse ends. It is probable that they represent Alfvén waves, but better observations are needed.

At supergranule boundaries, velocity fluctuations are of much reduced amplitude at the level corresponding to Hα ± 0.25 Å, quite the opposite of what Liu and Sheeley found for intensity fluctuations in K_{2v}. There is little correlation between intensity in Hα and direction of motion at this wavelength, a result in good agreement with Grossman-Doerth and von Uexküll's (1971, 1973) ffndings. Velocity films have not yielded obvious patterns of the behaviour with time; there is need for a much better study than any reported so far.

4. Active Regions

Fibrils near sunspots are almost certainly of the same basic structure as those in the quiet chromosphere. Bhatnagar and Tanaka (1972) have examined one region where the fibrils merged together to produce an almost structureless pattern, but failed to find any oscillations in Hα -0.5 Å. Giovanelli has recently been examining fibrils in active regions, finding that on the whole these too exhibit very few velocity fluctuations. However, these are occasional events which almost certainly represent Alfvén waves (Figure 3), a velocity pulse running along a fibril at a propagation speed of the order of 150 km s^{-1}. Much more study needs to be given to these events, including their connection, if any with changes in the appearance of fibrils at line centre.

Bhatnagar and Tanaka (1972) have studied intensity oscillations in plages at Hα -0.5 Å, finding periods of 282 ± 49 s. Subsequently Tanaka (private communication) was able to demonstrate that these were oscillations mainly of velocities. He found that

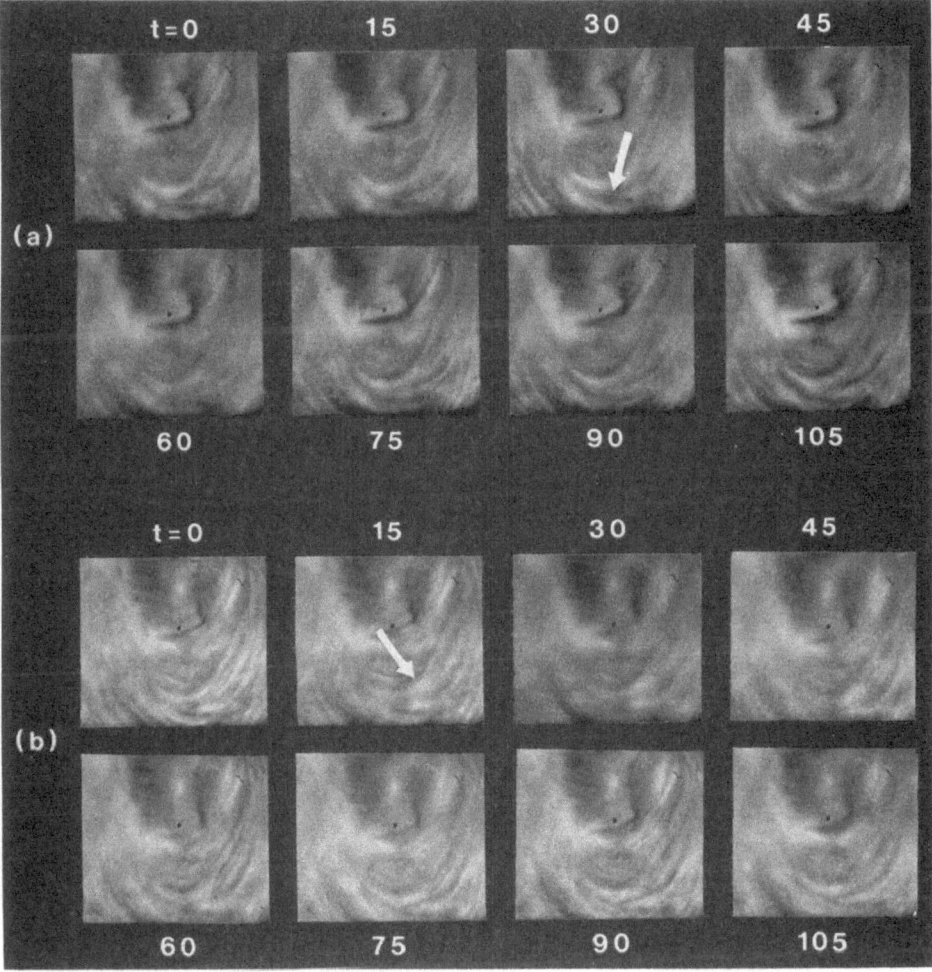

Fig. 3. Velocity pulses in active-region fibrils. (a) A downward displacement running from right to left (dark), (b) a similar upward displacement shortly beforehand (bright). Velocity of propagation is 150 km s^{-1} in both cases. Successive frames are at 15-s intervals.

the oscillations occur in individual points, 350–700 km across, spaced at intervals of about 2300 km.

An important feature of these oscillations is the phase coherency, which extends at times not only over the one plage but over plage regions up to 10^5 km apart (Figure 4). The propagation of this phase takes place at velocities of at least 1000 km s^{-1}. The duration of coherency of phase is typically 10–20 min, and the total fraction of the time during which phases are coherent has varied from 5–40%. This is an obviously intriguing phenomenon. It may also be of wider importance if further observation confirms Tanaka's suggestion that flares are associated with periods of phase coherence.

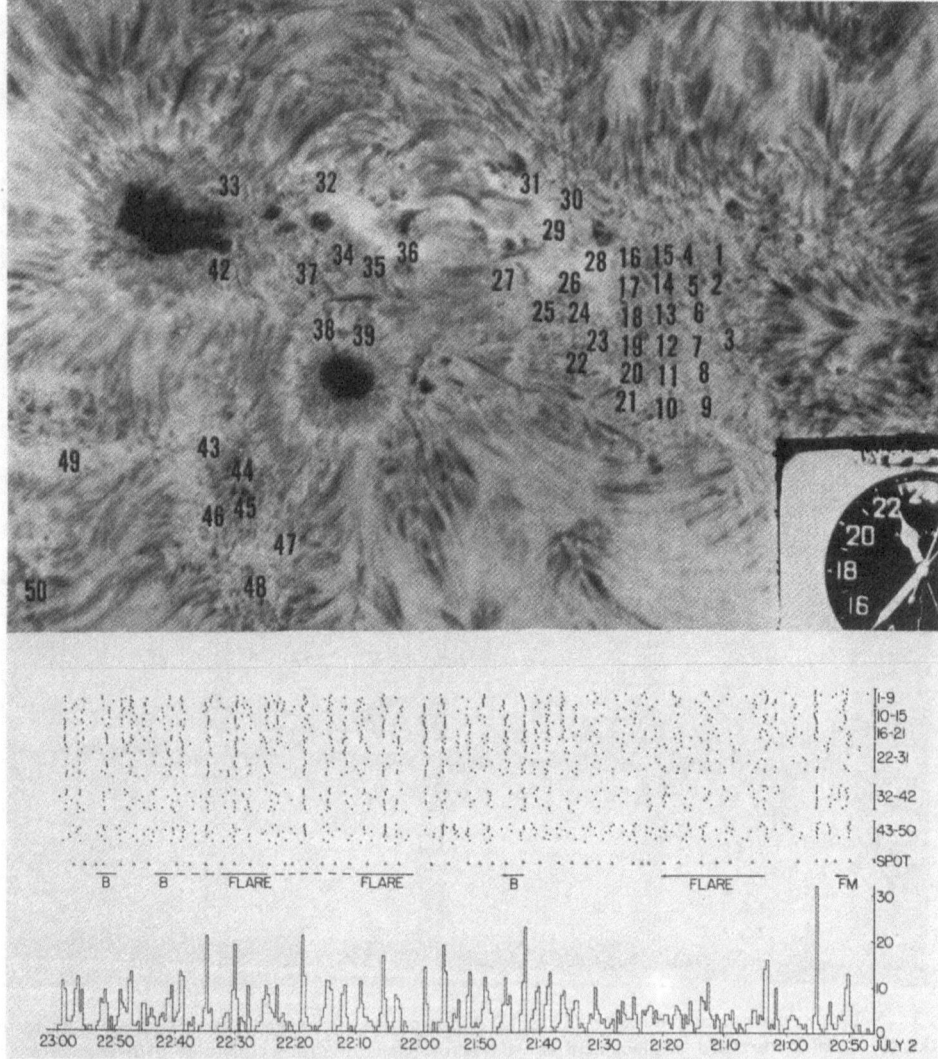

Fig. 4. *Top:* Active region photographed at Hα −0.5 Å. *Middle:* Times of maximum intensity at Hα −0.5 Å (maximum velocity downwards) at individual points in three plages (1–31, 32–42, 43–50) in above region. *Bottom:* Histogram showing for all these plages the number of points with simultaneous peaks in oscillation.

Times of flare activity, filament motion (FM) and filament brightening (*B*) are also shown (Tanaka, private communication).

5. Concluding Remarks

We now have direct evidence of Alfvén waves in penumbras and elsewhere in the chromosphere. Their significance in affecting the growth and decay of chromospheric structures is as yet unknown, but this should be a profitable field of study.

The well-known gravitational-acoustic waves are observed up to the highest levels accessible to ground-based optical observation, and it is highly desirable that methods be developed now for studying their propagation into the chromosphere-corona transition zone.

Observers have grand opportunities in both these areas, particularly if two-dimensional observations can be made in lines from different levels simultaneously.

References

Beckers, J. M. and Schultz, R. B.: 1972, *Solar Phys.* **27**, 61.
Beckers, J. M. and Tallant, P. E.: 1969, *Solar Phys.* **7**, 351.
Bhatnagar, A., Livingston, W. C., and Harvey, J. W.: 1972, *Solar Phys.* **27**, 80.
Bhatnagar, A. and Tanaka, K.: 1972, *Solar Phys.* **24**, 87.
Deubner, F.-L.: 1971, *Solar Phys.* **17**, 6.
Elliott, I.: 1969, *Solar Phys.* **6**, 28.
Giovanelli, R. G.: 1972, *Solar Phys.* **27**, 71.
Grossman-Doerth, U. and von Uexküll, Marina: 1971, *Solar Phys.* **20**, 31.
Grossman-Doerth, U. and von Uexküll, Marina: 1973, *Solar Phys.* **28**, 319.
Jensen, E. and Orrall, F. Q.: 1963, *Astrophys. J.* **138**, 252.
Leighton, R. B., Noyes, R. W., and Simon, G. W.: 1962, *Astrophys. J.* **135**, 474.
Liu, S. Y. and Sheeley, N. R.: 1971, *Solar Phys.* **20**, 282.
Orrall, F. Q.: 1966, *Astrophys. J.* **143**, 917.
Sheeley, N. R. and Bhatnagar, A.: 1971, *Solar Phys.* **18**, 195.
Stein, A. and Zirin, H.: 1972, *Bull. Amer. Astron. Soc.* **4**, 392.
Tanenbaum, A. S., Wilcox, J. M., Frazier, E. N., and Howard, R.: 1969, *Solar Phys.* **9**, 328.

DISCUSSION

Thomas: If there is a 1 km s^{-1} upward velocity corresponding to the outward propagation, why isn't there normal acoustic heating above that region? This is a bigger amplitude than one talks about in the old Biermann-Schwarzschild oscillation business.

Zirin: That's probably what causes the center-line brightening when we do see it.

Giovanelli: Zirin's results and mine are in conflict on this point. I've looked for brightening of the line center and I fail to see it.

Meyer: I was wondering to what Dr Thomas' question referred – to the umbral part or to the penumbral part. In the penumbral part Dr Giovanelli's suggestion was that this was an Alfvén wave which is a transverse wave along the magnetic field. This is practically a divergence free wave and would show no compression. It is quite a different wave mode than the ordinary compressional mode and would not give rise to the heating you asked about.

Zirin: I don't think that the divergence in observations between what we observe and what Giovanelli observes is real. First so far as the center line observations are concerned I will show tomorrow a film of plages which also has penumbral waves on it. There is one case where the waves can be seen clearly off-band and cannot be seen at all on-band, and we had good contrast. There are other cases where you see them on-band and not off-band. I think there are real differences from spot to spot and that there are cases where they are sometimes better seen as a velocity shift. In velocity they are much more visible than in brightness. I don't understand why they are sometimes seen better off-band than on-band except that it may just indicate where the lines of force are going. Second, our impression that the umbral flashes were exploding and producing an outward wave every second time has been based on observations made at the limit of resolution and should be taken with caution. This is simply an impression that we gathered from looking at films. Certainly the umbral flashes are more frequent than the penumbral waves. I think it is close to two to one but I would not argue strongly that every other one produces a wave. Finally I would add that we have not been able to detect any penumbral waves once the spot is more than four or five days from the centre of the disk. So I think basically we agree on most of the observations and the rest of it is just trying to understand these very subtle phenomena. I think we are just beginning to scratch the surface of what will turn out to be a much more interesting and involved phenomenon.

Beckers: My slide (see the figure in this discussion) shows observations taken with our universal filter in the Hα line. The time interval is too large to really show the motion of the waves but one can see the waves quite well. For example, in the second set of pictures one can follow the wave through the line centre clearly into the violet wing (Hα − 0.3 Å) but it can hardly be seen in the red wing (Hα + 0.3 Å). The waves are very well visible in the blue wing, marginally visible in line centre and not visible in the red wing. I think

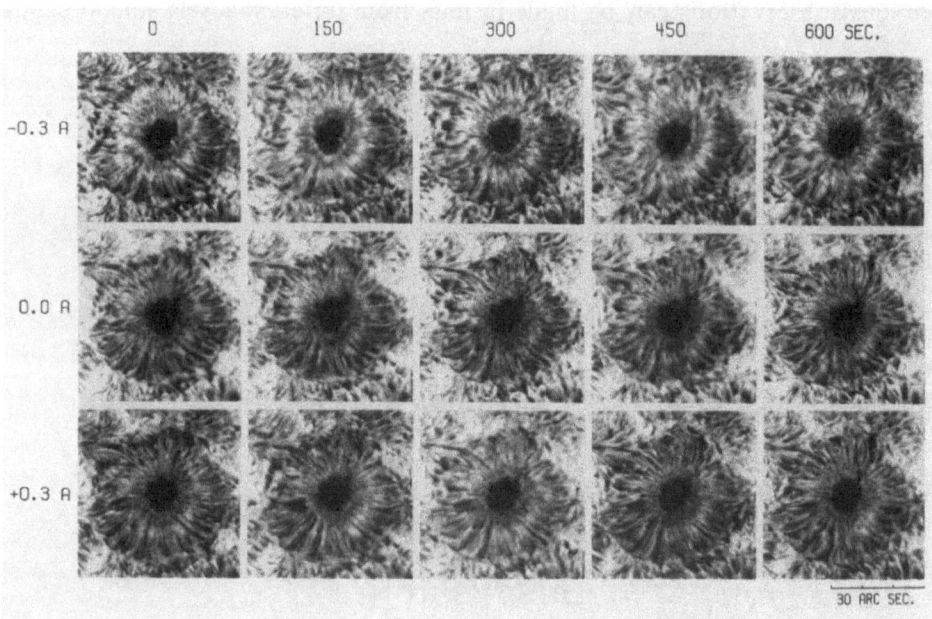

one sees a relationship between the upward motions and the brightening in the line centre. In other lines or further into the wings of Hα the wave becomes invisible. Waves cannot be seen in the sodium D-lines or the magnesium b-lines. They are best studied in the core of Hα.

Wilson: With regard to the invisibility of these waves in lines that lie deeper, as you go down deeper you expect a smaller transverse velocity because of the increased density. Would you expect to see these waves at deeper layers?

Giovanelli: I don't know what the propagation characteristics will be. The magnetic field may be more or less uniform but the densities will vary considerably and complicated refraction effects can be expected. On the other hand the wavelength is an appreciable fraction of the width of the penumbra. The waves disappear before the outer edge of the penumbra.

Sturrock: It is my opinion that any interpretation of waves in terms of an infinite, uniform medium is suspect because in this problem the scale height is small compared to the wavelength. Furthermore, when you compare the Alfvén velocity with the observed wave velocity by inferring a density from the velocity and field strength, you could probably get whatever you need by choosing the right layer. You must take account of the fact that this is a very highly stratified medium and you need to study the modes that exist in a stratified medium not in an infinite homogeneous medium. In this context I wonder whether your statement that the wave is transverse really rules out the acoustic wave. The same can be said about Athay-Moreton waves which are produced by flares and propagate on the surface of the Sun. One observes vertical displacements propagating horizontally with high velocities, but Dr Meyer showed that this could be interpreted as the response of the chromosphere to a shock wave travelling in the corona. It is possible that the waves you observe are a similar phenomenon. To sum up I think you must really take into account the stratified nature of the atmosphere and investigate the possible kinds of wave modes that can exist under these circumstances.

Uchida: I would think that there is a possibility that the vertical oscillation in the umbral portion is a secondary velocity induced by the horizontal velocity of the Alfvén wave in a cylinder with a node inside of it. The situation in such a case is fairly different from an Alfvén wave in an infinite medium. Each half of the cylinder moves towards each other, for example, with compression in the center, and then each half moves towards the outer boundary, and the material is compressed there, with induced vertical velocity each time. If we consider a situation like this, it may not contradict the fact that you do not observe systematic velocities at the limb. The Alfvén wave may thus be able to carry a considerable amount of energy without being detected except for a small induced velocity parallel to the field.

Giovanelli: It might be possible to test for such a possibility by making observation at different intervals from line center and thereby seeing to different optical depths.

Schmidt: I subscribe to the remark by Dr Sturrock but nevertheless I think that in the umbra the energy flux estimates for the vertical flux cannot be too wrong. In answer to the question by Dr Thomas I think that if you look up the model of Ulmschneider these are just the frequencies you need to escape dissipation in the chromosphere and dissipate in the corona.

Zwaan: Do you find the penumbral waves and the umbral flashes in all spots that have observed near the center of the disk?

Giovanelli: We have observed them in all that we have looked at. Zirin has not seen them in complex spot groups. I have looked in one complex spot group and seen them, but my statistics are not adequate to dispute Zirin's claims that they do not occur in the complex spot groups.

Zwaan: Then there seems to be a difference between the oscillations you see in Hα and the oscillations seen by Beckers in photospheric lines who found photospheric oscillations in a part of one umbra out of five spots observed, I believe.

Beckers: That was just a one time observation and not an observation of a large number of spots. We saw it in the umbra and an indication of an oscillation existing in the penumbra at a very much longer period. But that was only one spot out of six.

Martin: Ramsey and Phillis at the Lockheed Observatory have also been studying the Hα oscillations in sunspots. They find oscillations in essentially all spots. Their results completely agree with Dr Giovanelli's but they do have one additional interesting example of a sunspot which has an umbral oscillation with a period of 145 s and can be followed through 8 complete cycles. The variation is only ± 5 s from a completely sinusoidal wave.

Zirin: There was a misunderstanding of Zwaan's question. Zwaan asked if umbral flashes were observed in all spots. We observe umbral flashes in the K line in all spots. It is just the penumbral waves that we do not see in all spots.

Giovanelli: I think we should be careful not to call them umbral flashes – they are umbral oscillations.

Rosenberg: When you look at penumbral waves, it seems that they spread out with a rather wide opening angle. This contradicts your idea of Alfvén waves. They should be channeled. With regard to umbral oscillations, you said the periods were different for different positions within the umbra and this contradicts the remark by Uchida that the waves are possibly a standing oscillation in a tube. When you look at the ATM pictures, you see flux tubes which must be really very hot – it may be proof that the oscillations do heat the matter in the flux tubes and this is maybe an answer to Dr Thomas' question. There is apparently heating because we see these beautiful arches filled with hot material.

Giovanelli: On the first question I don't see any problem there. I regard an ideal sunspot as a monopole sending lines of force out in every direction. If you go on the hypothesis that these are Alfvén waves then you expect the Alfvén waves to go out in a more or less circular manner.

Rosenberg: The way you drew the picture is that one part of the umbra knocks against the field lines. I would expect that this would do something to just those field lines which were disturbed. It's not the whole umbra that goes up and down, is it?

Giovanelli: I see wavefronts which swirl around, expand and do all sorts of extraordinary things. This will cause a wave to propagate out, arriving at perhaps slightly different times at the penumbral-umbral border, but I see these umbral waves starting out and spreading over a very large part of the umbra. In some of the films you may have seen penumbral waves running out quite uniformly, but that is by no means the general rule.

Schatten: I don't see how you can calculate the energy flux associated with the waves without knowing the depth at which these waves are occurring. These waves could extend through the entire convective zone. How do you obtain the wave energy?

Giovanelli: The energy flux I gave is simply the energy flux per unit area at right angles to the propagation of the wave front at the depth where the waves are observed.

Athay: Then it is not valid to compare that with the radiation flux crossing the Sun per unit area in the radial direction?

Giovanelli: I think it is a valid comparison to compare the energy flux crossing unit area normal to the wave front with the radiation flux crossing unit area normal to the solar surface.

Wilson: But certainly if there is divergence of the magnetic field then you are reducing the amount of energy that you might calculate.

Beckers: Is the density of about 3×10^{16} atoms cm^{-3} for the depth where you observe these waves consistent with sunspot models? This seems rather high for the region where Hα is formed.

Giovanelli: Over the sunspot the temperature is low, the electron density is very low, and there is very low excitation. Under these conditions it is possible that Hα is formed quite deep in the atmosphere.

Wiehr: Do the penumbral waves run outside the outer border of the penumbra as seen in white light?

Giovanelli: No, they do not.

Wiehr: If they would do so it would be an argument against the Alfvén waves because the sunspot magnetic field apparently ends at the outer border of the penumbra. I have another comment concerning your estimate of the Alfvén velocity and the velocity of sound. You ruled out the possibility of interpreting these waves as a sound wave, as well as a possible explanation as gravitational waves. I am not at all convinced about your estimates of the sound velocity and the Alfvén velocity. You must deal with a very inhomogeneous material in the penumbra. It isn't known whether the penumbral waves occur in the bright filaments or the dark filaments in the penumbra. There might be a very large difference in both temperature and density between these two components of the penumbra. This could alter your estimates of the velocity of sound and the Alfvén velocity by at least one order of magnitude. Even if you take into account the magnetic field we don't know whether the magnetic field is stronger or weaker in the dark components of the penumbra as compared to the bright components.

Giovanelli: I would remind you that the wave is a transverse wave however.

Beckers: This is for an average penumbra and I don't think it is established yet whether the wave goes in the bright regions or the darker regions. Do you still want an answer to your first question as to why these waves do not penetrate outside of the penumbra? I would suggest that it is because the magnetic field just ceases to exist at that point.

Giovanelli: It's a bit of a problem. I can give an explanation but I'm not very happy with it. My explanation is that the magnetic field is oriented upwards and as the waves go upward along the field they reach a point where there is insufficient optical depth in Hα for the phenomenon to be observed. However I can see some rather nasty problems associated with this explanation.

Beckers: Do you see any change in velocity, either slowing down or speeding up as the wavefront approaches the border of the penumbra?

Giovanelli: I have tried to measure this but have not been able to convince myself that there is any acceleration. My visual impression from watching the films is that there is an acceleration but since I cannot measure it my answer has to be that there is no measurable acceleration.

Zirin: We have not been able to measure any acceleration in the waves we observe and this is one reason why I decided they must be sound waves. It would seem to me strange that the Alfvén velocity should remain constant over such a range although maybe it is. Perhaps this just proves that the ratio of B^2/ϱ is constant.

Giovanelli: The accuracy with which one can make this measurement is not good. It is difficult enough to measure the velocity and even more difficult to measure an acceleration, particularly when there is little more than one wavelength covering the penumbra. This is something that should be studied in a very large sunspot.

Schmidt: If the polarization is right and the velocity is locally right for Alfvén waves then it is up to us theorists to prove that they are not Alfvén waves if this is what we believe. The basic feature of a wave is its polarization and that is just right for the Alfvén wave. Also the local velocity of propagation seems to be right. Of course we could propose an explanation that this is a phase velocity associated with another type of wave but after doing some rough computations I am now convinced that the explanation of these waves as Alfvén waves is the correct one.

Meyer: May I ask you again. Do these waves end at the exact outer border of the penumbra or do they extend out into the border of the Hα penumbra in the fibrils that extend beyond the white light penumbra?

Giovanelli: The waves end at or before the edge of the penumbra as seen in the white light continuum.

Meyer: I would like to suggest that two of your observations already discussed might point to a somewhat different mode of propagation for your penumbral waves. Dr Wiehr has already commented on the observation that you see these waves in Hα only to a distance coincident with the white light boundary of the penumbra, and Dr Beckers pointed to a rather high value (10^{16} cm^{-3} for $B = 1500$ G and a propagation speed of 15 km s^{-1}) for the inferred density in the level of propagation. Dr Sturrock mentioned the problem of propagation in a medium of varying density. But it seems to me that all this loosely suggests the following picture. We know that quite a sizeable amount of magnetic flux leaves the penumbra and from this one has to infer that besides very horizontal penumbral field lines one must have there a large amount of field lines with strong vertical components. If this field configuration is pushed from a pressure disturbance in the umbra a compressional wave should radially travel outwards in which the more vertical field lines are horizontally compressed. Though the amplitude in the more photospheric penumbral level of $10^{16.7}$ cm^{-3} density might be rather small, still material would be vertically squeezed upwards along the field lines and in the decreasing density with height might obtain vertical velocities that make it observable in Hα. Still the main mass involved would be sitting in lower levels and provide the inertia indicated by the 15 km s^{-1} propagation speed. At the outer penumbral boundary one has run out of more vertical field and this might account for the end of the Hα wave pattern. Refraction of course will play a major role for such disturbances, and other difficulties might arise. But it is interesting enough to deserve a model analysis.

Athay: The last series of pictures you showed illustrated a type of motion that appears very similar to what the Sacramento Peak observers have called flagellant motion. Do you think this really is the same phenomena or is there something different?

Giovanelli: I believe the flagellant motions are primarily just the result of Doppler shifts, so the answer to your question is yes! I believe these are the same type of phenomena.

Wilson: Do you have any evidence as to the velocity amplitudes in these waves.

Giovanelli: I would be surprised if it were very large. The estimates given by other people of 1–1½ km s^{-1} might easily be about right.

Newkirk: I am uncertain about your identification of these motions with Alfvén waves. Are the velocities right for this type of wave?

Giovanelli: The velocities that are observed are consistent with Alfvén velocities as nearly as I can determine them. I choose fields and densities that seem to be reasonable and I get the right orders of magnitude for the observed velocities.

Schmidt: I think that the motions in the umbra are probably not Alfvén waves because they seem to be longitudinal. Here we do not have a well-defined field and we cannot make the checks that you have made in the penumbra. In the penumbra I think you have some good arguments that the penumbral waves are Alfvén waves. I say that because we are dealing with a phenomenon which is controlled from below. You cannot control the penumbra from the corona so the arguments applied to the Moreton waves do not apply here.

Giovanelli: I concede immediately that there is a fascinating problem here and that we have a non-uniform magnetic field surrounded by a plasma. I hope we can interest some of the theoreticians in this particular problem.

Deubner: There is another argument against the interpretation of these waves as Alfvén waves. You stated that the velocity in the fibrils was probably 80 to 100 km s^{-1} and that close to the spot where the field is much stronger the velocity is of the order of 12 km s^{-1} and I think this is a contradiction.

Giovanelli: When I use existing chromospheric models and sunspot models the models differ enough that I still get the right order of magnitudes for velocities.

CHROMOSPHERIC OSCILLATIONS IN PLAGES

KATSUO TANAKA

Tokyo Astronomical Observatory, Mitaka, Tokyo, Japan

Abstract. The 5-min oscillations in Hα plages as reported by Bhatnagar and Tanaka (1971, *Solar Phys.* **24**, 87) are discussed with respect to spatial – both horizontal and vertical – phase relation. Data for analyses include (a) Hα filtergrams at −0.5 Å (1050 min in total) (b) Hα filtergrams simultaneously taken at ±0.5 Å (60 min) (c) Simultaneous Hα − 0.5 Å and Fe I 5233 Å videomagnetograms (Doppler mode) (100 min in total). Projection (16 and 35 mm) -photocell method (Bhatnagar and Tanaka, 1971) was used for tracing the data with special attention to the guiding and reproducibility. Also photographic prints were used for comparison. The results are as follows:

(1) those showing oscillatory brightness changes are dark dot-like features (size ≲ 1″) in brighter regions of plage and slightly elongated features (1 × 1.5–2″) in less bright regions. The mean number density of these features is one per 2200 ± 500 km square.

(2) There are always a few regions with a size 8″ or more in one plage which show very regular oscillations with a period of about 5 min for a long time (at least 6.5 h, Figure 1). The distribution of the intervals between successive intensity peaks is much narrower than the Gaussian distribution and has a skewness towards longer period.

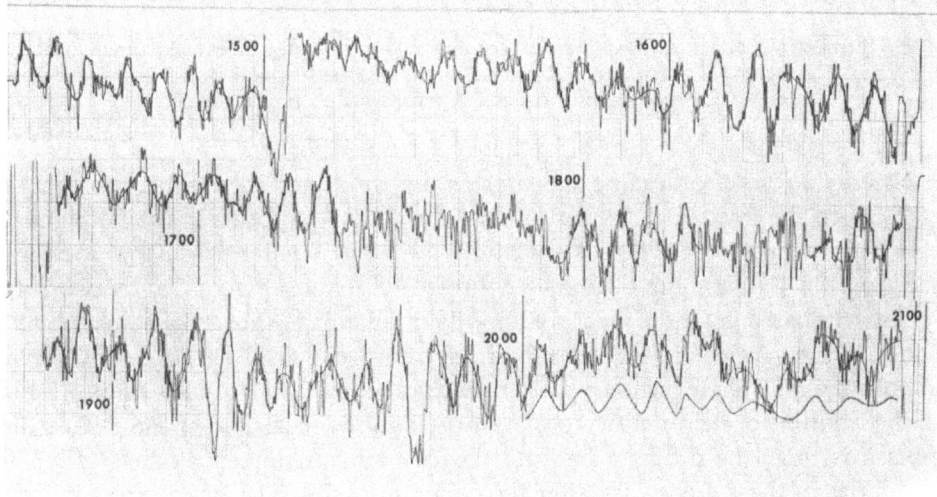

Fig. 1. Intensity variation in plage measured at Hα −0.5 Å with the 8″ aperture. Rather steady oscillations are seen for the observed duration (6.5 h) except the periods of bad seeing. The smoothed-out curve between 20:00 UT and 21:00 UT shows the record of the velocity oscillations measured in Fe I λ5233 Å at the same position.

R. Grant Athay (ed.), Chromospheric Fine Structure, 153–156. All Rights Reserved.
Copyright © 1974 by the IAU.

The other places in the plage show relatively irregular patterns, but with the mean interval of the fluctuations still being 5 min.

(3) In regular 5-min brightness oscillations the phase of oscillations in one wing is complementary to that in the opposite wing, indicating velocity oscillations. When the intensity records are relatively irregular there are many cases (76%) in which the blue wing patterns can be matched well for the whole time sequence with the red wing records by advancing the red wing records by 196 ± 18 s, (see Figure 2) indicating that

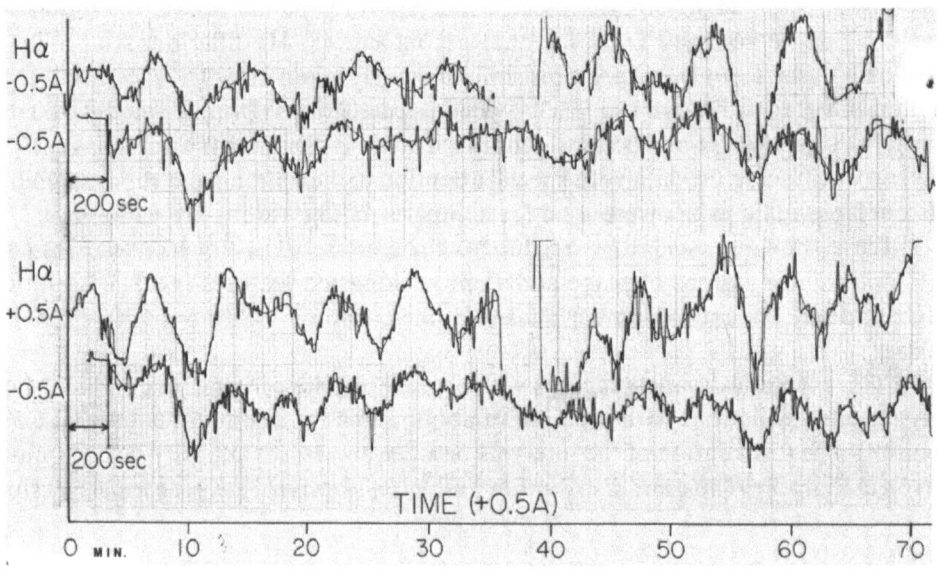

Fig. 2. Intensity variations in two different positions (8″) of the plage measured simultaneously at Hα ± 0.5 Å. The records of -0.5 Å are shifted by 200 s in time axis. Note the good correspondences of the $+0.5$ Å records with the -0.5 Å records with a time lag of 200 s.

the observable dark feature first appears in rising mode and then with a constant time lag (close to the free fall time from -0.5 Å to $+0.5$ Å) appears in the falling mode.

(4) The records of brightness variations are very similar when the aperture is changed from 4″ (minimum reliable aperture) to 10″ in most places. Comparison of the oscillation phases in all the measured points in one plage shows that a plage can be divided into several areas (for a certain time) with sizes from 7000 km to 30000 km in each of which the phases are very close to each other for 2 to 10 cycles. Figure 3 shows the distribution of the intensity peaks of oscillations measured at Hα -0.5 Å (the abscissa is time, the ordinate indicates space points arranged geometrically). The boundaries of these areas vary after the coherent cycles. The iso-intensity curves of the intensity-time records (see Figure 3) have slopes corresponding to horizontal phase velocity 100 km–600 km s^{-1}.

(5) The phase relation between Hα regular 5 min and Fe I 5233 Å 5-min velocity oscillations at the same place is almost constant for more than 5 cycles although the

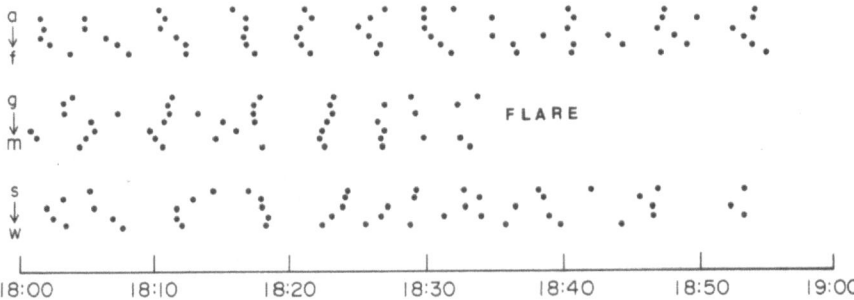

Fig. 3. Distribution showing only peaks of the intensity oscillations (dots) with respect to the various places (ordinate). All the measured places are divided into three groups ($a \rightarrow f$, $g \rightarrow m$, $s \rightarrow w$) in which phases of oscillations are correlated each other.

phase lag is different from place to place. (See, for example, the period from 20:00 UT to 21:00 UT in Figure 1. The smoothed-out curve denotes the record of the velocity oscillations in $\lambda 5233$ Å taken simultaneously.) The distributin of the phase lags is peaked at about 1 min. Assuming 1000 km for height difference of two lines the vertical phase velocity is equal to 17 km s^{-1}.

(6) There is an indication that the flare originating in the plage occurs about 20 min after the oscillations start to be in phase (1-min accuracy) in very large scale ($\gtrsim 20000$ km) although the samples (6 flares in three different plages) are not large enough for statistically meaningful discussions.

DISCUSSION

Pecker: I would like to discuss one of the slides that you showed. This is the one in which you showed the time lag of 200 s. The point I would like to make is related to the interpretation. This concerns the question of deriving velocities of propagation by comparing features seen in the red and blue wings. When there are strong velocity fields and two sides of the line are not formed in the same part of the atmosphere and the region where they are formed is not very clear. This has two consequences. One is that the 15–25 km velocity derived from the photometric considerations is subject to considerable criticism from this cause. The second point is that the 200-s delay might be a propagation time between the two layers in which you are looking, whatever they are.

Tanaka: I am not sure that filtergram measurements can give good velocities but 0.5 Å which is where we observe in the line corresponds to 23 km s^{-1}.

Pecker: Oh but you have to assume a lot of things about the profile to obtain this.

Deubner: If you try to do Fourier analysis of the velocity fluctuations in the core of Hα and the brightness fluctuations at the same place you arrive at two spectra which are not equal at all. They have entirely different mean frequencies, the peak of the velocity fluctuations being at about 150–180 s and the brightness fluctuations equalling those in the photosphere of about 300 s. This can be expected bécause the brightness in Hα pretty much reflects the brightness in the photosphere. When you look inside the profile you then have a combination of effects due to line shift and intensity fluctuations and that might explain the plages not having just one half of either period. In the second slide you showed the nice coherence between the brightness fluctuations in Hα and the velocity fluctuations in the photospheric lines. I agree that they are coherent but I would conclude from the slide that the phase differences are somewhere around 90° or 60° which again confirms that the intensity fluctuations in Hα go with the intensity fluctuations in the photosphere rather than with the local velocity fluctuations in Hα.

Athay: In that same connection I think it should be noted that when you are one half angstrom off-center in Hα about half of the observed intensity comes from chromospheric layers and half from photospheric layers. Current model chromospheres give a double peaked contribution function which at this wavelength is about equally divided between two regions. When you add the Doppler shift into the line you may see primarily chromosphere on one side and primarily photosphere on the other side of the line, so I think there are very large uncertainties in the height relationships between the two sides of the line.

Tanaka: Yes, I agree we do not know the height relationship. However in Hα I can identify dark granular features that have the characteristic appearance of chromospheric structures and we have measured the brightness fluctuations in these features so I think clearly we are talking about chromospheric features and that the brightness and velocity fluctuations are chromospheric in origin.

Athay: How can you be sure?

Tanaka: I have compared photospheric images with chromospheric images and although there is some similarity most of the features are quite different. I am convinced that the features we have looked at in the line wings correspond with those in the chromospheric images and not with those in the photospheric images.

Pecker: When looking at the 200-s delay shown on your (Tanaka) slide #2 between red and blue sides of the Hα line, and at Giovanelli's films, one cannot escape being terribly impressed by the quality of observation. But theoreticians should give a warning and some advice: the optical depth where the 'red' point and the 'blue' point are 'formed' is certainly not the same (a), and (b), it is very badly defined.

Hence I suggest (a) that the 200-s represent a propagation in the atmosphere between two levels, but that we cannot specify the two levels accurately enough. (b) that the velocities deduced by Giovanelli from photometry of his films are indeed very dubious, as they assume the profile to be displaced (from upwards going points to downwards going points) but not disturbed, perturbed, etc.

TRAPPED OSCILLATIONS IN THE CHROMOSPHERE
IN THE PRESENCE OF A MAGNETIC FIELD

Y. NAKAGAWA

High Altitude Observatory, National Center for Atmospheric Research, Boulder, Colo., U.S.A.*

Abstract. In the presence of a magnetic field, three types of magnetoatmospheric waves – magnetoacoustic mode, magnetogravitational mode, and hydromagnetic mode – can propagate in a stratified atmosphere, in contrast to the propagation of two types of atmospheric waves – acoustic mode and gravitational mode – in the absence of a magnetic field. The exact manner of propagation of the magnetoatmospheric wave is extremely complex, and most studies have been confined to certain specific circumstances, such as an isothermal atmosphere permeated by a uniform magnetic field (McLellan and Winterberg, 1968; Bel and Mein, 1971; Michalitsanos, 1973), and atmosphere in magnetohydrostatic equilibrium (Yu, 1965; Chen and Lykoudis, 1972) and the propagation across a density discontinuity (Stein, 1971).

In a previous paper (Nakagawa *et al.*, 1973) a general method of determining the trapped magnetoatmospheric waves was described, and in another paper (Nakagawa, 1973) possible interpretation in terms of such theoretical results of the observed modulations of chromospheric oscillations over the magnetically active regions of the Sun was discussed. Therefore we present a brief summary of the physically significant results of previous studies, with discussions focused on the validity of assumptions and the results obtained.

The main effect of a magnetic field is the modulation of the period of trapped oscillations, and this effect can be examined by means of the diagnostic diagrams. It was shown (Nakagawa *et al.*, 1973) that in a horizontal magnetic field except for the magnetoacoustic waves, most of the waves are trapped, and that in a general magnetic field most of the waves of frequencies below the local Brunt-Vaisala frequency are trapped (contrary to the non-magnetic case in which those waves are propagating gravitational waves). Further it was shown (Nakagawa, 1973) that a number of satisfactory agreements can be obtained between the theoretical and observational results. Notably, it was suggested that the short period oscillations of around 170 s reported by Elliot (1969) and Bhatnagar and Tanaka (1972) inside of supergranulation could be identified with the trapped magnetoatmospheric waves traveling along the magnetic lines, as such short period oscillations cannot result from the magnetogravitational waves traveling perpendicular to the magnetic field. Similarly, it was pointed out that the observed period of 600–700 s by Orrall (1966) and Blondel (1971) in bright facular regions could be identified with the trapped oscillations in the presence of a vertical magnetic field, together with the observed period of 900 s by Orall (1966), Elliot (1969) and Cha (1970) in the supergranulation boundaries.

* The National Center for Atmospheric Research is sponsored by the National Science Foundation.

In summary, the need of radiative-hydrodynamic both observational and theoretical studies for the understanding of the chromospheric oscillations is stressed, since most of the theoretical studies have been confined only to the examinations of adiabatic small amplitude perturbations.

References

Bel, N. and Mein, P.: 1971, *Astron. Astrophys.* **11**, 234.
Bhatnagar, A. and Tanaka, K.: 1972, *Solar Phys.* **24**, 87.
Blondel, M.: 1971, *Astron. Astrophys.* **10**, 342.
Cha, J. Y. M.: 1970, Thesis, University of Hawaii.
Chen, C.-J. and Lykoudis, P. S.: 1972, *Solar Phys.* **25**, 380.
Elliot, I.: 1969, *Solar Phys.* **6**, 28.
McLellan, A. and Winterberg, F.: 1968, *Solar Phys.* **4**, 401.
Michalitsanos, A. G.: 1973, *Solar Phys.* **30**, 47.
Nakagawa, Y.: 1973, *Solar Phys.* **33**, 87.
Nakagawa, Y., Priest, E. R., and Wellck, R. E.: 1973, *Astrophys. J.* **184**, 931.
Orrall, F. Q.: 1966, *Astrophys. J.* **143**, 917.
Stein, R. F.: 1971, *Astrophys. J. Suppl.* **22**, 419 (No. 192).
Yu, C. P.: 1965, *Phys. Fluids* **8**, 650.

DISCUSSION

Stix: Would you clarify one point. Is in each of your layers the Alfvén velocity and the density constant. If so, is this consistent with the stratification?

Nakagawa: The density falls off exponentially within each layer with a constant scale height. Thus the Alfvén velocity also changes. However, in the analysis the Alfvén velocity is assumed constant in each layer as we have found that superposition of such solutions gives an adequate approximation to the exact solution.

Souffrin: I am concerned about the identification of the slow mode with the magnetogravitational mode. In a homogeneous medium with a vertical magnetic field you already have Alfvén waves, a slow mode and a fast mode. In the presence of gravitation there is a new possibility for waves.

Nakagawa: No the dispersion relations still allow only three wave modes.

Souffrin: If you just suppress the gravitation you have three possible wave modes. Do you mean that when you add the gravitation you must modify these three modes?

Nakagawa: The dispersion relations still allow only three wave modes. The addition of a gravitational field modifies the wave mode but does not introduce a possibility for a new type of wave to exist. How you identify each of these modes with the modes in the non-gravitational case is of course somewhat open to question.

PART V

THE CHROMOSPHERE IN ACTIVE REGIONS

PART TWO

THE CHROMOSPHERE IN ACTIVE REGIONS

THE MAGNETIC STRUCTURE OF PLAGES

H. ZIRIN

*Big Bear Solar Observatory, Hale Observatories, Carnegie Institution of Washington,
California Institute of Technology, Pasadena, Calif., U.S.A.*

1. Introduction

Every sunspot group is surrounded by regions of weaker magnetic field; traditionally those parts which appear in emission in resonance lines are called plages. There is a more extensive region of enhanced magnetic field which is not bright because the magnetic field is horizontal; this region is easily discernible by dark horizontal fibrils in Hα and is included in our discussion. The plage includes most of the magnetic field in an active center and is the locus of most of the flares.

It is unfortunate that astronomers long viewed plages as westerners viewed orientals (or vice versa) as a number of indistinguishable individuals. Closer acquaintance shows that they have many distinguishing characteristics reflecting their history and magnetic structure.

For example, the spectroscopic data obtained so far on plages is hampered by the fact that there is no way of telling what part of the plage it refers to. In every case it is necessary to understand the kind of plage under discussion – whether old or new, *f* or *p*, closed or open. It is essential to remember that the plage structure is determined uniquely by the magnetic field structure and whether it is growing or stable.

For these reasons I shall place most emphasis on the evolution of plages and their structure as determined by magnetic fields. Since this structure depends strongly on the initial evolution, we shall discuss the latter in detail. Fortunately, although plages are most varied in nature, certain basic physical and developmental patterns constantly recur.

2. Development of Plages

Plages are formed in the inner parts of flux loops emerging from below. Figure 1 shows such an example, a bright plage crossed by arched filament systems. These were noticed by Waldmeier (1937) and their significance was pointed out by Bruzek (1967), who called them Arch Filament Systems. Zirin (1972) proposed the term EFR (Emerging Flux Region). In the early stages of active region growth the appearance of the group is symmetric, while a few days later the *f* spot may disappear, leaving an extensive plage. This gives rise to the most common type of spot group, *αp*, with a dominant *p* spot and a cloud of following plage, separated by a series of dark fibrils called field transition arches (Prata, 1971). The sunspots invariably form at the outer ends of the flux loops and the plages on the inside. The formation of a spot inhibits plage brightnees, probably because of the extensive horizontal field area immediately outside the penumbra, often called the superpenumbra. The formation of big *p* spots often pro-

R. Grant Athay (ed.), Chromospheric Fine Structure, 161–175. All Rights Reserved.

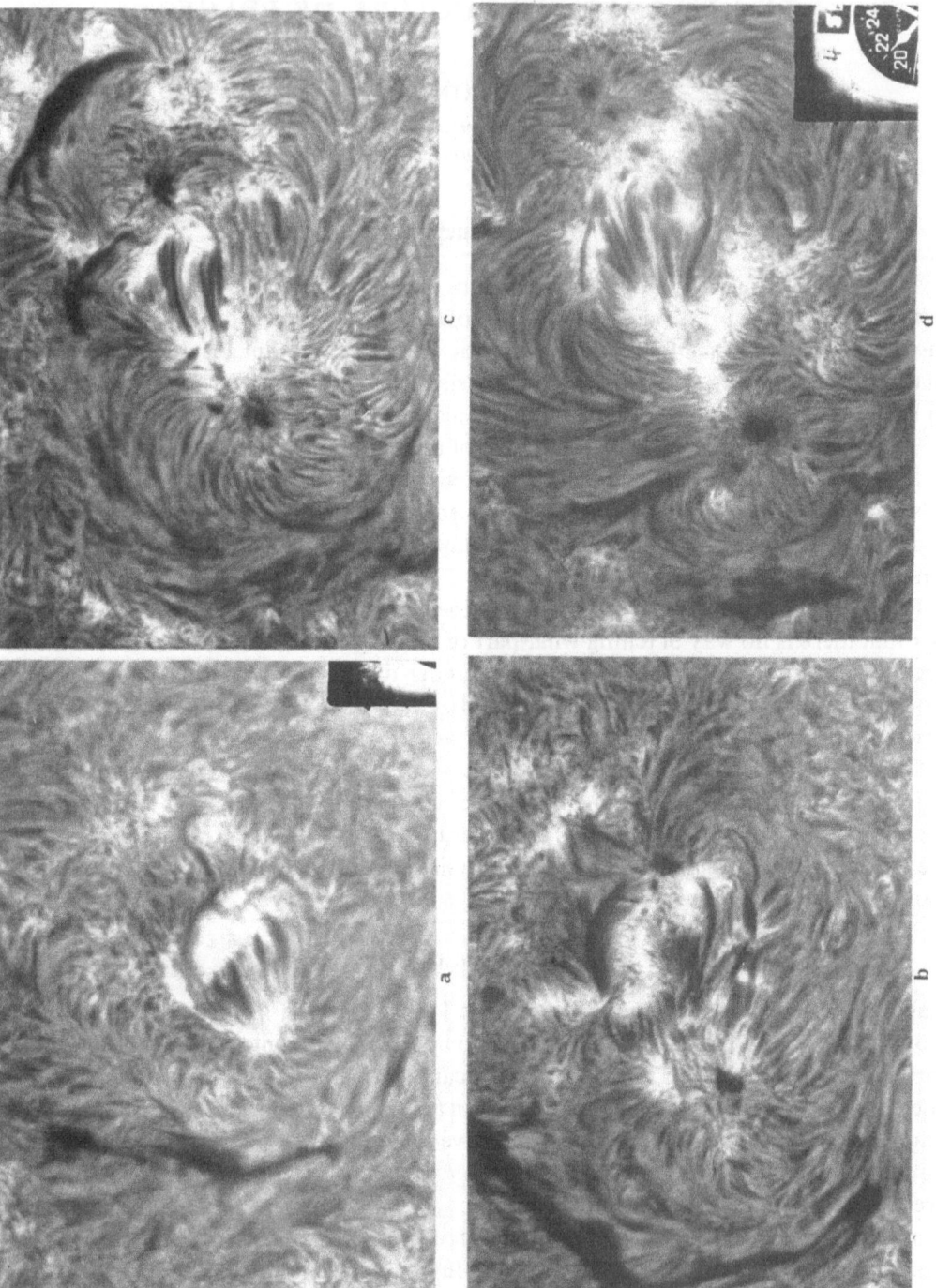

Fig. 1. Growth of an emerging flux region (McMath 11496) on successive days, 1971, September 1–4. The region arises in *f* polarity, so a boundary filament arose west and was gradually pushed away by the expanding spot and mostly destroyed after the 3rd, to be replaced by a broad system of horizontal field arches. Note how the arched filaments are confined to the inner part of the group where flux is emerging, while the sunspots rapidly move outward. (Clock is NE.)

duces a dramatic decrease in plage area, as the finely mottled plage region is replaced by horizontal field from the spot. There is also a sharp decrease in plage brightness as the field stops erupting. Further, the emerging plage is amorphous, while the mature plage is made up of fine granules. When sunspots break up a normal plage remains, so we may assume the total vertical field in umbra plus penumbra equals that in the plage. EFR's often occur in existing active regions or in close proximity to another. Since the number of EFR's per day is about two, the probability of two occurring close together is very small, and we must assume there is some connection between two or more EFR's in close proximity to each other.

The overall structure of an active region is determined both by the structure of the emerging flux loops and the sign and strength of the fields already present on the surface when it erupts. It is obvious that there must be some interaction between the emerging flux and the already existing field; although the erupting flux is closely tied together, it rapidly reconnects to nearby fields as may be judged from the fact that fibrils marking lines of force from most active regions go out to heighboring active regions. There is no evidence that existing surface flux is crowded out of the way by rising flux for the sizes of network cells on the edges of active regions are similar to those in the quiet areas.

No one has yet studied the question of the relative development of EFR's coming up in p or f polarity. It appears that emerging flux which is tilted to the rotation axis is more likely to produce high activity, but the smaller inclined EFR's were found by Weart to lie out earlier. If the EFR comes up in p polarity, it rapidly develops an open penumbral structure, with lines of force spreading out parallel to those already present to connect with distant f polarity. On the other hand, if the region emerges in f polarity, a boundary filament rapidly appears to separate the new p polarity from the old f polarity. It is interesting that ordinary connection with fibrils normal to the neutral line never occurs, but that the boundary is normally a filament channel, with lines of force parallel to the boundary. In some cases such a boundary may be very close to the p spot, even crossing the penumbra, and many flares occur at that point, but this only happens when the EFR emerges in a region of strong f polarity, i.e. part of an existing spot group.

As we noted above, flux loops always emerge in simple form; within the limits of resolution, no one has ever seen a bent or twisted arch filament; the lines of force appear to go cleanly and simply from one pole to the other. The material in the loops rises at the top of the arch and drains down at the edges. It takes about 7 min for a 15000 km arch to emerge. The complexity of active regions arises from the mutual interaction of new flux loops rising within and near old ones. As the sunspots separate, a region of emerging flux loops at the center of the active region may remain. There is a strong tendency for new flux to emerge in or near existing active regions; this is in fact the source of most flare activity. Therefore, the flux tubes in big emerging regions must have a large scale complexity in the sense that each has several loops or kinks, or is accompanied by other tubes. Once formed, the arched filaments are uniform and the direction changes only if new spots or flux loops emerge. If the flux loops were twisted ropes, we would see complex changes in directions as they emerged, and we do not.

I should point out that there appears to be no connection between either emerging flux regions or plages and the cells of the chromospheric network. Bumba and Howard (1965) found, on the basis of low resolution spectroheliograms, that EFR's started out as elements of the chromospheric network. Weart, on the basis of superficial analysis of limited data, decided that the AFS crossed over a network cell, i.e. just the opposite. Our data shows no connection whatever between EFR's and the elements of the network. A good example is the birth of the famous August 1972 region. A recent extensive study by Harvey and Martin (to be published) finds the same result. This is a very difficult determination, because one needs extensive observation before the emergence of the new flux. Once any flux emerges, bright areas appear which are *ipso facto* part of the network. Logic tells us that, since the EFR's are bipolar, at most one element could be part of the local network, which must be unipolar. Hence we cannot expect new flux to be produced from manipulation of the network itself, and in fact this is not observed. Sunspots emerge from below. Nor does the network cell size have any particular relation to the sizes of EFR's, sunspots, or active regions. Since the EFR starts from zero, young EFR's are small compared to the network; mature EFR's are about $\frac{2}{3}$ the size of a network cell, but much larger EFR's have been seen. Although a cellular pattern is seen on the edges of active regions, the plages themselves are several times larger than the network and show no connection. Finally, it has been suggested by Leighton and others that the large unipolar regions grow by a random walk process in which bits of active regions break off and spread over the surface of the Sun. In 1969 Richstone made an unpublished study of the size of the enhanced network region around various active regions, and found that this area tended to grow so long as new flux was emerging, but rapidly shrank as new flux stopped rising. Similarly, observations of detailed structures near sunspots (e.g. Figures 3–5 show little or no outward motion of the plagettes, despite the fact that Sheeley (1969) and others have observed outward motion of small magnetic elements around some sunspots.

3. Fine Structure of Plages

Figure 2 shows a videomagnetogram made in 5324 by A. Michalitsanos compared with a simultaneous K-line filtergram with 0.3 Å bandpass made by the author. Figure 3 shows the corresponding Hα centerline and −0.5 Å pictures. Dark corresponds to p polarity and light to f polarity in the magnetogram. We see (after allowing for slight misalignments) in almost perfect correspondence between the pictures. Even the faint K-line emission (except for hotizontal bright fibrils) corresponds to some longitudinal magnetic field, and we may expect that even weaker fields correspond to the weakest K emission. But we may be confident that most longitudinal magnetic fields are recorded by the videomagnetograph or by a K line or Hα picture.

We see in Figures 2 and 3 a mature αp type spot, with the f field broken down to a plage. A simple set of field transition arches (1) separate p and f polarity, although a

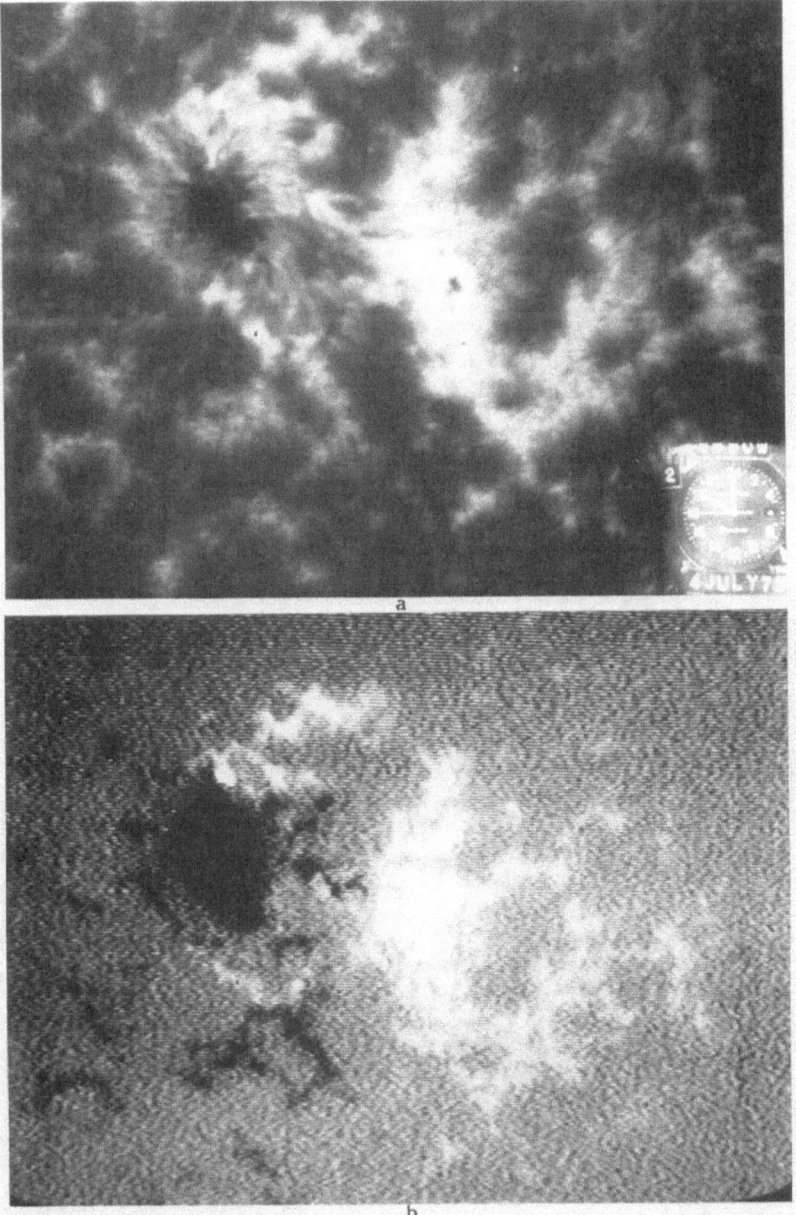

Fig. 2. (a) K line photo of McMath 12417, 1973, July 4, with 0.3 Å bandpass Halle filter. This photo shows how, with good alignment and proper exclusion of internal reflections, one may obtain K-line images comparable in quality to Hα. Careful comparison with 3(a) shows that all bright and dark K features in K correspond to similar features in Hα, except that the K contrast is stronger and dark features in K are lost against the background. We have not seen running penumbral waves in K. (b) Simultaneous videomagnetogram in 5324 made by A. Michalitsanos. The correspondence is very close. The bright rim of expanding field is marked (4). A rim of enhanced field is seen to the left of the penumbra, coinciding with a bright rim. (W is left.)

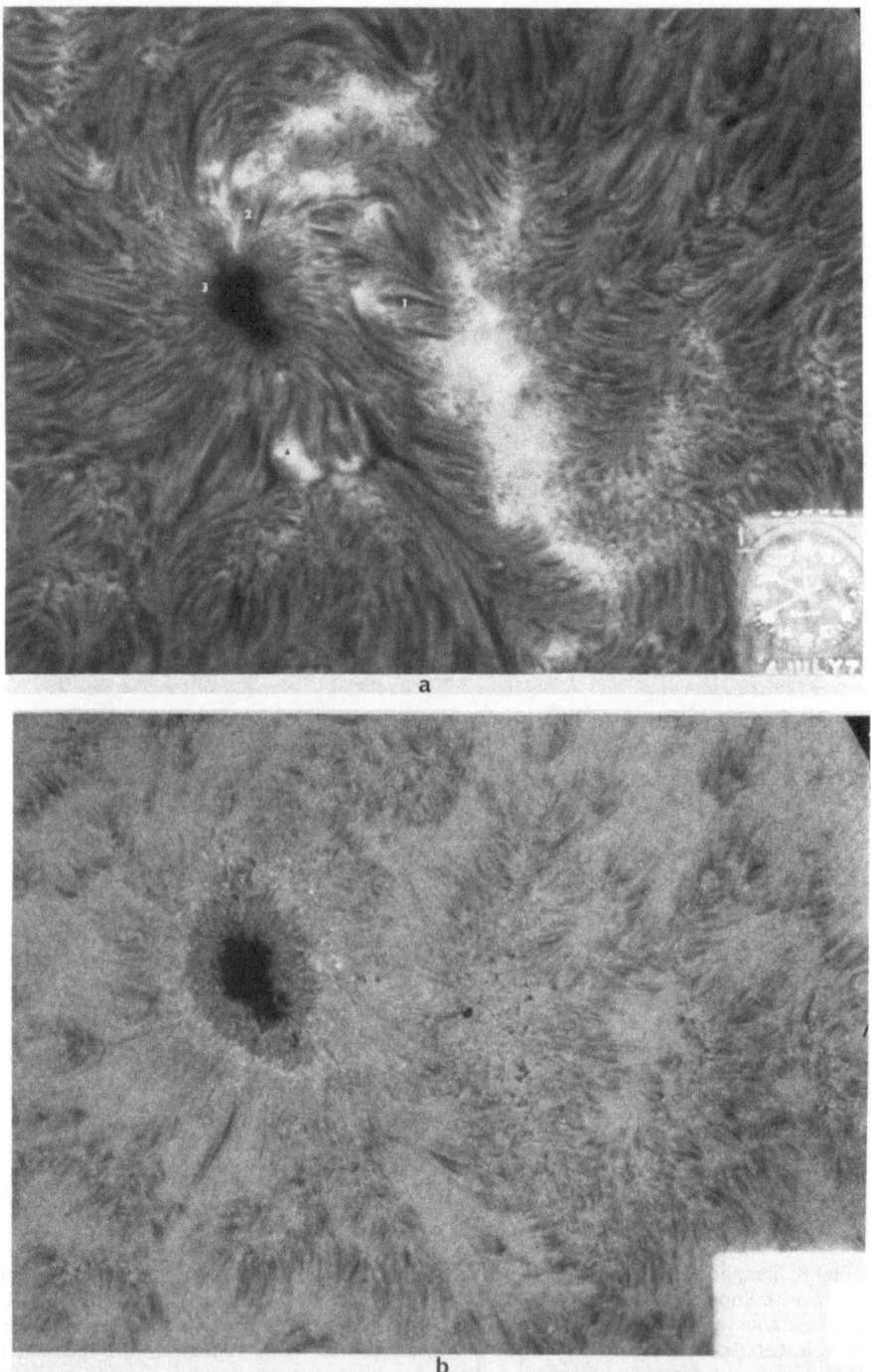

Fig. 3.   (a) Hα filtergram simultaneous with 2a and b. Numbers refer to features explained in the text:
(1) FTA (2) more complex FTA; (3) running penumbral wave (4) expanding front of flux. (b) Hα + 0.6 Å.
Note no running penumbral waves.

more complex neutral line (2) may be seen just above the spot. The dark fibrils here are considered to follow the lines of force in the chromosphere, and indeed they invariably are found to point to the nearest opposite polarity. The magnetic structures tilt sharply away from the vertical, and, as pointed out by Foukal (1971) the stronger magnetic structures are very shallow, as shown by regions seen on the limb. One can determine magnetic field polarities and distribution by using a few simple principles:

(1) *Regions of intense longitudinal fields (except spots) are bright in K and Hα centerline, but marked by fine, dark granular, structure off-band.* In centerline Hα some of these are obscured by overlying absorption, but all are seen in off-band on K pictures.

(2) *All dark fibrils in Hα or fine bright threads in the K-line mark the directions of lines of force;* neutral lines are marked by field transition arches (systems of dark fine parallel fibrils) where the lines are perpendicular to the neutral line, and filaments (which are larger and darker) when the field lines are parallel to the neutral line. It follows as was pointed out by Foukal (1971) and by Zirin (1972), that filaments only exist when there is a shear line between opposite polarity in which lines of force run parallel to the surface for some distance and can thus support the filament and also insulate it from the corona. It is a pity that none of the present filament models confront this apparent fact.

(3) *There is a rough proportionality of Hα or K brightness to field strength, except for newly emerging flux, which is brighter than normal.* Jansens (1972) and Frazier (1972) have taken issue with this proportionality, but others have supported it. However, all agree that the bright Hα regions mark the locus of enhanced vertical fields.

Our data, as illustrated in Figures 2 and 3, support the correspondence of brightness to field strength, with the exception of sunspots, which are too dark, and newly emerging regions, which are too bright.

We note that the salient characteristic of the plage is a fine granular structure which may be the intersection of flux loops with the surface. This region has been shown by Tanaka and Bhatnagar to oscillate in brightness with a 5-min period. In some cases (Zirin, 1972; Figures 5 and 1, Sept. 6, 1970) these may be seen near the limb to be somewhat curved. It is remarkable that there are no spicules or fibrils coming out of uniform plages; these phenomena are only found in plagettes (the small plages of the network) or at the edges of larger plages. The effect is probably due to two factors: first, because of the strong closed magnetic fields in active regions, the flux loops from the plages turn over steeply and return to the other polarity, and second, the temperature rises much more rapidly to coronal values in the region above the plage, thus evaporating any potential spicules. The horizontal field region under the field transition region (1) appears to be heated nearly as much as the underlying plage, since it appears bright in broad band K pictures. The darkness in only due to the overlying Hα absorption of the field transition arches and we may conclude that horizontal field regions are also heated, but not enough to evaporate the field transition arches.

Dunn (unpublished) has shown the existence of a fine filigree network in the upper photosphere which may possibly be the locus of the magnetic fields. The filigree is much finer than the Hα centerline structure, and this may indeed indicate that the magnetic fields spread out with height. A final determination awaits the obtaining of finer scale magnetograms.

Figures 2 and 3 show examples of other interesting phenomena. At (3), in the sunspot penumbra, (Figure 3a) we see a running penumbral wave. At (4), we see a band of p polarity which moved outward from the spot over a period of days, finally annihilating when it reached the nearby f polarity (Roy and Michalitsanos, submitted to Solar Physics). The magnetogram shows clearly the ragged distribution of flux. The (f) flux above resulted from a small EFR there.

It is instructive to compare the appearance of active regions in K, and various parts of Hα, because different heights are seen and different brightness contrasts. We note first that the contrast between excited regions and the background in K is much greater, presemably because the K line lies on the exponential part of the Planck curve. Although the K line is supposed to be formed higher, the narrowness of the K absorption makes certain features more transparent in K at the bandpass (0.3 Å). Thus we see a bright ring around the penumbra which is obscured by dark fibrils in Hα centerline but appears bright off-band.

One of the remarkable features in K is the network of fine bright fibrils extending over the surface, parallel to the dark Hα fibrils. At first these appear mostly bright, compared to the Hα fibrils. But careful comparison shows that this is only because the background is brighter in Hα; *dark features in Hα are dark in K*, and the bright K threads correspond to bright Hα fibrils which are barely visible onband and not visible at all off-band. Some of the Hα emission may be due to scattering of Doppler-shifted photospheric Hα (Zirin, 1969) an effect which cannot give off-band emission. The evidence is that spicules are both bright and dark, and that bright and dark spicules always occur simultaneously.

An important part of the apparent bright-dark structure must be due to the illumination of spicules by the photosphere below. Since the spicules are optically deep in Hα and K, their bottoms bask in the sunshine and are bright, while the tops are dark. These bright-dark pairings may be responsible for the doubling of spicules noted in off-band pictures by Tanaka (1973). Of course the comparison is not perfect because of quantitative contrast differences, but in principle all features correspond. Because of the contrast differences, the K pictures are best for determining the locus of bright features, while Hα is best for dark fibrils and the horizontal fields. Since the resolution of these filtergrams is better than magnetograms, they are very useful for mapping the magnetic structures.

An important part of the structure of active regions is the relatively low height of the flux loops. This is demonstrated by limb pictures and also by the beautiful X-ray pictures made by AS & E. On the other hand, the presence of higher loops is indicated by the more vertical spicule structures. The low, flat structure of active regions is a major puzzle. If it is force-free, the currents are large.

There has been considerable interest in the possible diffusion of flux away from active regions. In general our observations do not confirm such a migration. Figure 4 shows the region on July 5 and 6 in the K line and we see there has been very little change, with some small expansion to the right (E) but shrinking at the top (S). Certainly there is no definite pattern of expansion into unipolar regions as envisioned in Leighton's (1964) model. Yet unipolar magnetic regions must form, and flux is only separated on a sizeable scale in active regions. Although the bipolar flux in active regions emerges tightly coupled, p to f, the region rapidly achieves a state such as we see, in which all the flux at the outer fringes connects to somewhere else. This means that a sizeable amount of flux reconnects to somewhere else, i.e. distant plage. It may be that the unipolar magnetic regions are formed, not from the elements of the active regions, but their counterparts. Thus a large active region should be followed by a unipolar region of opposite polarity. The mechanism of reconnection of flux to distant poles is unfortunately not understood.

3. Changes in Plages and Active Regions

As noted earlier, major changes in active regions only take place in the following ways: (1) sunspot formation and break up; (2) flux outflow from sunspots; (3) new flux emergence; and (4) magnetic field reconnection.

In general there is no proper motion at all in the plage or the surrounding plagettes except for the latter two. The formation of sunspots is of course connected with EFR's, and large complex spots may be formed by the merger of p spots. When spots break up there is no particular enhancement of the plage left behind. This may be because the plage brightness is not so sensitive to the field strength for strong fields, or because the total field in the spot plus penumbra is not much greater than the surrounding plage.

We have mentioned the role of new emerging flux in producing complexity in sunspot groups. In my experience this is the principal progenitor of flares. A beautiful case is the August 1972 spot group (Zirin and Tanaka, 1973) which had completely reversed polarity from the start. The large activity in the group began when a new pair of spots began to develop and the leader, a p spot, plowed into the other flux. As we mentioned above, because the lines of force in such dipoles are tightly closed, the boundary with nearby flux becomes a shear boundary storing up the energy of the spot motions for release in flares. The ultimate source of this energy is of course the magnetic buoyancy of the flux loops.

In Figure 5 we see how a new emerging flux region developed in McMath 12417 on July 8, producing in a number of flares as it expanded, and eventually producing a second p spot as the region rotated off the disk. The likelihood of this new spot group emerging at random in the center of the one sizeable spot on the disk is negligible, so we must conclude it arose from some convolution in the same flux loop. The new EFR does come up in an area where some small emerging flux was seen June 30 and July 3, but otherwise there is no forewarning.

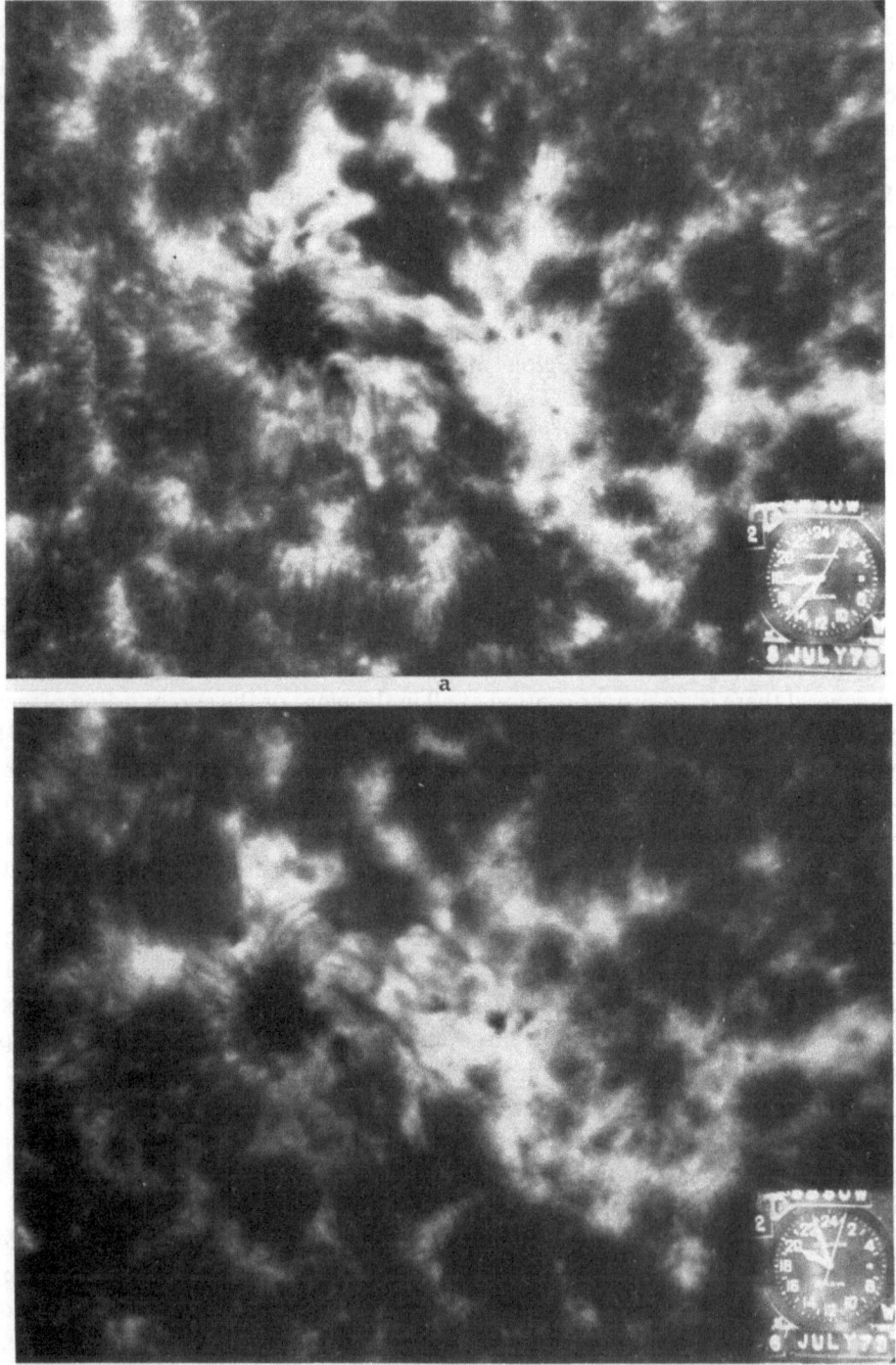

Fig. 4.   (a) K line on July 5. (b) K line on July 6, showing the stability of structures since July 4 (Figure 2a).
Brightness changes are due to printing variation.

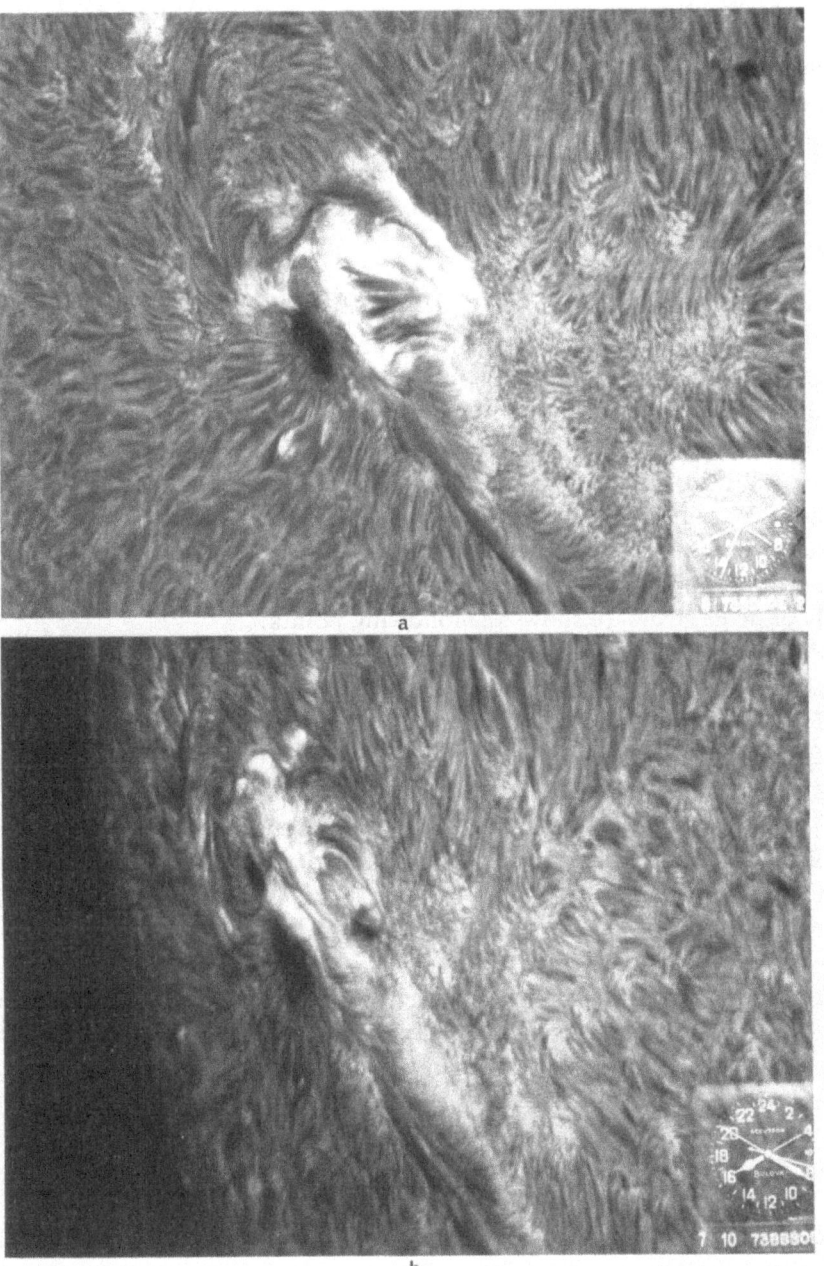

Fig. 5. (a) Development of an emerging flux region in McMath 12417 on 1973, July 8. The emerging flux is marked by bright Hα emission crossed by dark arched filaments following, in this case, the previous polarity orientation. (b) Same region, July 10, showing how two new sunspots formed from the EFR. The fact that this new flux emerged exactly in the only active region on the Sun shows that flux emergence is not random but connected with or focussed by existing sunspot groups. However, EFR's outside of sunspot groups are distributed at random (Glackin, 1972).

In many cases the emergence of the new flux does not involve a great deal of expansion, but because the flux does not arise exactly at the previous neutral line it produces a very jagged neutral line. A particularly impressive series of flares on 1972, February 15, was produced by such an effect. Field reconnection produced sizeable flares and left a smooth neutral line. The ultimate effect of reconnection and energy release is to leave an old plage consisting of two bright regions of opposite polarity connected by a fairly regular set of field transition arches.

4. Summary

I have tried to show that with modern high-resolution synoptic measurements and the understanding of the magnetic structure, a plage is transformed from an anonymous solar feature to a structure which reflects its history, development and future prospects. I should hope that those who carry out physical analysis of such plages pay more attention to the exact nature of the plage they are measuring and the field configuration at the point under study. Modern XUV results should enable us to learn exactly what the physical conditions are over various parts of the plage, but more visual spectroscopy is needed to define the chromospheric parameters.

Study of plage structure shows the following points:

(1) Plages emerge in the interior of emerging flux loops.
(2) Complexity is only produced by the emergence of new flux loops.
(3) There is no twist in the emerging loops.
(4) There are no spicules over plages.
(5) The plage structures are long lived.
(6) The magnetic structure of plages is quite shallow.
(7) The brightness of plages is uniquely determined by the magnetic field structure.

I have mentioned a number of important problems that remain to be solved. Among these are:

Why is a sheared magnetic boundary so characteristic and stable?

Why is the plage so much brighter when flux is emerging? (probably reconnection).

Why do plages form at the inside of flux loops, and sunspots at the outside?

Why are spicules suppressed over plages? (probably transition zone hating)

How do unipolar regions form? What is going on at the opposite end of the lines of force that we see?

Acknowledgments

This research was supported by NASA under NGR 05-002-034 and by NSF Atmospheric Sciences program under GA-24015.

References

Bruzek, A.: 1967, *Solar Phys.* **2**, 451.
Bumba, V. and Howard, R.: 1965, *Astrophys. J.* **141**, 1492.
Dunn, R. B.: 1972, unpublished.

Foukal, P.: 1971, *Solar Phys.* **20**, 298.
Frazier, E. N.: 1972, *Solar Phys.* **24**, 98.
Glackin, David L.: 1973, *Publ. Astron. Soc. Pacific* **85**, 241.
Janssens, T. J.: 1972, *Solar Phys.* **27**, 1972.
Leighton, R. B.: 1964, *Astrophys. J.* **140**, 1547.
Prata, S. W.: 1971, *Solar Phys.* **20**, 310.
Sheeley, N. R., Jr.: 1969, *Solar Phys.* **9**, 347.
Tanaka, Katsuo: 1973, BBSO #0115.
Waldmeier, M.: 1937, *Z. Astrophys.* **14**, 91.
Weart, Spencer R.: 1970, *Astrophys. J.* **162**, 987.
Zirin, H.: 1969, *Solar Phys.* **7**, 243.
Zirin, H.: 1972, *Solar Phys.* **22**, 34.
Zirin, H. and Tanaka, K.: 1973, *Solar Phys.* **32**, 173.

DISCUSSION

Beckers: I just want to make a comment about the supergranulation and the network. You intermix those two things. There is some correlation in the quiet Sun between network and the supergranulation but I am not convinced that at any point on the Sun there is such good correlation. I don't think that the correlation between brightenings in the image you have shown here and the supergranulation has been established in the active regions. I would prefer that we stick to the term network.

Zirin: I will show a slide later showing CN observations. I believe they are the same but so far as this discussion is concerned what I am talking about is the network as seen in K, Hα or in magnetograms. This picture is a magnetic picture and defines the rise of this new region relative to the magnetic field patterns.

Newkirk: Could you distinguish between what you call the field transition arches and the emerging flux loop. Are the field transition arches simply flux loops connecting regions of opposite polarity that have ceased to emerge?

Zirin: Yes, that's right. They also have a very narrow profile and connect old flux regions that are close together.

Mme Pick: What is the relationship between the appearance of new emerging flux and the appearance of opposite polarity around the sunspots?

Zirin: I do not have a systematic answer. I think they are a different phenomenon, that is the emerging flux region is a characteristic unique thing with a life of its own whereas the parasite polarity is something that is continually either erupting or appearing from a folding downward of flux loops. I don't understand parasite fields.

Altschuler: Two questions. First, do you always find from your observations that flares are preceded by emerging flux and with what time delay, and second, can you rule out once and for all models of flares which store the initial energy in the corona? Maybe in a neutral sheet?

Zirin: There is not a direct temporal relationship. Flux emerges and this produces a distortion in which the new flux has stressed the magnetic field which must reconnect, so this produces flares, but that doesn't happen in an hour or six hours or even twenty four hours after the flux emerges. However, if flux does not erupt there are no flares except perhaps for the small flares. There are lots of little things happening around the sunspots but larger flares occur when the flux erupts. In answer to your question, my thoughts are very much turned towards the chromosphere because this is what we look at. I cannot rule out anything that might be going on in the corona. You always tend to think that what you are looking at is the most important part. I would say this in the tight active regions. I think it is all stored down below because of the observations that Tanaka and I made in the August flares where you see a highly stressed magnetic field low in the surface and then afterward you see nice potential loops.

Athay: In your discussion of emerging flux regions outside of active regions you implied that they occurred with equal probability at all solar latitudes. Did you really mean this?

Zirin: No, they occur with equal probability in longitude but in latitude they occur primarily in the sunspot zones. The butterfly diagram is the end result of the EFR's and they do occur most frequently at the same latitude that has the sunspots still occurring.

Beckers: You raised the point a few times that the emerging flux loops and the field transition arches are horizontal structures. The same thing is true of the fibrils and the vortex structure that you see around the sunspots. I have often asked why it is that way and I like to give myself the following answer. If there is a bipolar magnetic region on the Sun and nothing but that bipolar magnetic region there will be lots of field lines that connect the two bipolar regions. Some of these will lie in the horizontal plane and some will rise above the horizontal plane extending into the corona. The ones in the corona will be emptied of their matter and will disappear, the only ones that will remain visible in Hα are the ones that are horizontal and lie at a low level in the atmosphere. If there is a much more confused magnetic field for example, a magnetic field concentration between the dipoles the picture would be distorted but I don't know in what way. I have been wondering whether the regions between these arches and below the arches may be completely void of field in the lower levels. I would just like to throw that out as an explanation.

Zirin: I would say that explanation is by and large right. However if one looks with high resolution at these regions near the surface the field lines seem to be nearly horizontal with very little vertical extent. I agree that our observations are very biased by the fact that it would be difficult to see enough neutral hydrogen to show the arches in the corona. When a flare occurs it normally illuminates those loops, but maybe that's just because we fill those loops up with material when a flare happens. Maybe those are real flux loops but I doubt it. However at the limb we still see these very low arches that are sharply bent over and are not potential fields they seem to require large current. Some people object to this but I have no control over it myself.

Giovanelli: I don't see any problem at all with the low arches. That is just what I would predict if there is a tight magnetic field below the photosphere; as it rises up it will expand because of the reduced gas pressure. As soon as it gets up above 1000 km it is surrounded by corona where the pressure is nearly constant over a considerable range of height. This means that a flux loop will become more or less uniform in diameter and the least energy is involved when the tube of force goes straight across between the two ends. According to this picture the tube of force rises up and expands until it is more or less of uniform diameter and then goes straight across to its other end.

Zirin: Someone needs to study these things, the observational material exists and the theoreticians have not really had a crack at it.

Martin: I would like to state a qualification about your statement about the number of active regions that form per day. Weart found that only two active regions form per day but this pertains only to active regions that attain the size of the supergranule cells or larger. For regions that don't attain this size and may not even form pronounced sunspots the number is very large. Karen Harvey and I found that the number is between 50 and 100 emerging flux regions per day.

Zirin: Yes, I was surprised at that big number that is on our docket to look at from our data. I agree that two per day is two active regions per day that reach supergranule size and form sunspots. We see the little ones but I don't see 50 per day.

Martin: The 50 per day was based on the count for the entire sun, and has been verified with Kitt Peak magnetograms taken since the beginning of the Skylab programs. On one magnetogram a day, which is insufficient to see all of the emerging flux regions we actually see 10 to 20 per day so that we figure that with more complete observations of the whole Sun we will be seeing the larger number.

Zirin: I think you will agree that there are only about 2 per day that attain large size, and therefore my conclusions about the probability of erupting within a sunspot group are not affected by that statement.

Martin: I agree with you but I think that the difference is that emerging flux is very common in active regions.

Newkirk: I am worried about your statement that flux emergence and field line reconnection accounts for practically all of the flux. Unless I misunderstood you this would imply that all of the flux is essentially annihilated locally in the active region. This is contrary to the picture that most of the flux in extended unipolar regions is flux that is just torn off from active regions. If you annihilate all of the flux in active regions you are then never going to get field lines connecting widely separating centers unless you can come up with some technique for preferentially destroying one field sign over the other.

Zirin: I thought I had addressed that question but maybe you missed it. I realize that we do not understand how flux loops connect to different places. It seems like filaments play an important role because they carry flux for such a large distance. There must be some reconnection which results in the connection of a new flux to some very distant flux somewhere else. But what I was trying to say is that one sees virtually no outward flow of magnetic flux in these plages. When new flux stops coming up the active regions shrink. The only thing I can guess is that when the plages come up, part of them gets connected to something farther out. Although we don't see the regions moving out each of the lines of force must connect back to the solar surface at some point in opposite polarity and that might conceivably account for the fact that we do not see flux expansion in the plages.

Schatten: I have recently come up with some ideas that might support Alfvén's old picture of Alfvén waves coming up from below. Can you rule that out on the basis of your observations?

Zirin: I don't understand how I would observe hydromagnetic waves.

Schatten: Isn't that what you are talking about?

Zirin: I shouldn't have used the word waves. What we saw was a sunspot with a large chunk of magnetic flux appearing and moving outwards. That is not a common phenomenon. The principle way that flux comes up is in the bipolar loops.

Schmidt: I have a nasty comment. I happen to be the guy who is worried about the Lorentz forces. I do not think the chromosphere and corona are equal footing when we consider the flux distribution. This is a geometrical question and in my own, old fashioned mind, this is a question of 4000 km height estimated for the fibrils versus my estimate here of 400000 km.

ON FACULAR MODELS

DAVID E. REES

Dept. of Applied Mathematics, University of Sydney, Sydney, N.S.W. 2006, Australia

Abstract. Weakenings of 14 Fe I lines observed in photospheric faculae by Chapman and Sheeley (1968) have been analyzed using Unno's (1956) LTE theory of Zeeman triplet formation. These observations can be explained by a facula-photopshere temperature excess of $\lesssim 250$ K in the layers ~ 100–400 km above $\tau_{5000} = 1$. The inferred magnetic field in the faculae is ~ 1000 G.

Such a temperature model is not consistent with the center-limb variation of the continuum contrast between faculae and the photosphere observed, for example, by Chapman (1970). In calculating the facular intensity it is necessary to take account of the small horizontal width of the faculae granules, especially near the limb where the light path traverses the photosphere as well as the facular granules. This has been done for a simple model of an isolated facular granule having vertical sides and a width of 760 km. The temperature excess required to explain the observed contrast near the limb is ~ 800 K.

The discrepancy between the facular temperature excesses derived from line and continuum data probably can be reconciled to some extent by varying the geometry and including the effect of clustering of the facular granules. Work on this is in progress.

The facular granules usually are not resolved. Coupled with the effects of scattered photospheric light, this indicates that the true weakenings and contrasts exceed those observed. Therefore higher temperatures would be favoured.

References

Chapman, G. A.: 1970, *Solar Phys.* **14**, 315.
Chapman, G. A. and Sheeley, N. R.: 1968, *Solar Phys.* **5**, 442.
Unno, W.: 1956, *Publ. Astron. Soc. Japan* **8**, 108.

DISCUSSION

Zirin: Since we know about the filigree in the photosphere and since we know that in off-band Hα the network spreads out by about 1000 km in the line wings; what do you think you would get if all of these things looked like cones?

Rees: This would tend to reduce the temperature for the line model.

Beckers: I completely agree with Dr Zirin's remarks. Also in the Magnesium b_1 line do we see the line center faculae to lie immediately above the photospheric filigree as seen 0.8 Å from the b_1 line center. The increase in size of the subgranular filigree elements ($\leqslant \frac{1}{4}''$) to the $\frac{1}{2}$–$1''$ line center facular elements really shows the faculae to have a conic structure which has to be taken into account in your modelling.

Rees: I think the geometry is the fundamental thing and until we do a more detailed calculation we can't put a number into this effect in the facula model.

Meyer: I would like to ask whether temperature structure in the vertical direction would have an influence on your contrast from center to limb. In the downward directed shock flow model one expects

R. Grant Athay (ed.), Chromospheric Fine Structure, 177–178. All Rights Reserved.

higher temperatures in the regions behind the shock front, and lower temperatures as the gas moves down-ward and cools by radiation. It would be interesting to know how this fits to the empirical models.

Wiehr: I think that the theoretical shock flow model is in good agreement with out model which requires a temperature excess exclusively in the high facula layers ($h > 250$ km) whereas the deeper layers show photospheric temperatures. The temperature excess in the high layers increases with height. Furthermore the downward flow is an observed fact, as I pointed out.

Rees: The temperature excess is greater in higher layers also in our models.

Thomas: I am curious as to what controls the line intensity – is it collisions or photo-ionization or what?

Athay: Which lines have you used in the analysis?

Rees: The line with the greatest change in central intensity is the 5250 line of Fe.

Athay: In my computations with Lites on Fe this line is collisionally controlled at these depths and should be in LTE.

Wilson: I am sure this would be important and this was my pont with Dr Wiehr if your excess tempera-ture occurs in a very shallow layer and of course increases upward then the contrast that you would see if you looked directly through this would be less. But you can't really answer the question until you play around with different geometries and trey to construct theoretical profiles.

Brueckner: In the ultraviolet continuum around 1600 to 1700 Å Tousey and Purcell placed an upper limit on the temperature excess of about 200°. But one has to keep in mind that these observations were done with quite low spatial resolutions so if you go to smaller structures this result may change. We have recently obtained a rocket spectrum which seems to fall on a facula area which seems to show that the temperature change was around 200° at 1700 Å, it also showed considerable change in the shape of the spectrum between the silicon triplet and silicon doublet continuum. In their words the opacity in this optical depth has changed considerably from the photosphere to the facula. I don't know how to explain this but I think photo-ionization needs to be taken into account. This coud modify the line profile calculations considerably and this may be a way out of the difficulty.

A FACULA MODEL AND ITS APPLICATION TO FACULA FINE STRUCTURES

E. WIEHR and G. STELLMACHER

Universitäts-Sternwarte, Göttingen, F.R.G.

Abstract. In a recent paper Stellmacher and Wiehr (1973) discussed in detail the discrepancies between 'continuum-' and 'line-profile facula models'. Their calculations show that continuum models (e.g. Kuz'minykh, 1962; Schmahl, 1967; Chapman 1970; Wilson, 1971) encounter the problem that the facula contrast *increases* towards the limb but *decreases* with wavelength λ. This discrepancy evidently cannot be due to NLTE effects since it concerns the continumm data.

Problems increase essentially when line profile data are taken into account. Measurements of several magnetically insensitive lines (Stellmacher and Wiehr, 1971) show that the normalized profiles are identical for facula and neighboring photosphere except for the line cores (rest-intensity effect). These results indicate a temperature excess in high facula layers ($\tau_0 \leq 0.03 \hat{=} h \geq 250$ km) increasing with height and a photospheric temperature stratification in the deeper facula layers. Such a (preliminary) model (A) also fits to the observed dependence of the rest-intensity effect on the line excitation potential.

The continuum contrast at the disc center including its λ-dependence ($\approx 2\%$ constant for $4000 \text{ Å} \leq \lambda \leq 8000 \text{ Å}$) requires in addition to the temperature stratification of model A, a reduction of the gas pressure by $\Delta \log P_g = -0.015$. The resulting model (C) is then able to represent all facula data observed at the disc center including the magnetic field (when taking $H^2 = 8\pi \Delta P_g$); however, it fails to represent the observed center-to-limb variation of the contrast (CLV). On the other hand, facula models which do fit the CLV strongly contradict the observed λ-independent contrast and the unchanged line wings.

Hence, we suggest assuming our 'disc center model' C and trying to explain the CLV exclusively by the facula geometry. Here we draw two possibilities:

(1) Elevation of the geometric facula scale by 200 km and a finite facula diameter of 2000 km easily yields the observed CLV. However, whereas the finite facula diameter seems reasonable, the elevation by 200 km contradicts our $\Delta \log P_g = -0.015$ which actually corresponds to a depression of 10 km.

(2) Applying the considerations on the Wilson effect in sunspots by Jensen *et al.* (1969) to the facula model and assuming a certain 'curvature' of the depression, one can obtain a CLV of the contrast depending on the assumed curvature.

None of these geometric models would limit the validity of our model C. In particular the data expected for facula fine structures remain compatible with our model: The continuum contrasts of single facula granules should reasonably be larger than those of the smeared-out facula region, but these larger contrasts would still be λ-

R. Grant Athay (ed.), Chromospheric Fine Structure, 179–181. All Rights Reserved.
Copyright © 1974 by the IAU.

independendent (because the smear-out by blurring is λ-independent). Furthermore, the normalized line profiles of a single facula granule can be expected to remain unchanged in the wings (otherwise our profile observations for smeared-out faculae would not contain one facula granule at all!). Model C can thus easily be adapted to single facula granules by applying an additional reduction to the gas pressure.

As an example we assumed a 'corrected' contrast of 6% for a single facula granule. Calculations with our model C yield this contrast including the required λ-independence and the unchanged line wings when applying $\Delta \log P_g = -0.06$. Since hydrostatic equilibrium requires a magnetic field $H^2 \sim P_g \Delta \log P_g$ the increase of $\Delta \log P_g$ by a factor of 4 yields an increase of the magnetic field by a factor of 2 as compared to the values for the smeared-out facula region deduced by Stellmacher and Wiehr (1973). It seems reasonable that the observed contrast and magnetic field for smeared-out facula regions are lower than the actual values for single facula granules. The resulting relation between contrast and magnetic field roughly agrees with the measurements by Frazier (1971).

Our model predictions of a temperature excess exclusively confined to the higher layers ($h \geq 250$ km) together with a pressure reduction relative to the adjacent photosphere fits well to the theoretical 'shock flow model' by Meyer and Schmidt (1968). This model proposes a temperature excess above the 400-km level increasing with height as well as a downward flow with subsonic velocities in the lower facula layers. Such a downward flow, however, usually is observed for facula regions and amounts to about 0.8 km s^{-1} (see e.g. Howard, 1972).

References

Chapman, G. A.: 1970, *Solar Phys.* **14**, 315.
Frazier, E. N.: 1971, *Solar Phys.* **21**, 42.
Howard, R.: 1972, *Solar Phys.* **24**, 123.
Kuz'minykh, V. D.: 1965, *Soviet Astron.* **8**, 551.
Jensen, E., Brahde, R., and Ofstad, P.: 1968, *Solar Phys.* **9**, 397.
Meyer, F. and Schmidt, H. U.: 1968, *Mitt. Deutsche Astron. Ges.* **25**, 194.
Schmahl, G.: 1967, *Z. Astrophys.* **66**, 81.
Stellmacher, G. and Wiehr, E.: 1971, *Solar Phys.* **18**, 220.
Stellmacher. G. and Wiehr, E.: 1973, *Astron. Astrophys.* **29**, 13.
Wilson, P. R.: 1971, *Solar Phys.* **21**, 107.

DISCUSSION

Pecker: How is your model with the reduction of the pressure in the photospheric layers by $\Delta \log P_g = 0.02$ compatible with an increase of the opacity that you suggest for the explanation of the center- to limb variation of the contrast? I would have placed the $\tau = 1$ line *deeper* in the faculae than in the surrounding photosphere. Now a second question. Your Figure No. 2 shows no variation in contrast with wavelength. That is a very intersting observational result. When you go near the limb what does the curve evolve to? Does it still show no variation with wavelength?

Wiehr: Yes, here we confirmed former measurements by Schmahl. I wonder why these observations are not better known, perhaps because they were published in *Zeitschrift für Astrophysik* in German. Concerning your first comment we are aware of the problem: If you evaluate the isolines from the pressure reduction $\Delta \log P_g = 0.015$, you actually get a *depression* of about 10 km in contradiction to the proposed *elevation* of 200 km. This discrepancy possibly disappears if you keep in mind that there is a downward motion observed in all faculae which might alter the pressure scale. But perhaps one should follow our second geometric suggestion.

Pecker: There is another solution which I don't like but which I will say and that is that the balance of pressure might be achieved by increasing the magnetic field above the value that you have computed.

Wiehr: Only a magnetic field of very complicated structure would be able to balance the *higher* facula pressure required for the elevation.

Wilson: Another discrepancy is the difference in the temperatures obtained from the continuum measurements and the line temperatures. What were the temperature values that you would get from the line data?

Wiehr: For the line weakenings we require a temperature difference of about 100° at $\tau = 10^{-3}$, for the center-to-limb variations you need larger temperatures but these definitely contradict the observed wavelength independence of the contract of the *continuum*.

Thomas: At which level does the temperature increase occur that you require for the representation of the central line weakenings?

Wiehr: At $\tau_0 \approx 0.03$ corresponding to $h \approx 250$ km above $\tau_0 = 1.0$. All these calculations, of course, are based on the assumption of LTE!

Wison: You mentioned the contradiction between continuum and line models, namely the temperature excess of 500–1000° required for Wilson's and for Chapman's models and that of 100–200° required for the line data. I feel that these contradictions clearly indicate that one-dimensional analysis are quite inadequate and that geometry is an essential ingredient in any analyst. In this respect geometry, i.e. a greatly increased surface area, may give rise to departures from LTE in lines which would certainly be in LTE at these levels of a one-dimensional atmosphere. A larger excess temperature will weaken the lines while departures from LTE will strngthen them. Thus the line data may be consistent with larger excess temperatures.

Wiehr: We agree with you that geometry is an essential ingredient, this clearly comes out from our discussion of the facula atmosphere and is the reason behind our proposal of two possible geometric models! On the other hand your NLTE considerations evidently are unable to remove the discrepancies between the continuum models and the observed λ-independent contrasts. Furthermore, I feel that the unchanged line wings give very strong indication for photospheric conditions in deeper facula layers ($\tau_0 > 0.03$) rather than for a non-photospheric temperature stratification the influence of which on the line profiles being exactly counter-balanced by NLTE. Last, but not least, the line strengthening by NLTE you mentioned would affect the line cores more than the wings, whereas the contrary would be required: The discrepancy between the continuum models and the observed line profiles is very strong in the wings and nearly absent in the cores (see Figure 4 of our paper).

THE RELATION BETWEEN CHROMOSPHERIC AND PHOTOSPHERIC STRUCTURES IN SUNSPOT GROUPS

V. BUMBA and P. AMBROŽ

Astronomical Institute of the Czechoslovak Academy of Sciences, Ondřejov Observatory, Czechoslovakia

Abstract. Using the high resolution photographs of the photosphere taken at the Ondřejov observatory a comparison of positions and forms of some chromospheric features with the distribution of photospheric fine structure morphological elements in the August 1972 active region has been made. The close relationship between active filaments, chromospheric threads, fibrils and photospheric small nuclei, dark interpenumbral and intergranular spaces or rudimentar penumbra is shown. Some considerations are demonstrated with the aid of computed models of the horizontal component of the magnetic vector simulating the real distribution of magnetic fields in the studied region.

1. Introduction

Recently, a relatively large number of papers studying the individual characteristic elements of chromospheric fine structure and their relation to the photospheric magnetic fields has been published. This work has been done mostly by the **Big Bear Solar Observatory** group (Zirin, 1970; Prata, 1971; Foukal, 1971a, b) as well as by others. In this paper, we study the relationship of the described chromospheric features to features seen in high resolution photgraphs of the photosphere, especially to those of sunspot structures.

For example, solar surges have been demonstrated to arise from certain regions of sunspots (Roy, 1973; Gopasyuk *et al.*, 1963; Gopasyuk and Obir, 1963). Although these regions have been characterized magnetically their morphology in the photosphere has not yet been examined. Also Frazier (1972) has shown that the ends of arch filaments are rooted in groups of 'knots' of the photospheric magnetic field.

These facts led us to an investigation of the relations of filaments, chromospheric threads, arches and fibrils to the underlying morphological details of sunspots, their penumbras and surrounding photosphere. An initial comparison of relatively low resolution chromospheric Hα photographs ($\sim 2''$) demonstrated that a great number of chromospheric features in an active region are related to certain peculiarities in spot penumbras or to small nuclei around large spots.

In the present note, we extend this study using good observations of the large August 1972 active region.

2. Observational Material

The great August 1972 sunspot group has been observed photographically at our observatory starting from July 31 till August 10. During this time interval a large number of photospheric photographs were obtained. The instruments and method of observation were described by Bumba *et al.* (1973). Each day many series of photo-

R. Grant Athay (ed.), Chromospheric Fine Structure, 183–191. All Rights Reserved.

graphs with different exposures, usually several tens of minutes apart have been made. The quality of photographs varied from day to day as well as during a given day depending on weather and seeing conditions. But in some exposure series photographs with resolution around 1″ or better may be found.

Good quality photographs of the chromosphere from the Big Bear Solar Observatory Flare Film 1972 (taken for the major part in the hydrogen Hα line) have been used.

3. Results

In the photosphere, the sunspot group was characterized morphologically by several large umbras (their development and magnetic field polarity distribution may be seen, for example in Bumba, 1973), and by an extensive penumbra, the greatest part of which was irregular. The penumbra enclosed many small umbral nuclei or darker, elongated spaces between the bright penumbral fibrils. Also, the photosphere penetrates into this penumbra in the form of gulfs, unusually bright filaments, chains of granules, light bridges etc. Only a small part of the penumbra was more or less regular, with radial fibrils and relatively sharp boundary with the photosphere. Among the more interesting features of the group were the light bridges. These had several different forms, some of them developing so vigorously that they appeared to divide the largest umbra and push its pieces apart. Some of these light bridges were in the form of a penumbra with very elongated, slightly curved fibrils coinciding with the magnetic field polarity boundary, which is very unusual. The special appearance of penumbral structures can be used by a skilled observer as an indicator of the presence of opposite polarity fields and as tracers of the magnetic field direction in the way that fibrils and filaments are used in the chromosphere. In the photosphere, of course, these associations may be checked by direct measurements of the magnetic field strength and polarity.

The general structural characteristics of the chromosphere above the active region were: a system of active region filaments (indicating the main polarity boundary) composed from individual fibrils connecting opposite polarities; regions formed by long fibrils or threads stretched out more or less radially from the group (magnetically characterized by presence of both polarities in islands); and regions without fibrils or threads. In the latter regions the dark space above the umbra and penumbra merges continuously into the quiet chromosphere far away from the spot. The quiet chromospheric structure around the spot group is characterized magnetically by a supergranular cell of predominantly one polarity which is the same as the polarity of the leader spot. There are also bright features and several chromospheric flares developed in the region.

Comparing the chromospheric and phtospheric structures, one sees that the chromospheric regions with long fibrils and threads are connected with the regions of disturbed penumbra, i.e., with those regions where changes in the distribution and form of penumbral structures and umbral nuclei take place (see Figures 1 and 2). On the other hand the chromospheric regions without distinct structures and with direct transition

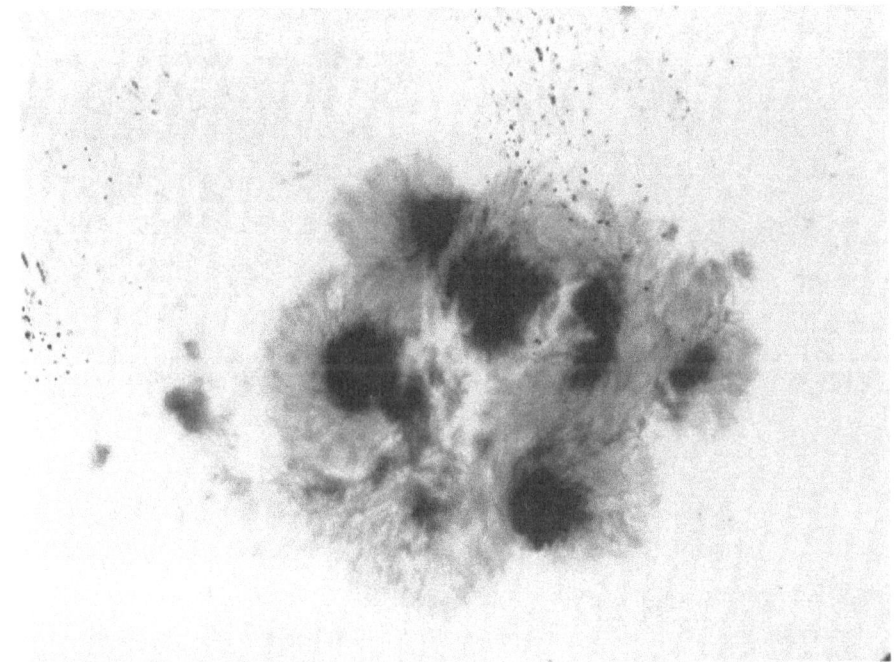

a

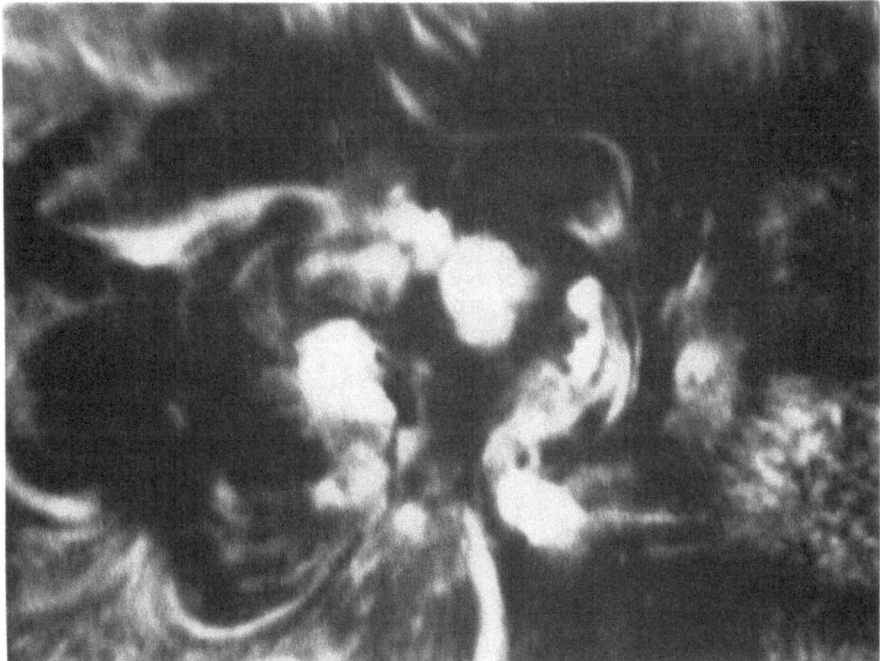

b

Figs. 1a–b. Photograph of the photosphere in the August 1972 active region taken in Ondřejov on August 4, at 14^{h}31^{m}34^s UT compared with the negative photograph of the same active region taken from the Big Bear Solar Observatory Flare Film in Hα at 19^{h}30^{m}50^s UT. Both photographs have the same scale and orientation. Using a sheet of transparent paper one can compare the mutual positions of details in the photosphere and chromosphere.

A chromospheric region with long fibrils and threads related to a photospheric region of disturbed penumbra may be seen, for example, in the lower parts of both photographs. Also, a region without distinct structures in the chromosphere and with a normal quiet penumbra is well visible on the right sides of both pictures.

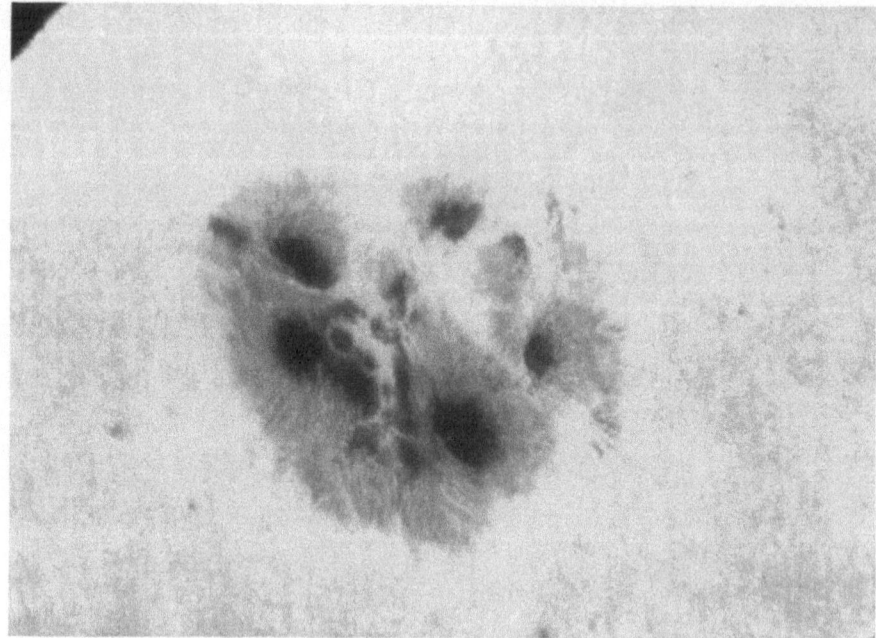

a

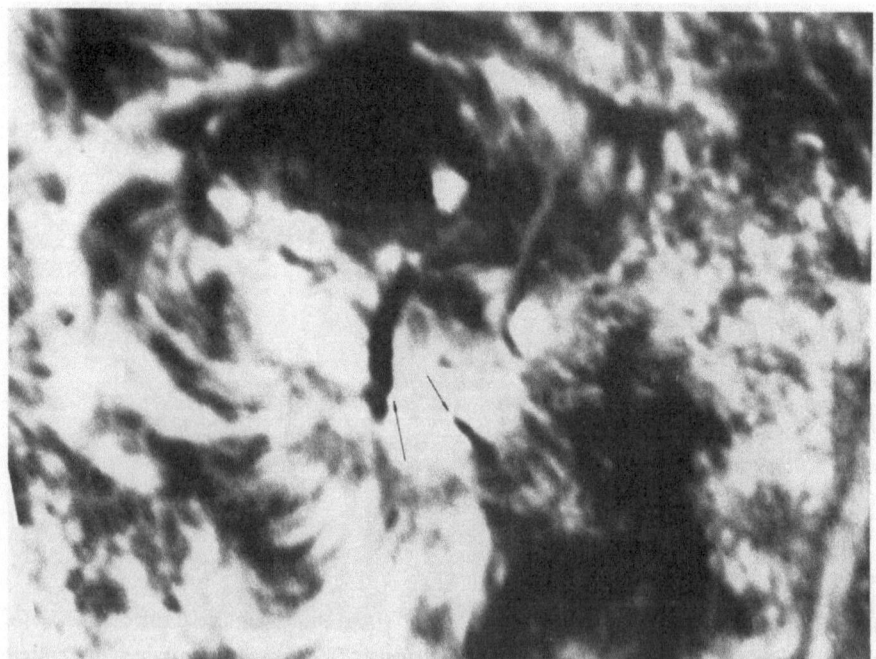

b

Figs. 2a–b.   Combination of the same photospheric photograph (15^{h}09^{m}36^s UT) and a chromospheric picture taken during the very first stage of large flare development (14^{h}59^{m}10^s UT) 1972, August 7. The brightening of emission above light bridges may be seen.

into the quiet chromosphere coincide with the regular umbras (see also Figures 1 and 2). The active region filaments and fibrils indicating the magnetic field line distribution in the active region and its vicinity coincide inside the group with the highest gradient of magnetic field across the boundary of polarities and therefore with the region of elongated, slightly curved bright fibrils of penumbra forming a special form of light bridges in this group (Figures 1 and 2). The individual fibrils of this filament join the groups of small umbral nuclei of opposite polarity placed close to the polarity boundary. In these latter regions there are fast changes due to growth as well as to disintegration.

Exact comparison of the positions of ends of individual fibrils, threads and filament fibrils shows that, for the most part, those in the nearest neighbourhood of the group are closely related to the darker elongated spaces between the penumbral bright fibrils, to small spots and nuclei without penumbra, to darker enlarged intergranular space or rudimentary penumbra in the vicinity of the great spots or to the secondary small spot groups developed around the main group (Figures a and 2). Due to the time differences between the best observing periods at Ondřejov and Big Bear, only a few of the photospheric and chromospheric photographs can be directly compared. (Pictures obtained at Ondřejov in late afternoon overlap the observing period at Big Bear.) By studying the development of those photospheric details in which the chromospheric features seem to be rooted during the whole days it is possible to say that, the fibrils are rooted in regions where the small spots or darker features are changing only slowly in position (Figure 2).

Chromospheric emission is usually very closely associated with underlying light bridges (Figures 1 and 2). During the first phase of flare occurrence the brightened regions of the chromosphere coincide with the light bridges even better (Figure 2). The flare brightening develop along the length of the light bridges and was still visible in these locations during the declining phase of the flare. The development of flare emission on these high resolution Hα photographs resembles the situation observed in the calcium chromosphere (Bumba and Howard, 1965).

Only few examples may be given concerning the dynamics of related chromospheric and chromospheric features. Westwards from the main group, a small regular spot of negative polarity and having, for a short time, a rudimentary penumbra was observed. This spot acted during the first few days as an attractive center for a system of fibrils and threads spiraling around it and with matter flowing into the spot. After August 5 the existence of this spot had no further influence on the distribution of chromospheric structures around and above the spot, although its photospheric appearance did not seem to change.

4. Discussion

What conclusion can be made from the fact that the greatest part of chromospheric features above an active region is rooted in small photospheric dark features on the periphery of a sunspot group or around it? The intensity of the magnetic field in such photospheric objects is at least of the order of 1000 G (Bumba, 1967). These objects

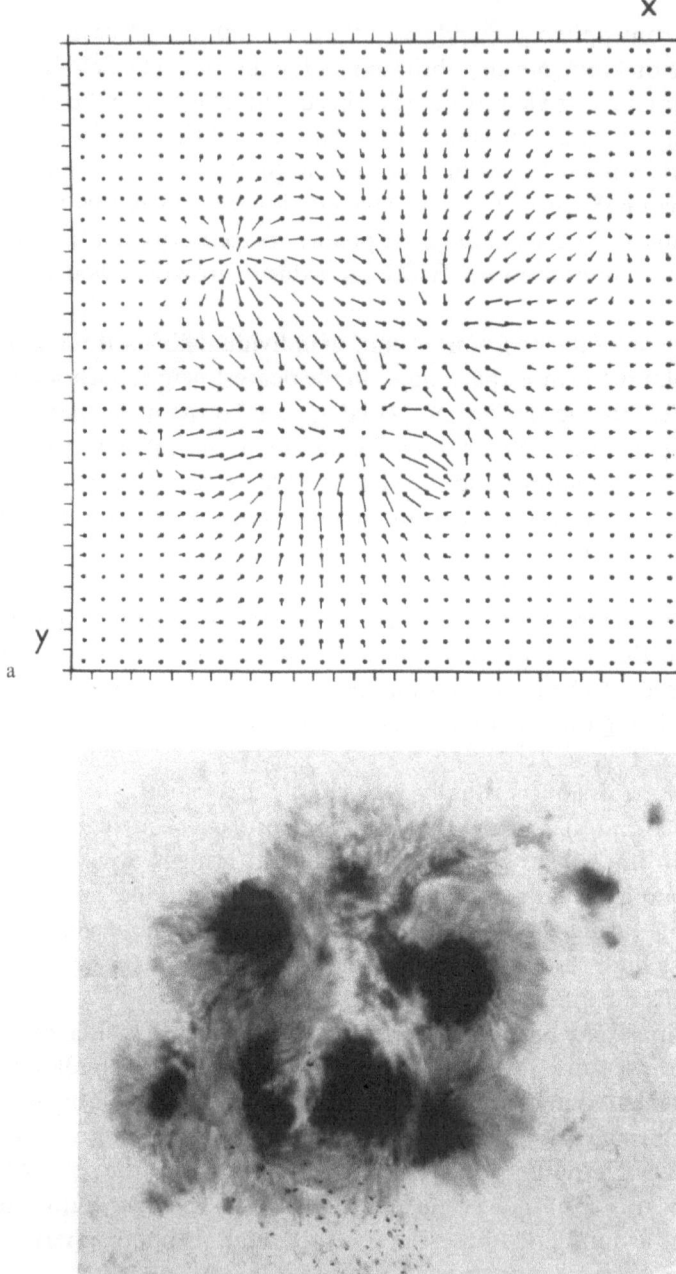

Figs. 3a–b. 1972, August 4: Combination of the photospheric photograph and the directions of the horizontal component of the magnetic vector slightly (the distance of two points in one line on the graph) above the studied group, calculated in potential approximation with the aid of a matrix corresponding to the photographically obtained magnetic field intensity and polarity distribution. The scale and orientation of both figures is the same.

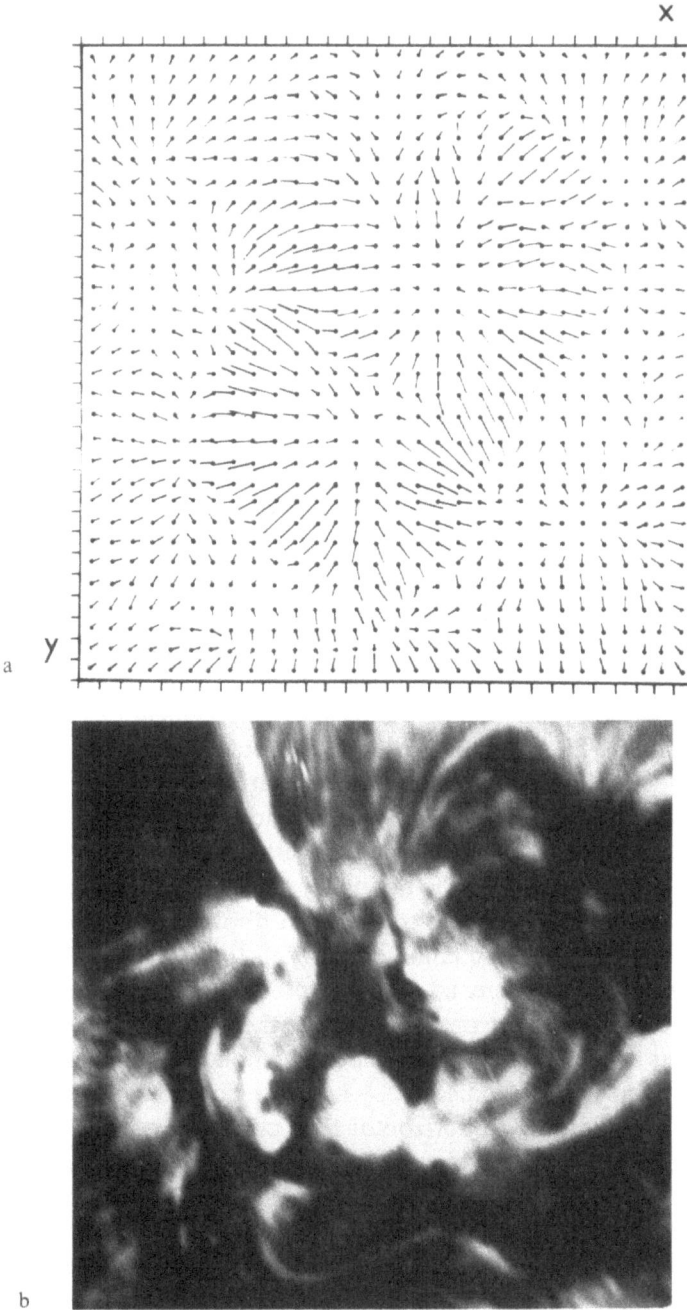

Figs. 4a–b. 1972, August 4: Comparison of the chromospheric photograph with the graph of distribution of directions of the horizontal component of the magnetic vector slightly (the distance of two points in one line of the graph) above the studied group, calculated in potential approximation with the aid of a matrix in which the magnetic field is represented by magnetic points with the same intensity, polarity distribution corresponding to that of a magnetogram (Livingston, 1973). The scale and orientation of both pictures is the same.

are relatively unstable. Their life-time seems to be of the order of hours, although the region in which they occur lasts for days. Usually one takes these features as of 'secondary' importance in an active region. As has been already demonstrated (Foukal, 1971a, b) individual fibrils join the opposite polarities of magnetic fields often lying along the polarity boundary and they lie very low. Thus, the structure as we envision it seems to be far from hitherto existing theoretical models. At the same time another question seems to arise: why does the morphology of the chromosphere above an active region seem to be influenced only by these 'secondary' eelements of photospheric structure and so little affected by the huge umbras with very high values of magnetic flux?

To study at least some aspects of this question we have calculated by a potential approximation (Schmidt, 1963) two models of the horizontal component of the magnetic vector slightly above the photospheric level. We have simulated the real distribution of solar magnetic fields for 1972 August 4, in two ways: (a) by a matrix in which the magnetic field is represented by magnetic points with the same intensity and polarity distribution as that of a magnetogram (Livingston, 1973); (b) by a matrix corresponding to the total magnetic field intensity and polarity distribution obtained photographically (*Roma Photographic Journal of the Sun*). Comparing the resulted field with photospheric and chromospheric structure (Figures 3 and 4), we see that in such a static model the magnetic fields of all magnetic features (small as well as big) are important in determining the field pattern. The agreement of the magnetic map (Figure 4), based on the fine simulation in which the role of smaller magnetic elements is included, with the photospheric and chromospheric morphology is better than in the case when only the larger spots are taken into account (Figure 3). (In our models the polarity boundary is demonstrated by the largest arrows representing the direction of the horizontal field component, the boundary being perpendicular to them.)

In the photosphere the direction of penumbral fibrils seems to coincide with the direction of the horizontal field component only in those region of spot penumbra bordered by normal photosphere, i.e., without other spots. In between larger spots, the penumbral fibrils seem to run perpendicular to the direction of the horizontal field component of our model. In the chromosphere the polarity boundary in closed field regions between large spots is indicated by active region filaments crossing perpendicularly to the direction of the horizontal field component. In the case when the polarity boundary lies on the periphery of the group, having on one side large spots and on the other side the photosphere with chromospheric network fields and secondary spots only, the chromospheric features seem to lie along the horizontal field component of our model, outlining the well known patterns of the chromospheric morphology.

References

Bumba, V.: 1967, *Solar Phys.* **1**, 371.

Bumba, V.: 1973, World Data Center A for Solar-Terrestrial Physics, Report UAG-28, Part I, July 1973, p. 82.

Bumba, V. and Howard, R.: 1965, *Astrophys. J.* **142**, 796.

Bumba, V., Ranzinger, P., and Suda, J.: 1973, *Bull. Astron. Inst. Czech.* **24**, 22.
Foukal, P.: 1971a, *Solar Phys.* **20**, 298.
Foukal, P.: 1971b, *Solar Phys.* **19**, 59.
Frazier, E. N.: 1972, *Solar Phys.* **26**, 130.
Gopasyuk, S. I. and Ogir, M. B.: 1963, *Izv. Krymsk. Astrofiz. Obs.* **30**, 185.
Gopasyuk, S. I., Ogir, M. B., and Tsap, T. T.: 1963, *Izv. Krymsk. Astrofiz. Obs.* **30**, 148.
Livingston, W. C.: 1973, World Data Center A for Solar-Terrestrial Physics, Report UAG-28, Part I, July 1973, p. 95.
Prata, S. W.: 1971, *Solar Phys.* **20**, 310.
Roy, J.-R.: 1973, *Solar Phys.* **28**, 95.
Schmidt, H. U.: 1963, *AAS-NASA Symp. on the Physics of Solar Flares*, p. 107.
Zirin, H.: 1970, *Solar Phys.* **14**, 328.

DISCUSSION

Zirin: I agree with you that there is a strong sheer in the magnetic field that is diverging and going to far away places, some very near the surface. This was our interpretation in the case of these big flares which is also very well seen in the case of the August 1966 flares. One sees that nearly all of the flux in these newly emerging spots in the active regions instead of connecting with something far away is channeled very close to the surface or to a nearby spot. The force that contains this new field is probably the existing magnetic field of the old spots. There is a whole class of sunspots either without penumbras altogether or with sharply sheared penumbras which characteristically produce big flares, and I think that they are characterized by this type of field that is sharply turned over and is completely force free and completely different from what it would normally be.

THE STRUCTURE OF THE
PROMINENCE-CORONA INTERFACE

F. Q. ORRALL

Institute for Astronomy, University of Hawaii, Honolulu, Hawaii, U.S.A.

and

R. J. SPEER

Dept. of Physics, Imperial College, London, England

Abstract. The relatively cool Hα emitting regions of solar prominences are evidently surrounded by an interface or structure of intermediate temperature that separates them from the surrounding hot corona. At present, very little is known about this prominence-corona interface (hereafter called the PC interface), because most of its distinctive radiation is in the EUV where observations of high spatial resolution have not been available. In this paper we report on studies made on photometrically calibrated monochromatic images of prominences and the surrounding corona recorded on slitless spectra made during the eclipse of 1970, March 7. These were obtained with a rocket-borne Wadsworth spectrograph flown by the consortium group (Speer *et al.*, 1970) and cover the range λλ 850–2159 with a spectral resolution of 0.17 Å and a spatial resolution better than 10″ over half the spectrum. They include spectral lines of a number of ionic species from H I through Ni xv. Because each species exists in a rather narrow temperature range, some of the monochromatic images show the low temperature prominence itself; some the coronal structure surrounding it; and some the intermediate structure we have called the PC interface.

At the time of the eclipse there were 10 prominences visible at the Sun's limb on Hα ciné observations made concurrent with the eclipse with the dual coronagraph at Mt. Halekala. Images of all ten prominences are visible on the EUV spectra in emission lines from a number of ionic species from H I through O vi ($T \sim 3 \times 10^5$ K). Images in lines formed at $T \geqslant 3 \times 10^5$ K do not show the prominences, although they may show other coronal structures nearby. This is consistent with the finding of Noyes *et al.* (1972) that the contrast of prominences against the coronal background drops sharply for lines formed with $T \geqslant 3 \times 10^5$ K.

Two prominences of the 'coronal rain' type (hereafter called prominences *A* and *B*, respectively) were especially well suited for a study of the PC interface. They have a simple geometry (apparently cylindrical) and the orientation of the magnetic field is known to be aligned parallel to the axis of symmetry of the cylinder.

Figure 1 shows the total corrected width at half intensity of these prominences as measured in several emission lines and plotted vs the temperature where the line is formed. These results suggest that these prominence threads consist of a cool Hα emitting core about 10^9 cm in diameter surrounded by a thin sheath in which the temperature climbs steeply to at least 3×10^5 K. At higher temperatures the width

R. Grant Athay (ed.), Chromospheric Fine Structure, 193–196. All Rights Reserved.

becomes uncertain because the contrast becomes low or because other features become visible which become confused with the prominence structure.

We have determined the absolute total intensities of all of the measurable lines in prominences *A* and *B*. The quantity measured for each line is the total intensity in the line from a slice of the prominence 1 cm thick, the slice being taken in the plane normal

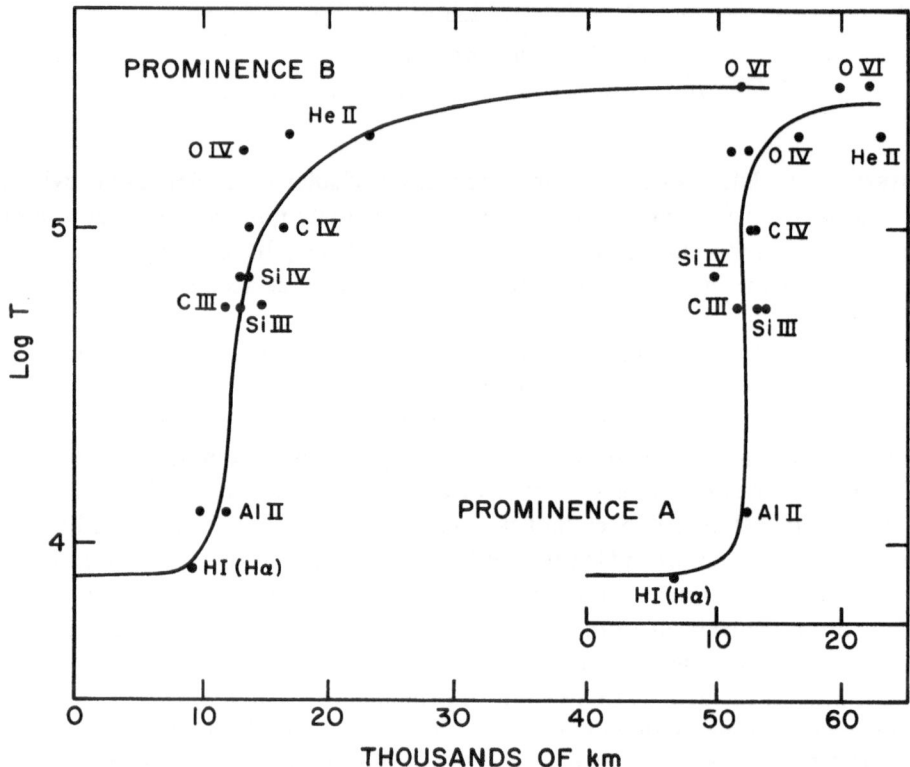

Fig. 1.   The total corrected width at half-intensity measured in several emission lines, plotted against the logorithm of the temperature where the line is formed.

to the axis of symmetry of the thread. The techniques that have been developed to study the EUV emission of the chromosphere-corona transition region (hereafter called the CC transition) are also applicable to prominences with slight modification (see e.g., Athay, 1971; Noyes and Withbroe, 1973). In the analysis we have used the EUV coronal abundances and (where possible, the atomic parameters given by Dupree (1972) and the characteristic temperatures given by Burton *et al.* (1971).

Among the measured lines are the density sensitive pair $\lambda 1176$ and $\lambda 1909$ of CIII. In prominence *B*, the intensity ratio of the CIII lines is observed to be $I(\lambda 1176)/I(\lambda 1909)=1.45$. From the calculations of Gabriel and Jordan (1972) this ratio implies $N_e = 3 \times 10^9$ cm^{-3}. Since this ion is formed at $T_e \sim 56000$ K, the product $N_e T_e \simeq 1.7 \times 10^{14}$ cm^{-3} deg. We shall assume that $\phi \equiv N_e T_e \sim 2 \times 10^{14}$ cm^{-3} deg in both prominences, corresponding to a total gas pressure of ~ 0.06 dyne within the PC interface. Analysis of the EUV resonance lines contains implicit information on the temperature gradient (Athay, 1966). Given the abundance, each measured line intensity yields the quantity

$$\phi^2 \left[\frac{1}{r} T^{5/2} \frac{\mathrm{d}T}{\mathrm{d}r} \right]^{-1},$$

where the bracketed quantity is appropriate to the temperature where the observed spectral line is formed. Here r can be inferred from Figure 1. The results of applying this analysis to the measured intensity is shown in Figure 2 where we have plotted

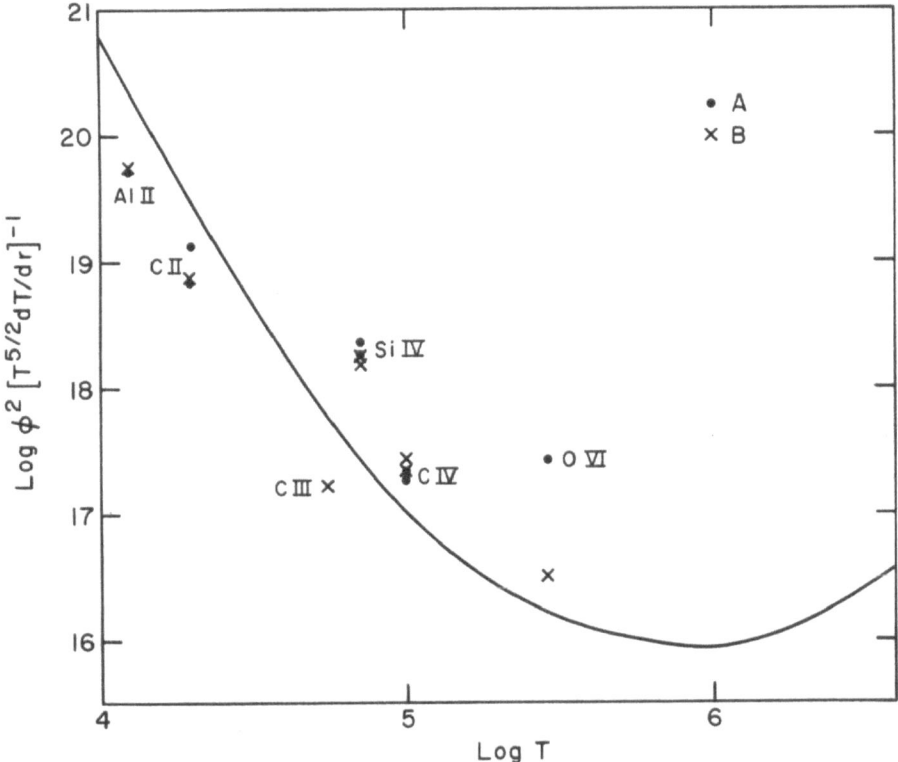

Fig. 2.   The temperature dependence of $\log \phi^2 [T^{5/2} \, \mathrm{d}T/\mathrm{d}r]^{-1}$ derived from resonance line intensities. The plotted points are for prominences A and B. The solid curve represents Kopp's (1972) results for the CC transition region, displaced downward to compensate for the lower pressure in prominences.

$\log \phi^2 [T^{5/2} \, \mathrm{d}T/\mathrm{d}r] -^1$ determined from each line vs the log of the temperature where that line is formed. The results from both prominences are plotted and are in good agreement, except for O vi. The solid line in Figure 2 represents the same quantity determined in the CC transition region by Kopp (1972) using a similar technique. We have displaced Kopp's curve downward to compensate for the fact that the pressure is a factor four higher in the transition region than in the prominence.

We do not wish to overstress the similarity between the PC interface and the CC transition region suggested by Figure 2. Nevertheless, it is worthy of note that within the scatter of both sets of data, the results are identical in the range $T = 4 \times 10^5$ and 10^5 K. Taken literally, this would imply that ∇T as a function of T is identical in the lower transition region and in the PC interface of these transient prominences. This is surprising. We expect the temperature gradient to be sensitive to the orientation and magnitude of the magnetic field, which is quite different in the transition region and in these prominences. However, these prominences are changing in time and may not even approximate the steady state, and thus may not be typical. We are making a similar study of quiescent prominences on the same plates.

This work was supported in part under NASA Grant NGL 12-001-011.

References

Athay, R. G.: 1966, *Astrophys. J.* **145**, 784.
Athay, R. G.: 1971, in C. J. Macris (ed.), *Physics of the Solar Corona*, D. Reidel Publ. Co., Dordrecht, p. 36.
Burton, W. M., Jordan, C., Ridgeley, A., and Wilson, R.: 1971, *Phil. Trans. Roy. Soc. Lond.* **A270**, 81.
Dupree, A.: 1972, *Astrophys. J.* **178**, 527.
Gabriel, A. H. and Jordan, C.: 1972, *Case Studies in Atomic Collision Physics*, vol. 2, North-Holland Publ. Co., Amsterdam.
Kopp, R.: 1972, *Solar Phys.* **27**, 373.
Noyes, R. W., Dupree, A. K., Huber, M. C. E., Parkinson, W. H., Reeves, E. M., and Withbroe, G. L.: 1972, *Astrophys. J.* **178**, 515.
Noyes, R. W. and Withbroe, G. L.: 1972, *Space Sci. Rev.* **13**, 612.
Speer, R. J., Garton, W. R. S., Goldberg, L., Parkinson, W. H., Reeves, E. M., Morgan, A. F., Nicholls, R. W., Jones, T. J. L., Paxton, J. H. B., Shenton, D. B., and Wilson, R.: 1970, *Nature* **226**, 249.

DISCUSSION

Jordan: I have seen your preprint and couldn't resist the temptation to put some of my own numbers into your computations. Using my CIII population and ionization equilibrium calculations I get a pressure which is a factor of two higher than the one you get. Also using the observed Fe XII, intensities in the corona and my own calculations for the populations you get a ridiculously low electron density, 10^7 cm^{-3}. Therefore I now modify my own calculations by about a factor of 2. This leads to electron densities in the quiet corona of about 2×10^8 cm^{-3}, and it raises the electron density in the region you have studied up to 10^9 cm^{-3}. Thus the pressure in the corona surrounding the prominence is about a factor of 3 larger than in the prominence and you would need a magnetic field of about 6 Gauss in the prominence to maintain equilibrium. Further you have really not compared your $\int Ne^2 \, dH$ curve with the best available data. We have far more points at the low temperature and of the curve between 10^4–10^5 K and this is much shallower than the curve that you showed from Duprea which was varying very rapidly. If you compared your results with our absolute intensities you would find that they agree much better and you will not have to scale your pressure so much.

Gabriel: The suggestion that the transition region should be thin because the magnetic field is parallel to the prominence is perhaps not valid because it is not a steady state phenomenon. As you pointed out the prominence is a falling or condensing coronal ring. It could have an intermediate temperature region surrounding it which is dragged on the magnetic field lines and has nothing to do with conduction from the corona to the cooler prominence material.

HELICAL FIELD LINES IN AN ERUPTING FILAMENT

JOAN VORPAHL

Dept. of Physics and Astronomy, Sacramento City College, Sacramento, Calif., U.S.A.

Abstract. An example of an erupting filament that exhibited helical structure during a flare on 1971, September 17, was discussed. This sort of event is not rare. Examples of rotational effects in flares, prominences and surges have been reported by others such as Öhman (1968). Accompanied by type II radio emission and the visible flow of Hα gas across $\frac{1}{2}$ the Sun's surface, the flare was very interesting in itself. However, here the emphasis is on possible ways in which the observed twist might develop. The latter is important because it yields information about: (1) the state of the subphotospheric field itself, and (2) the possibility of storing flare energy in the twisted filaments.

Possible reasons for observing twisted field lines include:

(1) Photospheric motion resulting in twisted flux tubes is attractive because various polarities or rotation of sunspots;

(2) minor field reconnection occurs in which new flux interacts with fields already present and produces intertwining filaments;

(3) flux tubes come up twisted and merely expand into the chromosphere.

Considering each of these in turn for the September event:

(1) Photospheric motion resulting in twisted flux tubes is attractive because various authors have shown that field lines can be twisted by a rotating sunspot (Barns and Sturrock, 1971) or by opposite polarity regions slipping past one another (Nakagawa and Tanaka, 1973). Two ways of determining shearing motion include: (a) sunspot motion or (b) high spatial and temporal resolution magnetograms. In the case of the September flare, no high resolution magnetograms exist. However, the twist did not seem to be due to shearing because: (a) no obvious motion (relative to a nearby larger active region) occurred in the bright plage region that constituted the flare site; (b) no surrounding Hα features such as fibril structures, appeared to move; and (c) the appearance of the twist was quite abrupt, i.e., it became obvious within a period of 1–2 h, on the day preceding the flare. Since opposite ends of some fibrils were displaced about 20 000 to 25 000 km, the necessary velocity of photospheric motions in explaining the helical filaments would be prohibitively high.

(2) Minor field reconnections could also result in filament twist. In the case under consideration, filaments encircling the bright plage did undergo oscillations, appearing first lighter then darker, several times on the day before the flare. On the other hand, it is not obvious as to why field reconnections would occur since they would result in a twisted field – and presumably a higher energy state.

(3) Due to subphotospheric activity, flux tubes might come up twisted. This is consistent with the September flare in that the twist became obvious over a short period of 1–2 h. Furthermore, no obvious photospheric or chromospheric motions were observed. On the other hand, no twist is noticeable in fibrils that make up arched

R. Grant Athay (ed.), Chromospheric Fine Structure, 197–198. All Rights Reserved.

filaments in a newly emerging magnetic region; however, this might be due to the inadequacy of current telescopes to resolve fine structure in tightly wound flux ropes.

In summary, helical structure in the September 1971 filament can be explained most easily by the fact that flux tubes came up twisted. However, since magnetograms of high quality resolution did not exist, photospheric motion that resulted in twisted filaments cannot be excluded.

References

Barns, C. and Sturrock, P.: 1971, *Astrophys. J.* **174**, 659.
Nakayawa, Y. and Tanaka, K.: 1973, High Altitude Observatory, preprint.
Öhman, Y. (ed.): *Mass Motion in Solar Flares and Related Phenomena*, Wiley Interscience Division, New York.

DISCUSSION

Giovanelli: The question of twisted field lines is connected with the question that has been worrying me for quite some time. Many years ago Bray and Loughhead observed some small sunspots that appeared very suddenly and migrated over to join with another sunspot thereby increasing its size. These observations were drawn to my attention when I saw Harold Zirin's observations this morning of the way flux regions emerged. How is it that flux comes up and gets gathered into sunspots? Is it because it comes up twisted or perhaps shredded apart in the photosphere but then when it spreads out these twists don't let it stay apart? It's a question I have to ask and I'd love to hear what somebody else has to say about it.

Meyer: This afternoon I am going to talk a little about the problem how sunspots come about and decay. It seems that the twist is not an essential feature in our model.

Newkirk: Can you make an estimate of how much energy is involved in the twist?

Vorpahl: The field lines in the twisted regions are about 20 000 to 30 000 km long and have about two or three twists in them. I have not tried to estimate the energy.

PART VI

EVOLUTION OF CHROMOSPHERIC FINE STRUCTURES

STREAMING MAGNETIC FEATURES NEAR SUNSPOTS

DALE VRABEC

San Fernando Observatory and Space Physics Laboratory, The Aerospace Corporation,
Los Angeles, Calif., U.S.A.

Abstract. Observations prompted by Sheeley's discovery of outflowing CN bright points from sunspots
have established that there occur small moving magnetic features (MMF's) near sunspots. These MMF's
exhibit a highly ordered pattern of movement directly related to the associated sunspot. The observations
are consistent with the concept of magnetic flux outflow (MFO), a process whereby net magnetic flux of
the same polarity as the sunspot is transferred from a decaying sunspot to the surrounding magnetic
network. Small magnetic flux concentrations are apparently convected outward by a velocity cell centered
on the sunspot. Doppler spectroheliograms have provided evidence for such systematic outward velocities
extending as far as 10000 to 20000 km beyond the outer edge of the penumbra of some sunspots, which
is comparable to the extent of MFO. While MFO is best observed by means of time-lapse movies of the
magnetic fields, it is also manifested morphologically on individual magnetograms by features that re-
semble moats (or bays) and wreaths around only those sunspots where MFO is present. Two examples
of magnetic features streaming toward and into rapidly forming sunspots are described, providing evidence
for the occurrence of magnetic flux inflow (MFI) associated with the growth phase of sunspot development.
It is, therefore, likely that MFI and MFO are basic aspects of the evolutionary development of sunspots.
Observational and instrumental aspects relevant to the investigation of MMF's are described in the
Appendix.

1. Background

Most of the phenomena that I shall review were unknown five years ago. In August
1968, Sheeley (1969) exposed a 2.3-h sequence of high-resolution CN 3883 Å band
head spectroheliograms of a sunspot region with the McMath Solar Telescope at the
Kitt Peak National Observatory (see Figure 1). When the 22 spectroheliograms were
later converted into a time-lapse movie, he could see a conspicuous horizontal outflow of bright points from the two principal sunspots. This outflow, which had an
average speed of 1 km s^{-1}, was present throughout much of the area surrounding the
sunspots to distances of 10000 to 15000 km, where it either encountered essentially
stationary fragments of incipient network forming from collections of stagnating
points or simply died out. Throughout this region of outflow, points moved along
approximately radial paths from the sunspots, some points fading or disappearing,
others forming or growing. Many of the outflowing points first appeared at the outer
edge of the penumbra of the sunspot, while points reaching the outer boundary of the
outflow zone merged with the network there, or (Sheeley, 1973) diverged around pre-
existing network fragments and stopped, thereby forming new network. Sheeley
correctly inferred that these moving CN bright points coincided with localized, outflowing magnetic field concentrations, but was unable to ascertain their polarities.
Subsequently, using Leighton's (1959) spectroheliographic technique for photographing the line-of-sight component of solar magnetic fields, Vrabec (1971) at the
Aerospace San Fernando Observatory obtained time-lapse movies of Ca I 6103 Å
Zeeman spectroheliograms, and confirmed that Sheeley's outflowing features were

R. Grant Athay (ed.), Chromospheric Fine Structure, 201–231. All Rights Reserved.

DALE VRABEC

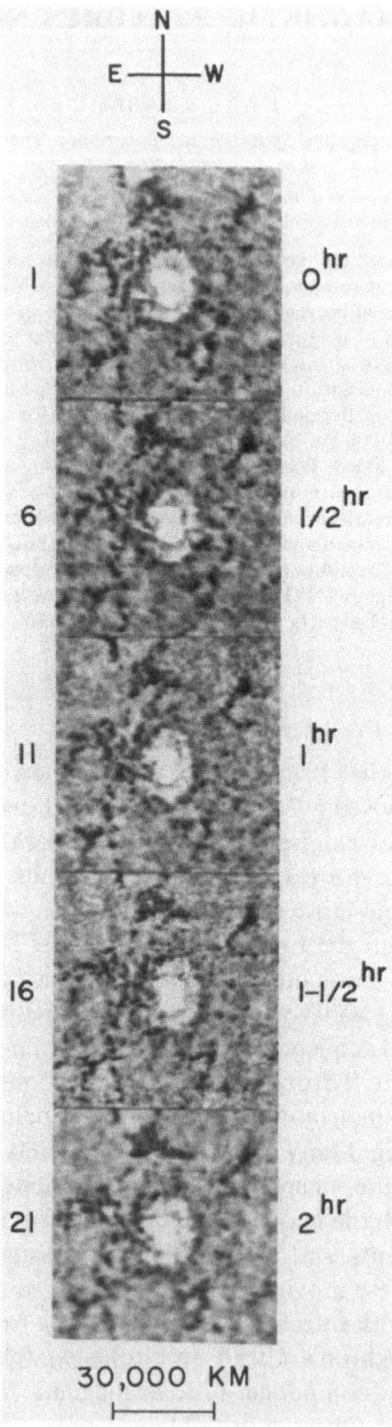

Fig. 1.   This sequence of spectroheliogram negatives exposed in the CN band head at 3883 Å is selected from a total of 22 taken on 1968, August 13. Sheeley first detected magnetic flux outflow from sunspots in a time-lapse movie made from these observations. CN bright points and the bright photospheric network appear as dark features. (Courtesy N. R. Sheeley and Kitt Peak National Observatory.)

indeed basically magnetic. In addition, he found that magnetic features of both polarities occur in the outflow around a sunspot of either polarity. (The technique of time-lapse movies has proven indispensible for discerning the ordered motion of these elusive features.) Recently, Harvey and Harvey (1972, 1973) have reported results of a study of a large sample of sunspots, observed with the Kitt Peak 40-channel photo-electric magnetograph, over half of which exhibited magnetic outflow. Due to their large sample and because quantitative measurements of the magnetic fluxes were obtained, their investigation is the most comprehensive of this phenomenon under-taken so far. In addition to confirming the earlier work, Harvey and Harvey obtained new and significant results which are discussed later. In view of these investigations, it can now be considered well-established that there occur around most sunspots, especially during their declining stage of development, numerous small magnetic features in a mixture of both polarities, which exhibit systematic outflow. This is a direct manifestation of magnetic flux transfer from sunspots to their surroundings, which has important implications regarding the evolutionary history of solar mag-netic fields. I will call this phenomenon 'magnetic flux outflow' (MFO).

While MFO is by far the best observed, it is important to note that we also observe examples of 'magnetic flux inflow' (MFI) in the form of instreaming magnetic features (Vrabec, 1971) that converge toward and actually enter sunspots. Published descrip-tions of MFI are scarce, although it appears to play an important role in sunspot formation and growth.

Both MFO and MFI involve a special class of solar features, which can be defined as follows:

(a) All members of the class are magnetic in nature, and this shall be regarded as their primary physical property. (As a consequence, it is necessary that any interpreta-tion of these features be consistent with the physics governing magnetic fields perme-ating a partially ionized fluid medium, where the appropriate magnetic field strengths, plasma densities, and velocities are properly taken into account. Throughout this review it will be assumed that the magnetic fields are essentially 'frozen' in the photo-spheric material, in accordance with the generally accepted meaning of this expression.)

(b) All members are associated in some direct way with a visible sunspot, or group of sunspots, and these sunspots are a dominant influence. (I suggest that in the future it may prove desirable in the case of MFI to broaden this definition by substituting for 'visible sunspot' the term 'localized magnetic disturbance,' such as occurs in a region during the incipient phase of sunspot formation whether or not this activity is accom-panied by visible sunspots or pores.)

(c) All members exhibit systematic (i.e., nonrandom) motion across the photo-spheric surface along paths that either diverge from or converge toward the associated sunspot.

A suitable nomenclature is needed for this class of features that are involved in both MFO and MFI. We will adopt the acronym MMF for 'moving magnetic feature,' a term introduced by Harvey and Harvey in their study of MFO. Of course, we must always bear in mind the suppressed prefix 'sunspot-associated' as well as the implicit distinction between outflowing and inflowing MMF's, which will be obvious from the context. It is likely that pores and small umbrae frequently fall in the latter category.

Most of the observations discussed in this review were made with instruments capable of sensing or measuring only the line-of-sight component of the photospheric magnetic field. To avoid tedious repetition, I will not explicitly refer to this each time these observations are described.

2. Magnetic Flux Outflow

2.1. CHARACTERISTICS OF DEVELOPMENT

It appears that magnetic flux outflow plays a very important role in the evolutionary history of a sunspot and manifests itself through the appearance of outflowing MMF's around the sunspot. According to Chapman (1972), Harvey and Harvey, and Allen *et al.* (1973), not all sunspots exhibit MMF's at the time they are observed. Rather, it was found that outflowing MMF's appear almost always in association with decaying sunspots. I have observed MFO in a sunspot that was near its maximum development, but was still growing in area. MMF's are associated with sunspots of all sizes and degrees of complexity. For example, the compact but very magnetically complex group of early August 1972 (Coffey, 1973; Zirin and Tanaka, 1973) exhibited exceptionally well-developed MFO when it was observed by Chapman (1973) on 5 August, while immediately adjacent to it an isolated small umbra independently provided an unusually well-defined example of MFO in its simplest form.

It is apparent that, closely related to the development of the sunspot with which it is associated, MFO must also undergo a characteristic evolution, progressing from its first visible onset through a stage of maximum activity and, finally, presumably dying out along with the dissolution of the sunspot. Unfortunately, our limited observations have afforded us only incomplete views of the overall phenomenon. Despite these severe limitations, I will try to sketch a simplified, but no doubt idealized, picture of MFO development.

During most of the growth stage of development of a sunspot, it is typical for active region magnetic fields surrounding the sunspot to extend right up to it. Active region fields are characterized by a high number density and compact morphology of the small magnetic field concentrations that constitute the fine structure of the fields. If the sunspot develops a penumbra, during this stage its outer edge remains nearly in contact with the surrounding fields, at least over a major portion of its periphery. Under these circumstances we do not observe any conspicuous MFO. As the sunspot approaches or attains its maximum development, the magnetic fields adjacent to it become increasingly fragmented and begin to be dispersed throughout an annular region centered on the sunspot, starting at the outer edge of the penumbra (see Figure 2). The outer boundary of this annular zone, within which the magnetic fields are becoming conspicuously thinned out compared to the fields in the adjacent active regions and network, expands outward up to distances of typically 10 000 to 20 000 km from the outer edge of the penumbra. Wherever extended areas of active region fields are encountered, an indentation in the latter is produced, resulting in a partially cleared area in the form of a 'bay,' the curved outline of which is concentric with the sunspot. Any network that is encountered assumes in time the circular shape of the boundary, producing the effect of a fragmented wreath of magnetic fields partially encircling the sunspot. This annular zone surrounding sunspots, first pointed out by

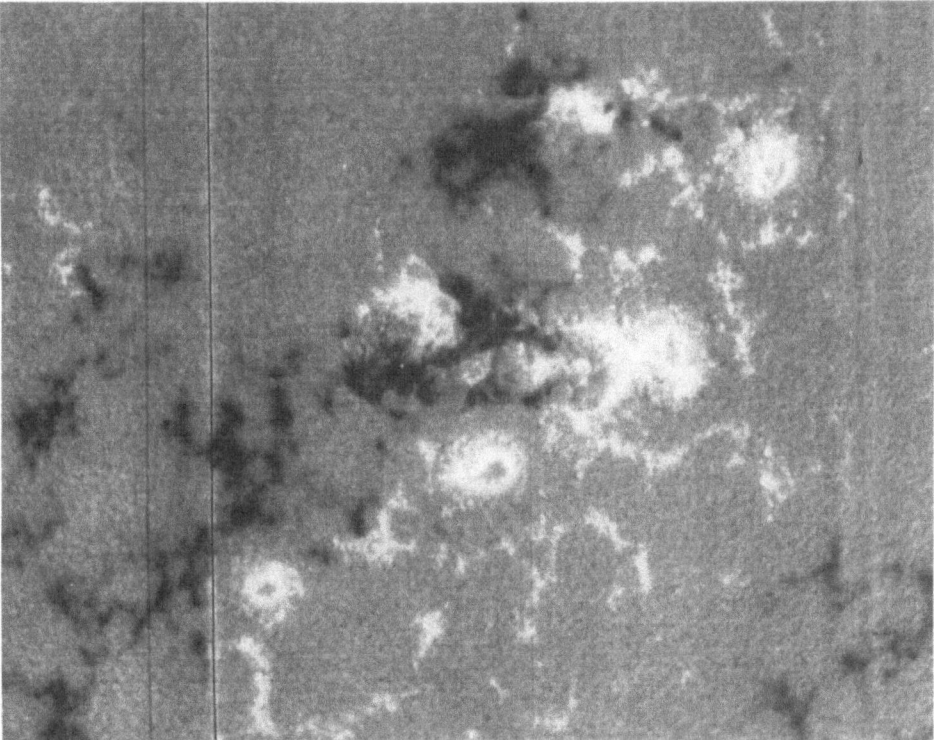

Fig. 2. Magnetic fields associated with McMath Region 10542 are depicted in this subtracted Zeeman spectroheliogram taken in the 6102.7 Å line of Ca I at 1735 UT on 1970, January, 26. White features are positive or north magnetic polarity, the normal polarity of the leading members of this southern group of sunspots. North is up and west is to the right. The height of the spectroheliogram corresponds to 237 000 km on the Sun. 'Moats' are conspicuous around the four prominent leading sunspots, and the small magnetic features in these moats are MMF's. Typical 'wreaths' partially surround the two sunspots at the right. The moat around the sunspot at the lower left produces an indentation or 'bay' in the black polarity network adjacent to it. (D. Vrabec, Aerospace San Fernando Observatory.)

Vrabec (1971) and Sheeley (1971), has been aptly coined the 'moat' by Sheeley. In some sunspots the moat characteristics do not develop in all azimuths around the sunspot. In the case of preceding sunspots, the undeveloped sector tends to point toward the following sunspot. Harvey (1973) has recently observed the formation of a moat during the course of a single day's observing run. If this single observation is indicative, moats may form very rapidly.

By the time bay and wreath forms have developed (Figure 2), the moat around the sunspot has become relatively cleared of magnetic fields with the exception of small fragments of each polarity distributed throughout it (Figure 3). In fact, on low-resolution magnetograms, the moat will appear almost field-free. However, it is these small fragments of field, typically less than 1500 km in size and always located within the moat, that exhibit the interesting property of outflow. These are the MMF's that reveal

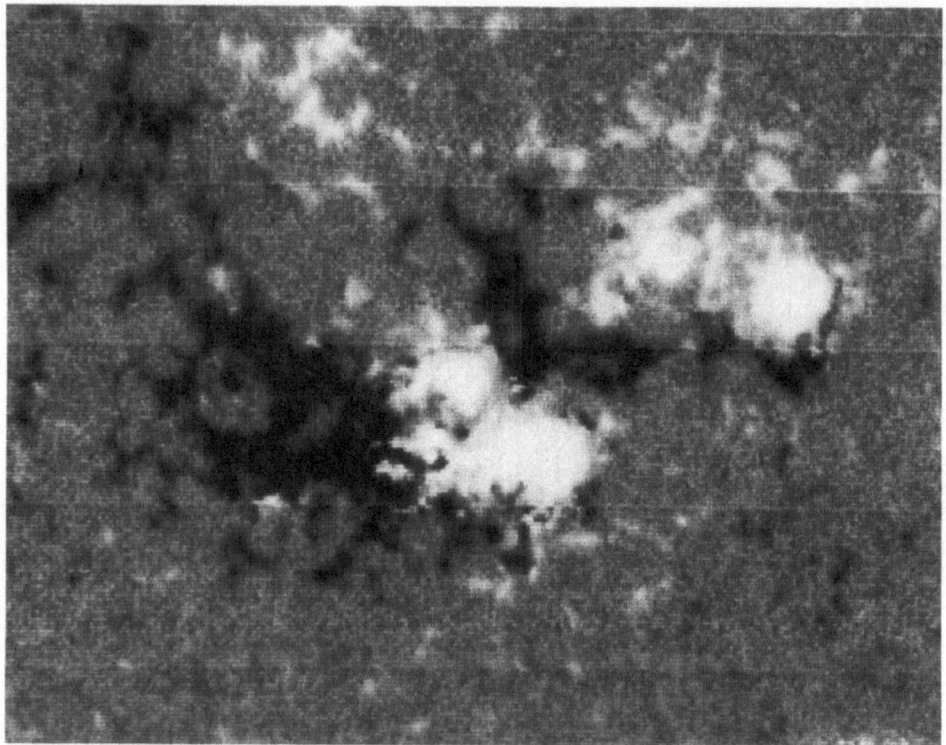

Fig. 3.   A conspicuous moat surrounds the black polarity sunspot at the left in this magnetogram taken on 1971, July 2, with the 40-channel photoelectric magnetograph. Note the many small MMF's of both polarities occupying the moat. The height of each horizontal swath produced by the magnetograph probe is roughly 100″. (Courtesy K. Harvey and J. Harvey, Kitt Peak National Observatory.)

to us the underlying phenomenon of MFO, namely, the direct transfer of magnetic fields from the sunspot to its surroundings.

From observations of the paths of these outflowing MMF's on time-lapse movies, it is evident that a highly ordered pattern of outward, radial movement from the sunspot occurs throughout the entire area occupied by the moat. Near the outer boundary of the moat, this ordered movement quite abruptly becomes ill-defined. Any magnetic fields at or beyond this boundary appear to remain essentially stationary, undergoing only natural evolutionary change unrelated to the outflow. Thus, these fields appear to be effectively isolated from the dynamic activity occurring immediately inside the moat.

During the declining phase of MFO development, we can anticipate a weakening of the outflow of MMF's accompanied by increased disordering of the characteristic bay and wreath forms in the surrounding magnetic fields as supergranulation encroaches into the moat. According to recent observations by Harvey (1973), termination of MFO can also occur abruptly.

2.2. OUTFLOWING VELOCITY FIELD

Thus far, I have described outflow only as the pattern of organized movement traced out by the proper motions of individual CN bright points or MMF's. However, these motions are so suggestive of the horizontal outflow of material envisioned to occur in a single supergranule cell that Sheeley (1969) interpreted the proper motions of the CN bright points he observed as evidence of horizontal outflow of photospheric material, in the form of a single supergranule centered on the sunspot. If the magnetic fields are indeed 'frozen' in the photospheric material and this material undergoes an ordered flow, then MMF's will appear, and their movements can be assumed to map the streamlines of this velocity field. The moat thus assumes the physical aspect of a single, outwelling velocity cell centered about the sunspot. As noted by Vrabec (1971), stable magnetic fields do not persist within moats. This is additional evidence of a velocity field similar to that in supergranules. Any magnetic field within the moat would be systematically transported with the material out to the boundary, so it would only be at and beyond this boundary that the fields could accumulate.

The evolution of MFO would thus be directly related to the evolution of a velocity cell around a sunspot, which must in turn be related to the development of the sunspot. The observed evolutionary changes in magnetic field morphology around the sunspot just described would to a large extent initially result from the action of this incipient velocity field on magnetic fields adjacent to the sunspot, forming the moat and bays. Subsequently, and to a progressively increasing extent, they would also be altered by the transfer of significant amounts of magnetic field from the sunspot to its surroundings.

Harvey and Harvey have suggested that a necessary condition for the occurrence of MMF's is that the sunspot be at least partially surrounded by a moat since they never observed MMF's around decaying sunspots without a moat. I believe that MMF's must accompany the formation of the moat, but it is likely that, at this incipient stage of MFO, their motions may be too slow and weakly ordered for them to be detected with the magnetograph. Hence, it is reasonable to assume that only when the velocity field has developed sufficiently to have formed definite signs of a moat would MFO be built up to a strength where MMF's are conspicuous.

The following dynamic properties of MMF's are among many described by Harvey and Harvey. MMF's exhibit a distribution of outflow speeds ranging from a few tenths to over $2 \mathrm{km \, s}^{-1}$. The speed of any single MMF remains essentially constant over most of its lifetime. There appears to be some indication of fast-moving channels imbedded in the main outflow field. There are no conspicuous differences between the dynamical behavior of MMF's of the same and of opposite polarity as the parent sunspot.

On Aerospace movies, this diversity of speeds is also evident, where many instances of faster moving, small MMF's overtaking slower moving, larger collections of magnetic field are seen as well as indications of high-speed streams. The MMF's exhibiting the highest speeds appear to be interlopers. In addition, the Aerospace movies show a clear example of a circulating eddy near the outer boundary of a moat and the occasional formation of a small network cell within a moat.

If these motions are actually due to an outflow of photospheric material, this outflow should be observable on Doppler spectroheliograms of sunspots located near the limb. Such spectroheliograms are known to show clearly the slower horizontal material outflow in supergranules which are typically only $0.5 \, \text{km s}^{-1}$ (Leighton *et al.*, 1962), or one-half the average speed of MMF's. Using a technique that eliminated velocity structure due to the 5-min oscillatory component (Sheeley and Bhatnagar, 1971a), these same investigators (1971b) obtained Doppler spectroheliograms showing definite, qualitative evidence of a long-lived, horizontal, photospheric outflow extending as far as 10000 km beyond the outer edge of the penumbrae of three sunspots located near the limb. The most conspicuous velocity features were the Evershed outflow in the penumbrae of these sunspots. One sunspot exhibited many spoke-like extensions of Evershed outflow beyond the average position of the outer edge of the penumbra. These velocity 'spokes' coincided with dark structures visible in the average (summed) wing intensity spectroheliograms, causing the outer edge of the penumbra to have a pronounced ragged structure. One exceptional spoke extended 12000 km beyond the average outer penumbral border and may correspond to the high-speed channels referred to previously. Sheeley (1972) subsequently verified extrapenumbral outflow in six sunspots to a maximum distance of 19000 km and with speeds ranging from 0.4 to $0.8 \, \text{km s}^{-1}$. He emphasized that this flow was clearly distinguishable from the much faster and more highly structured Evershed outflow, but on the other hand, very similar to the surface currents of a supergranule, suggesting that the sunspot was roughly centered in one. These Doppler observations strongly suggest that outflow of photospheric material accompanies the observed MFO. It is exceedingly important that this be put to the definitive test of whether both MFO and Doppler outflow occur together in the same sunspot and undergo related development. This will be extremely difficult to accomplish because MMF's are very difficult to observe toward the limb, which is where the observations will have to be made in order to detect the weak Doppler signals produced by the horizontal outflow and also because of the need for synoptic data. Notwithstanding these unresolved questions, it does now seem highly probable that what we observe and call MFO is indeed an outward convection of magnetic flux by a very large velocity cell centered on the sunspot.

2.3. PROPERTIES OF THE MMF'S

2.3.1. *Size*

MMF's appear in a wide range of sizes though they are most frequently $<2''$ (i.e., <1500 km) in extent. Thus, even on good magnetograms of 2.5" resolution, most are scarcely discernible, and it is only when time sequences of these magnetograms are converted into time-lapse movies, which average out seeing variations, that it is possible to prove from their persistence that some of these threshold features are real. These movies also reveal otherwise undetectable outflow, perceptible only by an unmistakable ordered movement of unresolved structure on the screen. On the very best magnetograms, which approach 1″ resolution, individual magnetic features appear around sunspots in increasing numbers as their sizes approach the limit of resolution. Beckers and Schröter (1968) carefully analyzed spectra showing 'magnetic knots' in the vicinity of a sunspot, and concluded that their average diameter was 1.3″. It is evident from the fact that these magnetic knots were distributed throughout the entire active region that many of them do not correspond to MMF's, since the latter are confined to moats. However, those magnetic knots that were located in close proximity to the sunspot almost certainly are MMF's. Simon and Zirker (1973), also using spectra, were unable to find magnetic field entities smaller than 1.5″ though their spatial resolution exceeded 0.75″. On the other hand, on CN spectroheliograms, the bright photospheric network characteristically exhibits a more 'point-like' structure than does the corresponding (i.e., co-spatial) magnetic network recorded on Zeeman spectroheliograms taken with the same instrument under similar conditions of seeing. Some CN bright points adjacent to sunspots appear to be less than 1″ in size. Harvey *et al.* (1972) have analyzed observations made with a line-profile Stokesmeter equipped with a $2.5 \times 2.5″$ probe, using a two-component model atmosphere to interpret their data. Their conclusion is that 0.6″ is the characteristic size of non-sunspot magnetic field fine structure. Thus, it is not presently clear whether the observational lower limit of 1 to 1.5″ for the size of the smallest magnetic field structures is a consequence of instrumental convolution combined with seeing degradation, or whether it is an intrinsic property of magnetic fields to fringe beyond the more compact boundaries of the bright, facular features.

Magnetic features in excess of 5″ are sometimes involved in MFO. Many of these large fragments of field first appear a few thousand kilometers beyond the outer edge of the sunspot penumbra, and then move outward. Some of the MMF's that appear in this fashion are of opposite polarity to the associated sunspot and are possibly remnants of the fields occupying the same region before development of the sunspot, which lay hidden below the photosphere. Invariably, if the resolution permits, these larger fragments exhibit a fine structure which suggests that they are collections of smaller magnetic entities. In general, these larger components of MFO move conspicuously more slowly than the average speed of MMF's, and are frequently overtaken by smaller, faster moving MMF's. Frequently, they appear in the form of extended arcs centered on the sunspot. On lower-resolution magnetic movies, barely discernible wavelike features sometimes appear to emanate from the sunspots. Whether these 'waves' are intrinsic or simply the extended arcs forming in the moats has not been determined.

2.3.2. *Magnetic Field Strength*

The only direct measurements of magnetic field strengths in features positively identified as MMF's are those made by Harvey and Harvey. They found the strongest *measured* longitudinal field to be 300 G, but this unquestionably underestimates the true field strength because of the limited spatial resolution of the magnetograph. They also believe that the fields are primarily vertically oriented with respect to the photosphere since they were unable to detect MMF's around sunspots near the limb. Beckers and Schröter (1968) measured longitudinal magnetic field strengths between 250 and 400 G in the magnetic knots they found scattered around a sunspot located near the center of the disk, and calculated, after corrections for scattered light and inclination of the field lines, field strengths up to 1400 G. As previously noted, some of these features were very likely MMF's.

Recently, Simon and Zirker (1973) measured maximum magnetic field strengths up to 1300 G at the roots of spicules and fibrils in both quiet and active regions of the Sun. The applicability of these related measurements to MMF's remains to be determined.

Clearly, quantitative measurements of magnetic fields in MMF's are both meager and extremely difficult, but it is upon these measurements that the very important determinations of magnetic flux of MMF's are based. Thus, there is a great need for more measurements of the maximum magnetic field strengths in MMF's as well as the distribution of field strengths and the inclinations of the field lines. The fact that the measured strength of a magnetic feature on the Sun is affected by the spatial resolution of the observations as determined by the instrument, by seeing conditions, and by any changes in the profile of the line used for the measurement resulting from differences in physical conditions existing within and external to the magnetic regions is, of course, what makes these measurements so difficult to obtain. The limitations on measuring the transverse magnetic field components are also well known.

2.3.3. *Magnetic Flux*

Harvey and Harvey measured the magnetic fluxes of 34 MMF's of each magnetic polarity. The resulting two histograms are remarkably similar, the fluxes per MMF ranging between 6×10^{17} and 8×10^{19} Mx (G cm^2). The average magnetic flux of an MMF was found to be about 10^{19} Mx, independent of the magnetic polarity. This large range of measured fluxes no doubt reflects the tendency for small magnetic field entities to become clumped together in various numbers, forming a hierarchy of sizes of features that collectively are the MMF's that were observed and measured. On closer examination, these histograms seem to indicate a minimum flux per MMF of about 2×10^{18} Mx if we neglect features whose fluxes are at the indicated noise level of the measurements, a value about three times smaller.

Let us independently calculate the magnetic flux of an MMF whose diameter is the previously noted value of 1.3″ and whose average vertical field component is 400 G. The flux of this idealized MMF is 3×10^{18} Mx. Clearly, this value is subject to considerable uncertainty. For example, an assumed vertical field component of 1400 G would increase this to 10^{19} Mx, the average value obtained by Harvey and Harvey. Later we will use the smaller of these values to estimate the total magnetic flux of MMF's in the moat around a single sunspot.

2.3.4. *Lifetime*

Lifetimes of MMF's are very hard to determine because MMF's are difficult to observe with ground-based instruments and their evolutionary histories and morphological properties are complex. As previously noted, MMF's appear and disappear in the moat in transit, and also merge with one another. Thus, it is extremely difficult to identify a particular MMF when it first makes its appearance near the sunspot, and then to follow it throughout its subsequent history.

From magnetic movies it is obvious that MMF's exhibit a wide range of lifetimes. Some last less than 1 h while the slower-moving, large fragments tend to persist for at least 6 to 8 h, corresponding to the longest intervals of continuous observation of MFO obtained to date. I find that only a small fraction of the total number of MMF's that can be individually seen on the best magnetic field movies obtained at Aerospace persist long enough to complete their transit across a typical moat. This apparently conflicts with the findings of Harvey and Harvey who note that most MMF's persist long enough to reach the network fields at the boundary of the moat. Since the Harveys worked with lower resolution data, it is likely that this discrepancy can be attributed to a bias resulting from their having measured lifetimes of only the largest MMF's, which according to the Aerospace movies, tend also to be the longest-lived. The best time-lapse movies show these largest magnetic fragments to be eroding away at their edges, apparently by the action of the outflowing velocity field. Small pieces appear to be swept away by the faster-moving outflow while, along their trailing side, overtaking MMF's continually merge with them.

At the other extreme, the smallest MMF's tend to be the least persistent. This is very likely an intrinsic property, but it may be partly an observational effect since their visibility depends entirely upon the instantaneous quality of the seeing. In fact, there is always a background of ephemeral, unresolved threshold magnetic features that collectively exhibit MFO as well as the most rapid apparent changes.

Provided one bears in mind the complexity and disparity of the behavior of MMF's, a characteristic lifetime between 2 to 3 h may be assumed, but this should not be taken too literally. Thus, in addition to their systematic outflow, a very distinctive characteristic of MMF's associated with MFO is their continual and generally rapid change as is evident in Figure 4.

2.3.5. *Morphology*

The most distinguishing characteristic of MMF's is their organized motion. However, even if we disregard their dynamic and evolutionary properties, MMF's can be distinguished from all other non-sunspot magnetic fields on the basis of their unique morphology. The basic characteristics of this morphology are: (a) a high degree of fragmentation and reduction of the fields into small-size entities; and (b) the intimate mixing of features of both polarities on a fine scale, producing the effect of 'salt-and-pepper' on magnetograms as shown in Figure 5 (Vrabec and Janssens, 1972).

When we carefully examine the fine structure of the two-dimensional spatial distribution of non-sunspot magnetic fields recorded on high-resolution magnetograms,

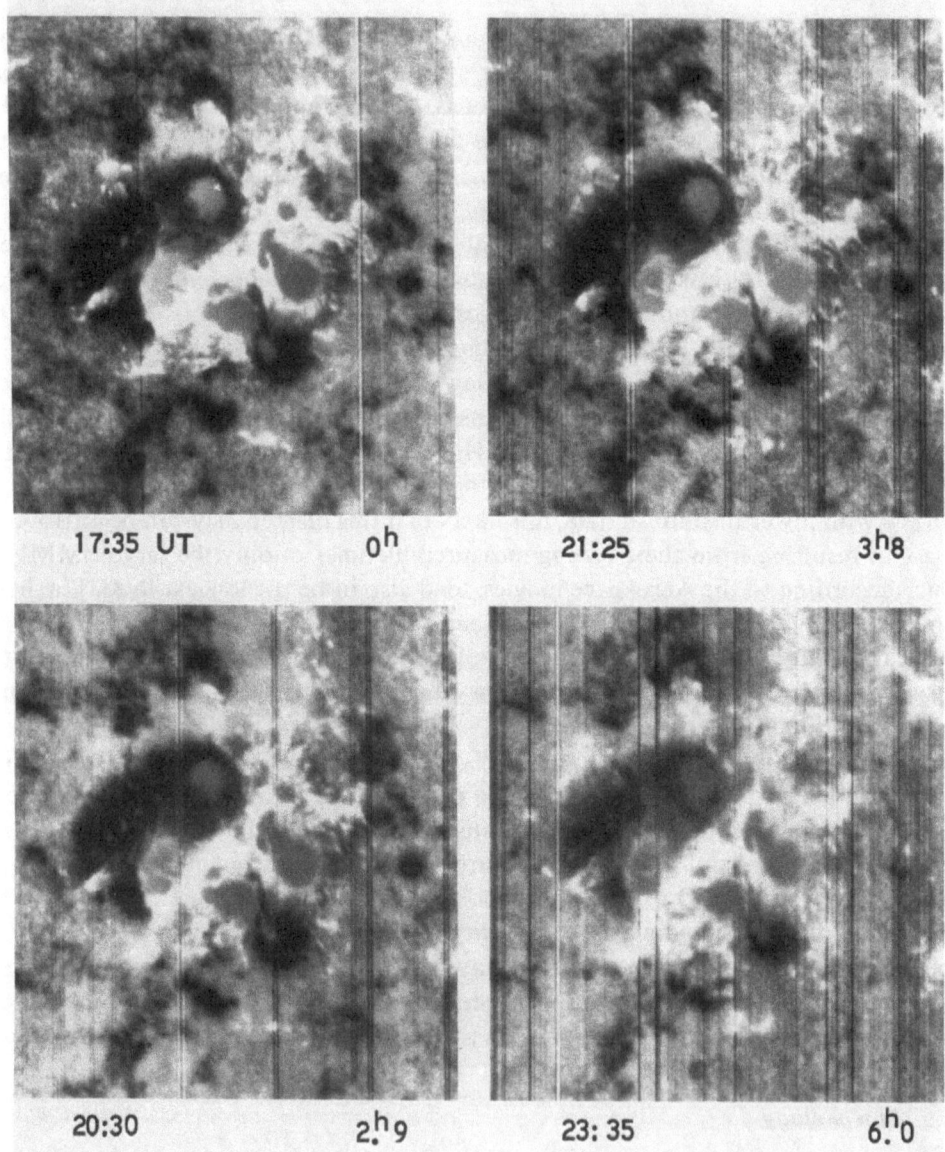

Fig. 4. This sequence of subtracted Zeeman spectroheliograms depicting magnetic fields of McMath Region 11976 was taken on 1972, August 5. The changes occurring in the magnetic fields just outside the sunspot are associated with well-developed magnetic flux outflow in progress during the observations. This MFO is much better seen in the time-lapse movie referenced in the text. (Courtesy G. Chapman, Aerospace San Fernando Observatory.)

we see that this structure is ultimately formed from small magnetic field concentrations of various sizes, most of which are at the limit of resolution (i.e., 1″). Outside of these resolved concentrations, the photospheric surface appears to be essentially field free. Within these concentrations the magnetic field strengths are high, probably of the order of many hundreds to over a thousand Gauss (Sheeley, 1967; Livingston and

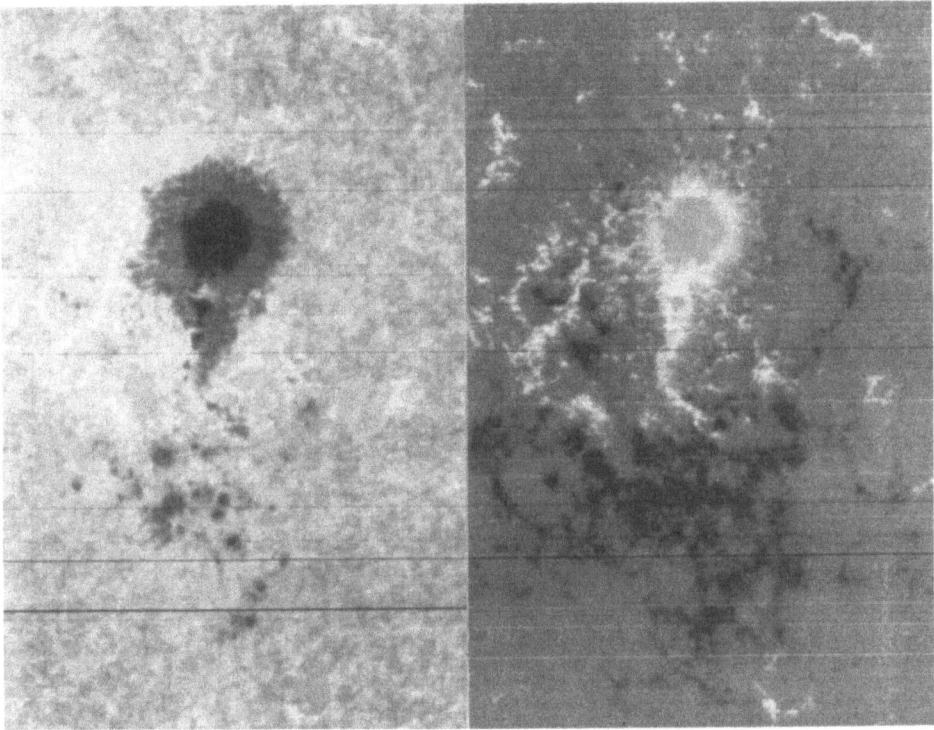

Fig. 5.   The subtracted Zeeman spectroheliogram (right-hand photograph) of magnetic fields associated with the sunspot in the left-hand photograph illustrates the fine structure and mixing of both polarities on a fine scale that is a unique characteristic of MMF's. These are seen immediately surrounding the radially oriented, elongated magnetic structures corresponding primarily to the middle portion of the penumbra of the sunspot. The sunspot's underexposed umbra is featureless in this magnetogram taken at 1843 UT on 1971, January 9. (Top is west, left is north.) (D. Vrabec and T. Janssens, Aerospace San Fernando Observatory.)

Harvey, 1969; Harvey *et al.*, 1972; Simon and Zirker, 1973). There is now considerable evidence for magnetic fine structure beyond the limit of resolution of magnetographic instruments operating under the best seeing conditions (Harvey *et al.*, 1972; Howard and Stenflo, 1972; Frazier and Stenflo, 1972). Hence, these instruments directly record the *average* magnetic field strength (not the true field strength) over the effective resolution element projected upon the photosphere, or more correctly, upon the level of line formation. This quantity is the magnetic flux threading the resolution element. Consequently, magnetograms are records of the surface distribution of magnetic flux, and the small entities that are just resolved on them correspond to the cross-sectional intercepts with this surface that contain the measured flux. When describing the fine structure recorded on magnetograms, it is necessary to refer to these small concentrations of magnetic flux. I shall therefore call these small-size magnetic field entities 'fluxules,' which emphasizes the fact that magnetic flux rather than magnetic field strength is the physical quantity actually recorded. (Other terms are, of course, more appropriate when referring to the true three-dimensional filamentary structure of the fields not depicted on magnetograms.)

Fluxules clump together in various numbers and with varying degrees of compactness to form different-size aggregates, which in turn collect in a similar manner into larger complexes that comprise the network and active region fields. This results in a clustering hierarchy which probably continues well below the limit of resolution. If this proves to be true, then the smallest fluxules we presently observe are likely to be similarly structured.

In contrast to those observed everywhere else on the Sun, the fluxules occurring within the moats show little tendency to group together to form larger magnetic features. Instead, they exist more or less as individual and independent entities, which we now know are MMF's. Thus, MMF's tend to represent the simplest and most elementary magnetic field structures found anywhere on the Sun's surface. Could this be evidence that the magnetic fields in sunspots are already inhomogeneously distributed, and are gathered into spaghetti-like strands that are swept up by the outflowing velocity field? Presumably, within the moat, the diverging property of the outflow would not favor their collecting into larger associations.

A marked characteristic of magnetic fields in both the active regions and quiet network is that the two polarities occur separated to a high degree. Invariably, we observe large regions of one or the other polarity exclusively, which typically occupy the boundaries of many contiguous supergranules. It is only in the moats around sunspots that we observe the unique distribution of roughly equal numbers of separately identifiable fluxules of both polarities intimately mixed together on a fine scale, which are MMF's.

Notwithstanding these morphological distinctions, MMF's exhibit one remarkable property in common with all other non-sunspot photospheric fields; namely, the fluxules that make up the MMF's are virtually indistinguishable from those that comprise the fine structure components of the fields in general, wherever else they occur outside of sunspots. In other words, magnetic fields imbedded in the photospheric plasma exhibit a propensity to become concentrated into fine flux threads. Apparently this is true all over the Sun, at least wherever the fine structure of the fields is directly observable, which for the time being excludes sunspots. Also, the morphology of this fine structure exhibits a remarkable stationarity in time, evidenced by the similarity between the fluxules that make up MMF's, where the surface fields are just starting their existence outside of sunspots, and those that make up the quiet network fields, which we presume are old remnants of former sunspots. What agency can act upon these diverse fields in a common way to produce and maintain such globally uniform fine structure, and on a sufficiently short time scale to impress its effect upon MMF's when they first appear in the moats? In my opinion, this must be the normal granulation, which has just the necessary spatial, temporal, and dynamic properties. The effect of granulation velocity cells upon the fields can be readily pictured by drawing an analogy with the familiar action of supergranulation on fields to form the large-scale network, applying, of course, the appropriate transformation of length and time scales (Harvey, 1971). The magnetic fields near and including a sunspot are shown in Figure 6. Let us assume, however, that we have at our disposal a 'supermagnetograph' and,

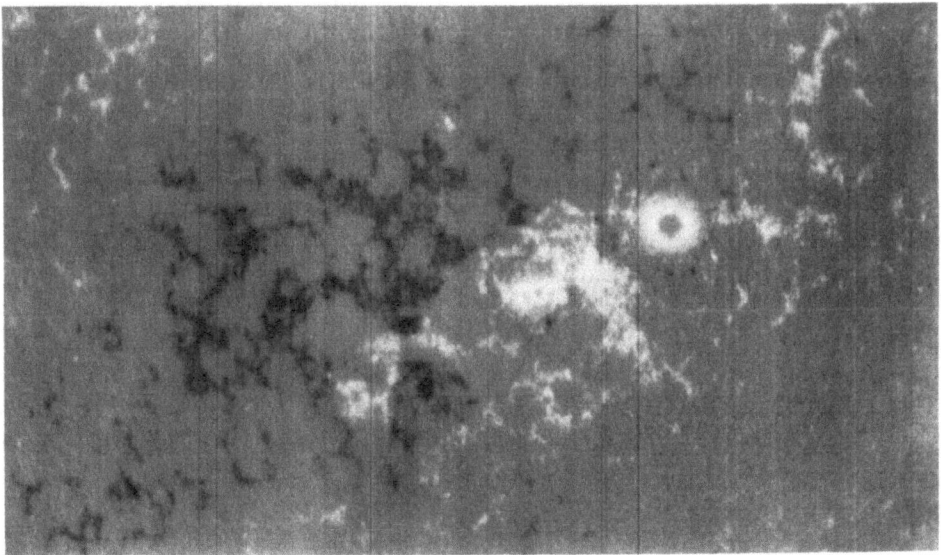

Fig. 6.   This subtracted Zeeman spectroheliogram depicting the magnetic fields of McMath Region 10584 (return of McMath Region 10542 shown in Figure 2) was taken at 1915 UT on 1970, February 21. The tendency of network fields to be exclusively of one magnetic polarity over large areas occupying the boundaries of many contiguous supergranule cells is evident. (See text for the suggested analogy drawn between this magnetogram and the fine structure of solar magnetic fields.) (D. Vrabec, Aerospace San Fernando Observatory.)

instead of a regular sunspot, imagine it to be a very small pore. I expect that the magnetic field fine structure would appear to be distributed in a network pattern very similar to the fields in this picture, cospatial with the intergranular regions, where now the dimensions of these network cells correspond to the sizes of ordinary granules. This network, with its granulation-size cells, would be reshuffled on a time scale of the lifetime of the granulation. Only the most extended areas of field such as those occurring at the cell vertices might be expected to persist over several granulation lifetimes. These may be the fluxules that we are presently barely able to resolve with magnetographs, and which exhibit the wide range of lifetimes and ephemeral characteristics of MMF's and Sheeley's CN bright points. Some evidence of field-free, open cell structure presumably corresponding to the largest granules can be discerned on high-resolution magnetograms such as Figure 6.

2.3.6. *Trajectories*

The time-lapse magnetic field movie made from the Aerospace spectroheliograms taken by Chapman (1973) of McMath Region 11976 on 1972, August 5, is so far the only one with sufficient resolution to enable the outflow trajectories of individual MMF's to be studied. Just before leaving for this meeting, I plotted the paths of over 75 individual MMF's distributed around this complex sunspot group. In each case the path of the feature was traced for the entire length of time it could be identified. Several excellent photographs of this group taken on the same day (Bumba and Suda, 1973; Zirin and Tanaka, 1973) were then examined to determine the orientations of the penumbral filaments, which were found to be very complicated (Pfister, 1973). For example, in some places the filaments are highly inclined with respect to a radius drawn from the associated sunspot umbra, and in others their orientations abruptly change.

A comparison between the MMF trajectories and the penumbral filaments revealed the interesting fact that the trajectories appear to match the orientations and even, in many places, the sign of the curvatures of the penumbral filaments when they were projected out into the moat beyond the points where they actually terminated. This strongly suggests that, at this advanced stage of development of a sunspot, the Evershed outflow in the region of the sunspot penumbra is very closely related to the outflow of photospheric material throughout the entire moat surrounding the sunspot, despite the apparently very real difference in speeds of the two phenomena found by Sheeley and Bhatnagar (1971b) and Sheeley (1972).

2.4. MAGNETIC FLUX TRANSFER

The systematic outflow of large numbers of magnetic field concentrations across the moats surrounding sunspots represents a cumulative horizontal transfer of magnetic flux easily shown to be comparable to the flux of the sunspot itself and of the entire active region associated with it. If we assume the previously estimated value of 3×10^{18} Mx to be the flux $\phi_{+,-}$ of an MMF just resolved on the best spectroheliograms, the total magnetic flux $\phi_T = \sum \phi_+ + |\sum \phi_-|$ of the 50 to 200 MMF's typically seen in the moat around a sunspot is 1.5×10^{20} to 6×10^{20} Mx. In approximately 4 h, which is the time an MMF takes to move across a typical 15 000-km wide moat, all this flux is replenished, so the corresponding rate of magnetic flux transfer by MMF's is 4×10^{19} to 1.5×10^{20} Mx h^{-1}. At this rate, in 4 days a total flux $\phi_T (t=4d) = 4 \times 10^{21}$ to 1.5×10^{22} Mx has crossed the moat, compared with 10^{21} to 10^{22} Mx, which is considered to be representative of the fluxes of individual sunspots and also of active regions. The actual total flux transferred is probably significantly higher than this estimate, because the signals of unresolved MMF's of opposite polarities will tend to average out to produce no observable flux, and because 4 days is probably an underestimate of the duration of typical MFO.

In order to test the validity of the concept that MFO is the agency by which the magnetic field of a decaying sunspot is transferred into the surrounding magnetic field network, as originally suggested by Simon and Leighton (1964), it is necessary to measure separately: (a) the rate of decrease of flux in the sunspot; (b) the rate of outward transfer across the moat of net flux by MMF's of mixed polarity; and (c) the rate of increase of net flux in the active region magnetic network external to the moat. If this concept is valid, all three of these rates should be equal throughout the duration of MFO. Clearly, these three critically important flux measurements are exceedingly difficult to make. Consider, for example, the measurement of (b). Whether we base our determinations of magnetic flux on counts of individual MMF's or, more accurately, on actual magnetograph measurements, it is quite common to find that the flux involved in MFO is roughly equally distributed between MMF's of positive and those of negative polarity. Since $\sum \phi_+ \approx |\sum \phi_-|$, the net flux transferred $\phi_N = \sum \phi_+ - |\sum \phi_-|$, being the difference of two nearly equal quantities, is only a small fraction of their sum, which is the total flux transferred. The measurement of this net transferred flux is, therefore, very sensitive to errors made in determining the total transferred fluxes of each polarity separately. In many instances, uncertainties in the observations make it difficult to determine even the polarity of the net flux transferred, which is, of course, a very vital piece of information.

In the light of these considerations, a comparison of actual measurements reported by Harvey and Harvey with these estimated values is very interesting. In the case of two sunspots, it proved possible to measure the fluxes of MMF's crossing a boundary established within the moats at a fixed distance from the sunspots. Harvey and Harvey found values for the rate of transfer of total flux $|\dot{\phi}_T| = 3 \times 10^{19}$ to 10^{20} Mx h^{-1} and for the rate of transfer of net flux $|\dot{\phi}_N| = 10^{19}$ Mx h^{-1}, the latter being of the *same polarity* as the associated sunspot. Moreover, they estimated the rate of decrease of the magnetic flux of each of these sunspots by calculating the total flux of the sunspots from measurements of their maximum magnetic field strengths and their areas (see Tandberg-Hanssen, 1967). The decay rate of sunspot magnetic flux was also found to be 10^{19} Mx h^{-1}. This value was further substantiated by the later observation of the dates of final dissolution of these two sunspots. In both cases these dates were in agreement with the dates extrapolated on the basis of the measured rate of net magnetic flux transfer by MMF's (Harvey, 1973). Thus, for these two sunspots, Harvey and Harvey were able to verify the equality of (a) and (b) above; namely, that at the times they made their observations, the rate of loss of flux in each of these sunspots was equal to the rate of transfer of net flux by MMF's moving outward across the surrounding moats. Unfortunately, though they attempted the measurement of (c), they were unable to obtain definitive results.

These important and difficult observations made by Harvey and Harvey add considerable weight to the hypothesis that MFO is the primary process by which sunspots decay. It would, however, be a serious mistake to assume too much on the basis of the presently limited data. Actual observations indicate that MFO is a considerably more complicated process than the idealized one of piecemeal fragmentation of a sunspot into MMF's that are convected away by the velocity field surrounding them. We have simply to note the conspicuous occurrence of MMF's of both polarities in virtually all observed examples of MFO. Many MMF's disappear during their transit through the moat, while others of either polarity first appear not at the outer edge of the sunspot penumbra but, rather, at some distance in the moat from it. This greatly complicates any attempt to measure the magnetic flux transferred out of sunspots by MMF's. Also, all observers have encountered difficulty in keeping track of individual MMF's after they have reached the outer boundary and have merged into the magnetic structure already present there. Virtually nothing is known about the interactions that take place between MMF's of opposite and of the same polarity. In addition, it is not yet clear whether MFO plays as important a role in the dissolution of following sunspots as it appears to for preceding ones. It is interesting to note that in the case of the magnetically complex sunspot group observed on 1972, August 5, where both preceding and following polarity members were compactly grouped together in an unusual polarity configuration, the MFO was overwhelmingly dominated by MMF's of following polarity, which is an exception to the previously noted fact that MMF's tend to occur in roughly equal numbers of each polarity. Clearly, very much more quantitative, as well as synoptic data must be obtained on MFO before its effects can be considered really well-established and its basic nature understood.

3. Magnetic Flux Inflow

Magnetic flux inflow (MFI) is a dynamic, sunspot-associated phenomenon manifested by moving magnetic features that stream toward the sunspot along paths converging upon it. In contrast to MFO, many of the MMF's involved in MFI, but not all, are pores easily observed in integrated light. Some of the MMF's, including pores, move directly into the umbra of the sunspot and coalesce with it. It is almost a certainty that MFI is associated with the growth phase of sunspot development and with the emergence of new magnetic flux at the photospheric surface.

In some respects, there is little new about what I have elected to classify as MFI, since it is well known that in many cases the growth of a sunspot may involve the coalescence of smaller sunspots and pores in the manner just described (McIntosh, 1967 and 1969). However, it is important to recognize the distinction between the MMF's associated with this form of activity and those involved in MFO. Also, I wish to call attention to some dynamic and morphological features of MFI that are certainly relevant to the underlying processes involved in sunspot formation and growth.

I will illustrate MFI with a set of observations made on 1970, January 26, of Mc-Math Region 10542, which contained four major southern hemisphere sunspot groups. These data consist of time-lapse movies of the magnetic fields, produced from Zeeman spectroheliograms, and time-lapse movies of the region in various wavelengths within the profile of Hα, exposed through a birefringent filter which was cyclically tuned during the 6-h observation interval. This combination of concurrent magnetic field and wavelength-tuned Hα data in the form of time-lapse movies has proven to be very effective. The latter permit the full spatial extent of Doppler-shifted features to be traced, provided they fall within the wavelength interval scanned.

Two examples of MFI were noted by Vrabec (1971) to occur in one of the sunspot groups classified $\beta\gamma$. (See Figure 7.) The sunspots comprising this group developed rapidly during the preceding two days, especially the dominant leading member of

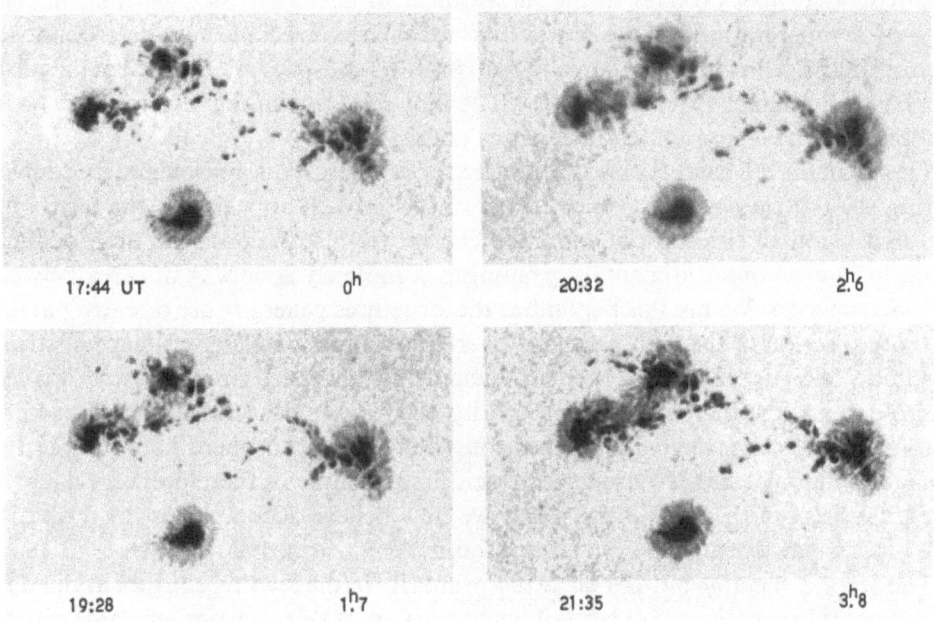

Fig. 7.   Magnetic flux inflow is illustrated in this sequence of photographs of sunspots occupying the center of Figure 2. (See Section 3 for description.) (D. Vrabec and W. Mott, Aerospace San Fernando Observatory.)

north magnetic polarity seen at the right in Figure 7. During the observations, the most rapid growth took place in a newly forming bipolar sunspot pair in the following portion of the group, seen in the upper left of Figure 7. The leader of this bipolar pair emerged at the following edge of some small south polarity sunspots probably paired with the leading sunspot at the right. (The magnetic fields associated with this sunspot group can be seen at the center of Figure 2, which covers most of the active region.)

The Hα filtergrams clearly show that this rapid growth of the new bipolar pair was accompanied by a very conspicuous arch filament system (AFS) (Bruzek, 1967 and 1969) covering the area between the two sunspots. The opposite ends of the arch filaments terminate in magnetic areas of opposite polarity (Vrabec, 1971), in some of which pores formed that were conspicuous on filtergrams exposed approximately $+6.5$ Å from line center. These filtergrams show only the sunspots, pores, and granulation as they would appear in integrated light.

The magnetic field movie shows a conspicuous instreaming toward each of the two sunspots of MMF's of the same polarity as the sunspot. The paths of these MMF's show marked curvature, corresponding to a pronounced spiral structure somewhat resembling a tilted comma, exhibited by both the umbra and penumbra of the following sunspot. These MMF's exhibit a diversity of horizontal speeds ranging from 0.25 to 1 km s^{-1}. Frazier (1972) has interpreted these oppositely moving, opposite-polarity MMF's to be the photospheric intercepts of the footpoints of arched magnetic flux tubes brought to the surface near its center by the upwelling velocity field of a supergranule. These footpoints are then convected laterally to opposed vertices of the cell where they come to rest, forming sunspots after sufficient magnetic flux has accumulated (McIntosh, 1969). The flux tubes rise, lifting material with them, and become the observed dark structures of the AFS.

Viewed in the off-band Hα movie, the areas of both sunspots approximately doubled during the 6 h they were observed. Most of this growth appeared in the form of *in situ* expansion of the sunspot umbrae. The movie also records the dramatic and rapid formation of a rudimentary penumbra completely spanning the area between the two sunspots. Within this penumbra the structures generally are oriented parallel to the projection of the arch filaments overlying them, providing evidence of strong magnetic fields tightly linking the two sunspots. Under these circumstances, it is reasonable to assume that, at least temporarily, the newly surfaced fields exhibited predominantly horizontal components throughout the photosphere between the two sunspots. The type of MFI observed in this first example is, therefore, very likely an integral aspect of the AFS activity that we now believe accompanies the formation of a bipolar pair of sunspots (Weart and Zirin, 1969; Zirin, 1970; Weart, 1970, 1972).

The second example of MFI occurred primarily in the leading sunspot at the right of Figure 7. Emanating from the following portion of this sunspot are three distinct tail-like chains or aligned strings of pores and small umbrae. The appearance of a three-pronged tail seen in the off-band Hα filtergrams is considerably enhanced in the Zeeman spectroheliogram of the magnetic fields (Figure 2) where the gaps between the pores delineating the prongs are almost continuously filled in with magnetic features. All prong features, including the pores, are of the same north magnetic polarity as the leading sunspot.

MFI is conspicuous in both the magnetic field movie and the off-band Hα movie in the form of highly organized movement localized in these prongs. All MFI is confined to the prongs, and all magnetic features comprising the prongs stream toward the leading sunspot at speeds averaging 0.25 km s^{-1} relative to it. Most noteworthy of the observed characteristics of this second example of MFI is that, within each prong, all instreaming MMF's follow a single path that very nearly coincides with the axis of the prong, producing the effect of threaded beads sliding on a string stretched along this axis.

During the 6-h span of the movie, three conspicuous pores or small umbrae are observed to enter the sunspot, two feeding in from the lower prong and the third from the middle one. A small area of the sunspot penumbra preceding the leading pore and separating it from the following portion of the sunspot umbra, together with a small area of photosphere directly behind it and separating it from the following two pores, moves with the pores into the following portion of the umbra of the main sunspot. In so doing, each of these two confined areas is deformed into a thin light bridge that partitions the umbra of the main sunspot. Both of these light bridges assume a curvature resembling a 'bow wave' corresponding to the direction of intrusion.

The morphologies associated with the dynamic processes just described are strikingly similar to those described by Bumba (1965) in a discussion of the forms and evolution of light bridges in sunspots. It is very likely that the forms Bumba described were produced in the same manner, namely, by the intrusion of pores into the main sunspot umbra, a phenomenon that has also been described by McIntosh (1972). Evidence for both this type of MFI and the formation of a similarly curved light bridge can be seen in a sequence of videomagnetograms and continuum filtergrams of a sunspot group discussed by Schoolman (1973). According to McIntosh (1973), strings of pores frequently occur near sunspots.

It is interesting to note that, at the same time that MFI was actively taking place along the three azimuths of the prongs in the following quadrant of the sunspot, well-developed MFO extended uniformly throughout the remaining three quadrants, producing a well-developed moat and wreath. As a consequence, pores were streaming into the sunspot along the outer prongs immediately adjacent to regions occupied by outflowing MMF's. The latter, as is characteristic of MMF's associated with MFO, were restricted to features visible only in the magnetic field movie. Thus, at this particular stage of its development, and while magnetic flux was being fed into it by instreaming pores and MMF's, this sunspot was already being acted upon by the process of MFO by which it would presumably decay.

It should be noted that there also occurred an analogous, but less conspicuous, oppositely directed movement of south polarity features toward the south polarity following member of the new bipolar sunspot pair. All of these magnetic features, including pores and one small sunspot, were similarly organized into an extended linear strand pointed toward this following sunspot. Some of these streaming pores and the small sunspot grew appreciably in area as they approached the following sunspot, evidencing the emergence of new magnetic flux. The observations ended before it could be determined whether or not these streaming features actually entered the following sunspot.

This second example of MFI differs from the first in at least four respects: (a) the MMF's include a large fraction of conspicuous pores, (b) the loci of these MMF's and pores are in the form of 'tails' or 'strings' converging upon the sunspot, (c) MFI is strictly confined to the features comprising these tails or strings, and (d) the motion of MFI toward the sunspot is directed roughly along the axis of these tails or strings. This last characteristic appears to be inconsistent with convective transport of magnetic fields by velocity fields which appears to underly both MFO and the first example of MFI. I am inclined to believe that the explanation should instead be sought in terms of the configuration and dynamics of the subphotospheric magnetic fields of the sunspot.

The two types of MFI just described are schematically illustrated in the upper half of Figure 8. The characteristics of the first example of MFI are consistent with the movement of the magnetic footpoints of upwelling magnetic flux tubes that first appear near the center of a supergranule cell outlined by the dashed lines, and then are transported by the material outflow to the cell boundary where the sunspots also form. A possible explanation for the occurrence of the tails, or aligned strings of pores, characteristic of the second example of MFI is depicted in the second and third diagrams where the magnetic footpoints including pores are shown to be ac-

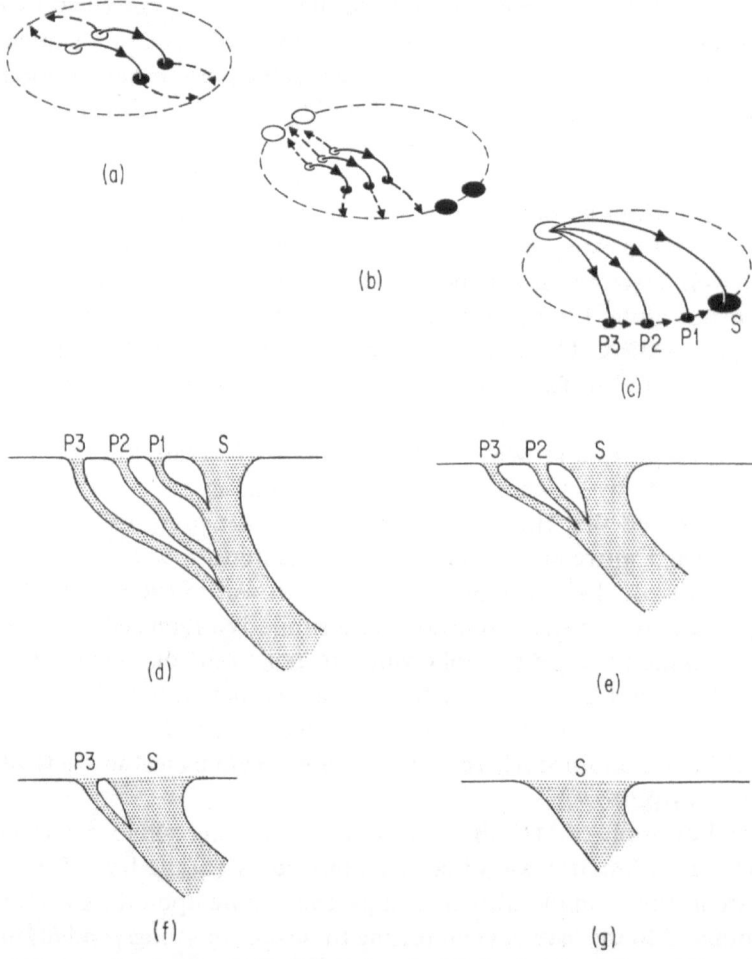

Fig. 8.   The two examples of magnetic flux inflow described in Section 3 are shown schematically. In the upper three diagrams, continuous lines denote magnetic field lines, and dashed arrows show the direction of movement of magnetic features. The dashed ellipse delineates the boundary of a supergranule cell. *P* and *S* denote pores and a sunspot, respectively. In the lower four diagrams, the shaded areas represent vertical cross sections of a frayed magnetic flux tube depicted as rising.

cumulating along the same supergranulation cell boundary on which the new sunspots formed. The movement of MMF's and pores into the sunspot along the axis
of the 'string' is illustrated in the third diagram. This movement may result from the
simple geometry of a rising magnetic flux tube that is frayed just below the photosphere into a number of fine strands that individually produce MMF's and pores
where they intercept the photosphere. These strands are assumed to be moved by
the material, so they can be transported to the cell boundary to form the observed
tails. As schematically illustrated in the remaining four diagrams of Figure 8, if the
main flux tube is assumed to rise, the initially separated strands will successively
coalesce with the main flux tube when the points at which they join it reach the photospheric surface.

On the basis of these two well-observed examples of magnetic features moving into
growing sunspots, we may presume that the same types of MFI accompany the formation and growth of many sunspots. Although we do not know at present what
fraction of the total they constitute, it is probable that these two forms of MFI are
quite common inasmuch as the first example represents the magnetic aspects of AFS
activity generally believed to play a basic role in sunspot formation, and provided
we broaden the second example to include the coalescing of small sunspots to form
larger ones. It will be interesting to learn what other forms of MFI may occur.

4. Concluding Remarks

Magnetic flux inflow affords us the opportunity to observe directly the surface streamlines of a highly ordered velocity cell in the photosphere and also to observe the
convective transport of magnetic fields over the Sun's surface. Because the outflow
speed is roughly twice that of a normal supergranule, the moat around a decaying
sunspot is an optimum place to repeat Simon's (1966) extremely difficult measurements of a proper motion component of granulation systematically directed toward
the boundary of a supergranule cell, as Schmidt has just urged us to do.

The outflow velocity field associated with MFO has been likened by various investigators to supergranulation. I wish to emphasize that we must be very careful to
distinguish between the two, since in MFO we are dealing with a velocity cell with
unique properties. First, the existence of this velocity field depends upon the presence
of a sunspot (presumably, actually upon the magnetic field associated with the sunspot), which completely determines the geometry of the outflow. In contrast, in the
case of supergranulation, any magnetic fields present appear to play a passive role.
Second, the systematic outward velocities are at least twice as high as those associated
with supergranulation. Third, the diameter of a typical moat is roughly twice that of
a typical supergranule. Finally, the lifetime of MFO is at least several to tens of days,
compared to one day for a supergranule.

The theoretical aspects of MFO and MFI are the subject of the paper by Meyer,
Schmidt, Weiss, and Wilson, which will be presented at this Symposium. These investigators have taken care to incorporate much observational data into their theo-

retical treatment. According to them, magnetic flux is first expelled by supergranule eddies, and is then concentrated by the converging velocity field to form a sunspot. At this point, the concentrated magnetic fields of the sunspot modify the velocity field and, in fact, reverse the direction of circulation, resulting in outflow. The observed slow decay mode of persistent sunspots is explained by turbulent diffusion of flux tubes within the sunspot toward the moat, where they are swept up by the outflow, producing MMF's.

Note: This review concluded with a film that included time-lapse magnetic field movie sequences of MFO occurring in the sunspot groups shown in Figures 2 and 4. In addition, a time-lapse sequence was shown of the sunspot group in Figure 7, illustrating the two examples of MFI described in Section 3.

Acknowledgments

I am indebted to Dr G. Chapman for generously placing his excellent spectroheliograms of 1972, August 5, at my disposal, and to W. Mott who assisted in the observations of MFI. These observations were skillfully converted by R. Maulfair into a time-lapse movie of magnetic fields and MFI, for which J. Paul provided photographic support. Dr P. Wilson provided valuable insight into various theoretical aspects. Karen Harvey, Dr J. Harvey, and Dr N. Sheeley kindly reviewed the manuscript and clarified many points. P. McIntosh contributed detailed information on sunspot growth and development. The preparation of this review was accomplished under The Aerospace Corporation Company-sponsored research. An NSF-USNC-IAU Travel Grant made it possible for me to attend the Solar Symposia in Australia.

Appendix. Observational and Instrumental Aspects

As described by their acronym, MMF's are fundamentally magnetic features that move (and change with time). They are small (typically $<2''$), so the largest can be individually resolved by only a few magnetographs currently in operation, these being of the type designed especially for high-resolution observations. Though MMF's can be observed with excellent resolution as components of the bright photospheric network through a simple filter (Chapman, 1970), such nonmagnetic observations cannot supplant (but may certainly complement) direct magnetic field observations that reveal the polarities of the individual features. Therefore, to observe MMF's, the first basic requirement is a capability for obtaining magnetograms depicting the two-dimensional distribution of photospheric magnetic fields (i.e., of the line-of-sight component) with a resolution of at least $3''$, polarity discrimination, and a magnetic flux sensitivity better than 3×10^{18} Mx. Also, because MMF's are dynamical, evolving entities, it is an essential second requirement to be able to repeat these observations a number of times, spaced over an interval of time long enough for significant changes to be recorded. I would consider three to four such magnetograms spanning 2 h to constitute a minimum requirement for simply detecting MMF's. However, if one intends to study their movements and evolutionary changes, considerably better data are needed. By far the most effective method for both detecting and observing MFO has proven to be the transformation of magnetograms into time-lapse movies (Vrabec, 1971; Sheeley, 1971; Harvey and Harvey, 1972 and 1973; Schoolman, 1972), but these require large numbers of magnetograms and longer time coverage. For example, a recent movie of MFO prepared from observations made by Chapman (1973) with the Aerospace spectroheliograph reveals many interesting details regarding the fine structure, polarities, and patterns of flow of MMF's as well as the manner in which they first appear, subsequently change, interact, or finally disappear. In this instance, a sequence of 92 Zeeman spectroheliograms spanning 7.8 h were obtained on 1972, August 5, of an active sunspot group rich in MMF activity. (See Figure 4.) The resolution ranged from 1.5 to 4″, and the magnetic flux sensitivity was approximately 3×10^{17} Mx. Actually, to study the evolutionary development of MFO, we need observations of this quality repeated over many successive days spanning the development of a sunspot group. Clearly, a major limitation of ground-based efforts is the infrequency of occurrence of sustained good seeing prevailing over extended intervals of time.

To date, it has proven extremely difficult to achieve the resolution of solar magnetic fields adequate to observe MMF's. The instruments used to produce magnetograms are, unfortunately, presently complex and costly, since they must sense or measure the Zeeman effect and transform this information into a pictorial record. Excellent summaries of magnetograph techniques are available for reference (Bray and Loughhead, 1964; Evans, 1966; and Beckers, 1971). With the exception of videomagnetographs (Janssens and Baker, 1971; Smithson and Leighton, 1971), which employ TV cameras and display the magnetic fields in real time, the raw output data of magnetograph systems, whether photoelectric or photographic, must undergo subsequent processing before a 'picture' of the fields is obtained. This has a major impact through the greatly expanded effort required to produce the many individual magnetograms that constitute a single MMF observing record, e.g., a time-lapse movie. To date, the major sources of MMF data have been two spectroheliographs and one photoelectric magnetograph.

The spectroheliographic method of photographing solar magnetic fields was first developed and used by Leighton (1959). Subsequently, MMF's have been observed with the Aerospace spectroheliograph (Vrabec, 1971; Chapman, 1973) and with the Kitt Peak spectroheliograph (Sheeley, 1971). Examples of high-resolution Zeeman spectroheliograms appear in Livingston (1972), Vrabec and Janssens (1972), and Chapman (1972). Because this method involves the photographic subtraction of a pair of spectroheliograms to produce each Zeeman spectroheliogram, a special cine optical-printer facility has been found to be indispensible for producing the magnetic field movies used extensively at Aerospace. When quantitative results are not required, film offers many advantages for recording and storing magnetograms, especially in the large numbers required for making movies.

Currently, the only photoelectric scanning magnetograph capable of resolving MMF's is the 40-channel Kitt Peak magnetograph (Livingston and Harvey, 1971), soon to be supplemented with a 512-channel linear array of diode pairs (Livingston, 1973). Examples of the magnificent results obtained with this instrument can be seen in Harvey (1971), Livingston (1972), and Figure 3. Harvey and Harvey utilized computer-generated movies of solar magnetic fields recorded with the Kitt Peak magnetograph in their investigation of MMF's. These calibrated, digitized magnetograms remain to date the only source of quantitative measurements of the field strengths of MMF's and of the magnetic fluxes involved in MFO. As we have seen, these quantitative data are vital for ascertaining to what degree MFO transfers the magnetic flux of a sunspot into its surroundings. The extraction of quantitative magnetic field measurements from the spectroheliographic data has been impeded by the well-known difficulties encountered in photometrically transforming two-dimensional photographic images, compounded by the large number of images involved, but these are being overcome. In a number of ways, these two basically different instruments complement each other.

In both the spectroheliograph and the photoelectric magnetograph, resolution is degraded by the necessity to use a finite-aperture slit or probe and to scan the solar image over an extended interval of time during which the seeing may vary considerably, leading to inhomogeneity in the quality of the data from point to point in the magnetogram. These variations constitute 'noise', in the presence of which the elusive MMF's must be detected. Eventually, both these currently effective methods will be supplanted by magnetographic instruments that utilize narrow-band filters and, therefore, are limited only by seeing and the resolution of the telescope. Moreover, instantaneous records can be made of the entire field of view without scanning.

The earliest instrument to exploit these advantages was the Culgoora magnetograph (Ramsay *et al.*, 1971), with which time-lapse movies showing evolutionary changes in magnetic fields have been obtained (Schatz, 1971). The tunable narrowband filter upon which this instrument is based consists of three automatically controlled Fabry-Pérot interferometers in series, and has been described by Ramsay *et al.* (1970). 'Hybrid' filters consisting of a Fabry-Pérot interferometer blocked by means of Lyot-type birefringent elements (Zirin, 1966; Title, 1970) are used in both the Aerospace and Caltech-Big Bear videomagnetographs referred to previously. Smithson (1972, 1973b) has reported observations of evolutionary changes in the magnetic field network of quiet regions seen on time-lapse movies obtained with the latter instrument. In principle, the real-time videomagnetographs are ideally suited for making time-lapse movies because it is a very simple matter to photograph the magnetic fields automatically directly off the monitor screen.

Preliminary, high-resolution, nonmagnetic observations obtained with a 'universal,' continuously tunable, birefringent filter have been presented by Beckers (1973). A straightforward modification will convert this into an extremely high-resolution photographic magnetograph, particularly in consideration of the fact that this filter is being used in conjunction with the high-resolution Sacramento Peak vacuum telescope. The interacting (or coherently-phased) Fabry-Pérot interferometric narrow-band filter (Title, 1970) is also a new development in narrow-band filter technology with direct application for improving magnetograph resolution while also greatly simplifying the design of the instrument. Multichannel diode arrays (Dunn and Spence, 1973; Smithson, 1973a) are ideally suited for the real-time differential photometry involved in generating quantitative magnetic field data, and will no doubt play a significant role in a new generation of magnetographs that will far outperform the present ones.

Measurements of the total vector magnetic field within MMF's will be necessary for determining whether or not their appearances and disappearances are a result of changes in the orientation of their fields. For this, the line-profile Stokesmeter (Harvey *et al.*, 1972) offers considerable promise. For obtaining the most complete spectroscopic data on which to base in-depth astrophysical analyses of the physical conditions, velocities, and the complete magnetic vector field in MMF's, the classical spectro-enregistreur des vitesses, as utilized by Michard *et al.* (1961) or its modern equivalent, the spectra-spectroheliograph (Title and Andelin, 1971), both photograph spectra of a two-dimensional area on the Sun, from which the spectrum line profiles and mean wavelengths of all individually resolved features within the area can later be retrieved.

As was stated previously, one of the most outstanding needs of MMF research is to obtain continuous, synoptic magnetic field observations over extended intervals of many days so that the complete evolutionary history of MFO and MFI, as they relate to the evolutionary development of sunspots, can be observed. Ultimately, such observations will come from space observatories, but it should be realized that it is also feasible to obtain continuous solar data with a suitable network of ground-based observatories distributed in longitude, or from a single installation at a sufficiently high latitude (Janssens, 1970; Rogers, 1970).

While direct observations of MMF's and MFO of the types described will always prove indispensible, these must also be combined with concurrent observations of a nonmagnetic nature if we are to understand what relationships exist between them and other solar phenomena. For example, we need to relate MMF's and MFO to chromospheric fibril structure, Ellerman bombs, and Hα bright points observed around sunspots. Thus, it is very important to obtain, in addition to the direct magnetic field data, simultaneous, tuned Hα filtergrams. For investigating the velocity structure of MFO, we need Doppler spectroheliograms or filtergrams. CN and various other bright photospheric network observations in addition to white-light or continuum data are clearly essential for locating precisely where in sunspot penumbrae many MMF's first appear as well as for studying MFI and the role of pores. These are but a few examples of the wide range of data constituting a third, general requirement which must be satisfied by any serious observational program directed toward improving our present meager knowledge of MMF's and the underlying processes of magnetic flux inflow and outflow.

References

Allen, R., Edberg, S., LaBonte, B., and Sheeley, N. R.: 1973, *Bull. Amer. Astron. Soc.* **5**, 268.

Beckers, J. M.: 1971, in R. Howard (ed.), 'Solar Magnetic Fields', *IAU Symp.* **43**, 3.

Beckers, J. M.: 1973, *Bull. Amer. Astron. Soc.* **5**, 269.

Beckers, J. M. and Schröter, E. H.: 1968, *Solar Phys.* **4**, 142.

Bray, R. J. and Loughhead, R. E.: 1964, *Sunspots*, Chapman and Hall, Ltd., London, p. 186.

Bruzek, A.: 1967, *Solar Phys.* **2**, 451.

Bruzek, A.: 1969, *Solar Phys.* **8**, 29.

Bumba, V.: 1965, in R. Lüst (ed.), 'Stellar and Solar Magnetic Fields', *IAU Symp.* **22**, 305.

Bumba, V. and Suda, J.: 1973, in H. E. Coffey (ed.), Report UAG-28, Part 1, 'Collected Data Reports on August 1972 Solar-Terrestrial Events', World Data Center A for Solar-Terrestrial Physics, U.S. Department of Commerce, National Oceanic and Atmospheric Administration, Environmental Data Service, Asheville, North Carolina, U.S.A., p. 155.

Chapman, G. A.: 1970, *Solar Phys.* **13**, 78.

Chapman, G. A.: 1972, *Solar Phys.* **26**, 299.

Chapman, G. A.: 1973, in H. E. Coffey (ed.), Report UAG-28, Part 1, 'Collected Data Reports on August 1972 Solar-Terrestrial Events', World Data Center A for Solar-Terrestrial Physics, U.S. Department of Commerce, National Oceanic and Atmospheric Administration, Environmental Data Service, Asheville, North Carolina, U.S.A., p. 85.

Coffey, H. E.: 1973, *ibid.*

Dunn, R. B. and Spence, G. E.: 1973, *Bull. Amer. Astron. Soc.* **5**, 271.

Evans, J. W.: 1966, in *Proceedings of the Meeting on Solar Magnetic Fields and High Resolution Spectroscopy, Rome, 14–16 September*, p. 123.

Frazier, E. N.: 1972, *Solar Phys.* **26**, 130.

Frazier, E. N. and Stenflo, J. O.: 1972, *Solar Phys.* **27**, 330.

Harvey, J.: 1971, *Publ. Astron. Soc. Pacific* **83**, 539.

Harvey, J., Livingston, W., and Slaughter, G.: 1972, in 'Line Formation in the Presence of Magnetic Fields', High Altitude Observatory, National Center for Atmospheric Research, p. 227.

Harvey, K.: 1973, private communication.

Harvey, K. and Harvey, J.: 1972, *Bull. Amer. Astron. Soc.* **4**, 384.

Harvey, K. and Harvey, J.: 1973, *Solar Phys.* **28**, 61.

Howard, R. and Stenflo, J. O.: 1972, *Solar Phys.* **22**, 402.

Janssens, T. J.: 1970, *Solar Phys.* **11**, 222.

Janssens, T. J. and Baker, N. K.: 1971, in R. Howard (ed.), 'Solar Magnetic Fields', *IAU Symp.* **43**, 44.

Leighton, R. B.: 1959, *Astrophys. J.* **130**, 366.

Leighton, R. B., Noyes, R. W., and Simon, G. W.: 1962, *Astrophys. J.* **135**, 474.

Livingston, W. C.: 1972, *Sky Telesc.* **43**, 344.

Livingston, W. C.: 1973, private communication.

Livingston, W. C. and Harvey, J. W.: 1969, *Solar Phys.* **10**, 294.

Livingston, W. C. and Harvey, J. W.: 1971, in R. Howard (ed.), 'Solar Magnetic Fields', *IAU Symp.* **43**, 51.

McIntosh, P. S.: 1967, oral communication presented at *IAU Symp.* **35** on the 'Structure and Development of Solar Active Regions', Budapest, Hungary, 4–8 September.

McIntosh, P. S.: 1969, in A. C. Stickland (ed.), *Annals of the IQSY*, Vol. 3, 'The Proton Flare Project, (The July 1966 Event)', The MIT Press, Cambridge, Massachusetts, p. 40.

McIntosh, P. S.: 1972, in H. E. Coffey and J. V. Lincoln (eds.), Report UAG-24, Part II, 'Data on Solar-Geophysical Activity Associated with the Major Ground Level Cosmic Ray Events of 24 January and 1 September 1971', World Data Center A for Solar-Terrestrial Physics, U.S. Department of Commerce, National Oceanic and Atmospheric Administration, Environmental Data Service, Asheville, North Carolina, U.S.A., p. 303.

McIntosh, P. S.: 1973, private communication.

Michard, R., Mouradian, Z., and Semel, M.: 1961, *Ann. Astrophys.* **24**, 54.

Pfister, H. J.: 1973, in H. E. Coffey (ed.), Report UAG-28, Part 1, 'Collected Data Reports on August 1972 Solar-Terrestrial Events', World Data Center A for Solar-Terrestrial Physics, U.S. Department of Commerce, National Oceanic and Atmospheric Administration, Environmental Data Service, Asheville, North Carolina, U.S.A., p. 35.

Ramsay, J. V., Kobler, H., and Mugridge, E. G. V.: 1970, *Solar Phys.* **12**, 492.

Ramsay, J. V., Giovanelli, R. G., and Gillett, H. R.: 1971, in R. Howard (ed.), 'Solar Magnetic Fields', *IAU Symp.* **43**, 24.
Rogers, E. H.: 1970, *Solar Phys.* **13**, 57.
Schatz, D. L.: 1971, in R. Howard (ed.), 'Solar Magnetic Fields', *IAU Symp.* **43**, 243.
Schoolman, S. A.: 1972, *Bull. Amer. Astron. Soc.* **4**, 390.
Schoolman, S. A.: 1973, *Solar Phys.* **32**, 379.
Sheeley, N. R.: 1967, *Solar Phys.* **1**, 171.
Sheeley, N. R.: 1969, *Solar Phys.* **9**, 347.
Sheeley, N. R.: 1971, in R. Howard (ed.), 'Solar Magnetic Fields', *IAU Symp.* **43**, 310.
Sheeley, N. R.: 1972, *Solar Phys.* **25**, 98.
Sheeley, N. R.: 1973, private communication.
Sheeley, N. R. and Bhatnagar, A.: 1971a, *Solar Phys.* **18**, 195.
Sheeley, N. R. and Bhatnagar, A.: 1971b, *Solar Phys.* **19**, 338.
Simon, G. W.: 1966, in K. O. Kiepenheuer (ed.), *The Fine Structure of the Solar Atmosphere*, F. Steiner Verlag, Wiesbaden, West Germany, p. 47.
Simon, G. W. and Leighton, R. B.: 1964, *Astrophys. J.* **140**, 1120.
Simon, G. W. and Zirker, J. B.: 1973, *Bull. Amer. Astron. Soc.* **5**, 280.
Smithson, R. C.: 1972, *Bull Amer. Astron. Soc.* **4**, 392.
Smithson, R. C.: 1973a, *Bull. Amer. Astron. Soc.* **5**, 280.
Smithson, R. C.: 1973b, *Solar Phys.* **29**, 365.
Smithson, R. C. and Leighton, R. B.: 1971, in R. Howard (ed.), 'Solar Magnetic Fields', *IAU Symp.* **43**, 76.
Tandberg-Hanssen: 1967, *Solar Activity*, Blaisdell Publishing Co., Waltham, Massachusetts, p. 181.
Title, A.: 1970, 'Fabry-Pérot Interferometers as Narrow-Band Optical Filters, Part 1, Theoretical Considerations', TR-18, Harvard College Observatory, p. 75.
Title, A. M. and Andelin, J. P.: 1971, in R. Howard (ed.), 'Solar Magnetic Fields', *IAU Symp.* **43**, 298.
Vrabec, D.: 1971, in R. Howard (ed.), 'Solar Magnetic Fields', *IAU Symp.* **43**, 329.
Vrabec, D. and Janssens, T. J.: 1972, *Solar Phys.* **25**, 2.
Weart, S. R.: 1970, *Astrophys. J.* **162**, 987.
Weart, S. R.: 1972, *Astrophys. J.* **177**, 271.
Weart, S. R. and Zirin, H.: 1969, *Publ. Astron. Soc. Pacific* **81**, 270.
Zirin, H.: 1966, *The Solar Atmosphere*, Blaisdell Publishing Co., Waltham, Massachusetts, p. 33.
Zirin, H.: 1970, *Solar Phys.* **14**, 328.
Zirin, H. and Tanaka, K.: 1973, *Solar Phys.* **32**, 173.

DISCUSSION

Wilson: You mentioned that the lifetime of the MFO's is about four days and yet these are found in association with slowly decaying spots which may last for weeks. If a spot lasts for several weeks as it may well do, will the MFO's persist for that time or for something like four days?

Vrabec: That is a very good question and I knew I would get in trouble if I gave you too many numbers. The problem in citing numbers is that there is such a large dispersion in the observations. I chose four days to illustrate that even if we make a conservative estimate for the duration of MFO, a lot of flux is transferred by this process. Four days is probably a minimum value. I know that moats persist for at least this long, but by the time the sunspot has become dispersed, the whole phenomenon weakens and is difficult to observe. Karen Harvey has informed me that she has seen moats disappear very suddenly – during the course of a single day.

Sturrock: One of the questions in my mind is whether the fluxules are being swept out by a general motion of the photospheric gases, or whether they are being moved through a comparatively stationary gas. In the former case one would expect fluxules of different strengths to move with the same velocities, whereas in the latter case one would expect the more intense fluxules to move faster than the weak fluxules.

Vrabec: Sheeley and Bhatnagar and Sheeley have measured, by means of Doppler spectroheliograms, an outflow of material from the sunspot. These velocities turn out to be somewhat less than those of MFO, ranging from 0.4 to 0.8 km s^{-1}, whereas the moving magnetic features have velocities averaging 1 km s^{-1}. However, the velocity of the magnetic features and the gas velocity have not been measured simultaneously so it is only by inference that I have assumed that the two are associated. We need more observations to answer this question. In answer to the second part of your question, we do observe a large spread in velocities of the magnetic features in the moats, and I think I see the large features acting as a sort of barrier in the moat. They tend to be sluggish in their motion. The smaller features (and by 'small' I mean smaller in area, not smaller in the field strength, because most of the fluxules, I believe, have about the same field strength) are the most mobile and include the fast interlopers.

Sturrock: What you say suggests that the outward force moving the gas out is the magnetic force on the fluxules, and that the moving fluxules tend to drag the surrounding gas along with them.

Giovanelli: Have you any model for the appearance of fluxules of opposite polarities in the decaying stages of sunspots?

Vrabec: A number of people have worried about this. The Harveys suggested a model in which the lines of force going out of the sunspot get vertically kinked, and these kinks move outward away from the sunspot. Then, depending on how many kinks are in the magnetic field lines, you may have an excess of one polarity over the other. For one kink you can have a two-to-one excess, and for many kinks essentially an equal number of each of the two polarities. The other popular idea is that we have an umbrella of field lines which leave the sunspot and return in the moat. Another possible picture is that the subphotospheric fields are tangled, and fields of opposite polarity existing below the surface are carried upward to the surface by the velocity fields. This could bring up polarity opposite to that of the sunspot and result in MMF's of opposite polarity.

Athay: One feature of the magnetic field pattern that seems to stand out is the persistence of rather small features of the field through a great height range. The network extends from the upper photosphere through the chromosphere without really diverging very much. Sunspots are observed in the center of the Hα line and are not very much larger than sunspots seen in the photosphere even though the height difference is 1000 km or more. In view of this, isn't it quite likely that the magnetic features extend more or less intact into the subphotospheric layers and, if this is the case, isn't it likely that the guiding forces that move around the magnetic field elements are associated with subphotospheric motions that may have nothing directly to do with the types of motions that we observe at photospheric levels such as the supergranulation.

Wilson: A short answer to that question is yes, but we will have more to say about this in a later paper presented by Meyer.

Vrabec: The concept we have is analogous to seaweed that is anchored to the bottom at some depth below the surface where it is observed. The seaweed tendrils move about with the motion of the water, but they are still anchored down below. The same may be true of the magnetic flux tubes observed at the photosphere in that they may be anchored to some more permanent feature of the field at deeper levels.

Zwaan: Could you say anything about possible differences in the velocity distributions of the MMF from one sunspot to another?

Vrabec: I don't think we have enough data to answer that question, but the movies you just saw were the ones used for the initial estimate of 1 km s^{-1} for the magnetic features. Sheeley had previously estimated 1 km s^{-1} for the bright CN features. After studying 34 sunspot groups, the Harveys came up with 1 km s^{-1} for the average velocity, but with a wide range of velocities from 0.2 km s^{-1} extending up to 2 km s^{-1}. The average seems rather well determined but there is a large dispersion. I should remark that, as the resolution of our observations improves, we see more and more of these features not visible in lower resolution observations. The features that are near the limit of resolution tend to have the shortest lifetimes. They come and go faster than the larger, more conspicuous ones and I have the impression that they move faster. The smallest features may be moving with the fluid velocity, whereas the slower moving large features appear to act somewhat as obstructions to the flow.

Grossmann-Doerth: I wonder whether these moving fluxules can be connected in one way or another to the Ellerman bombs? Could it be that the Ellerman bombs are related to two fluxules of opposite polarity that possibly annihilate each other, and can it be shown that the Ellerman bombs occur in the same areas and are associated with the moving fluxules?

Vrabec: Zirin showed examples in his talk of what was described many years ago by Bruzek – that when we have an arch filament system and emerging flux loops, these are regions where we see bright points in Hα. It's very difficult, however, with only a filter to distinguish between Ellerman bombs and other bright features in Hα. There are many bright features that extend for only about an Ångstrom into the line wings whereas the true Ellerman bombs show brightening many ångströms beyond the line. So the Ellerman bombs are very distinctive and exceptional bright points. We do see, as Zirin showed in his pictures, that the bright points also ring sunspot penumbrae. Aside from the fact that they occur where the fields have dynamic properties, such as near the feet of the surfacing arch filaments, we do not have an explanation for the bright points.

Meyer: I wanted to come back to this point of the velocity of the moving magnetic features with respect to the gas velocity in the moat. Following Harvey, the four authors of the paper to be presented on the theory of sunspot growth and decay suggest that in the subphotospheric layers the flux tubes are carried along more or less horizontally and that the kinks developed in the magnetic field, possibly by granular action, constitute the moving magnetic features. The velocity of the moving magnetic features in this picture should be a little bit higher than the gas flow because they should travel with the Alfvén speed super-posed on the speed of the moat flow. In a very general estimate, the numbers seem about right and the magnetic features should move a little faster than the gas flow itself.

Vrabec: Is there anything in the theory that suggests that if they are moving faster than the material they may cause a local excitation such as the bright points?

Meyer: No, I did not intend to imply that.

Zwaan: There is one observation that has a bearing on Grossmann-Doerth's suggestion and that is that Ellerman bombs tend to show a recurrence for a couple of hours. As far as I know this observation reported by early observers is still valid.

Vrabec: I will make the point that in the few cases that I, as well as others, have looked into, a necessary condition for an Ellerman bomb is that there be a concentrated field at the same place. However, there are many concentrated fields constituting MMF's, and we have not found what circumstances are sufficient for bombs.

Wilson: Does the MFI show both polarities or only that of the spot to which they move?

Vrabec: Only the latter.

Wilson: Does the moat lifetime of four days which you quote apply to the spots with lifetimes of several weeks?

Vrabec: No. It is just a typical figure limited by the length of our observing runs.

ON THE RELATION BETWEEN MOVING MAGNETIC FEATURES AND THE DECAY RATES OF SUNSPOTS

CORNELIS ZWAAN

Astronomical Institute, Utrecht, The Netherlands

Abstract. Bumba (1963) inferred from plots of areas of sunspot groups against time that stable spots may show a phase of slowest decay which is independent of the spot's area. From the Greenwich Photoheliographic Results we analysed the areas of individual spots. Indeed there is a clear lower limit in the decay rates close to Bumba's value $dA/dt \simeq -4.0 \times 10^{-6}$ visible hemisphere/day $\hat{=} -5.0 \times 10^{15}$ cm^2 h^{-1}. However, only a small fraction of regular spots and an even smaller fraction of the irregular spots pass through the phase of slowest decay.

A slow erosion by some supergranular velocity field cannot explain why the rate is independent of area. Rates independent of area are compatible with a model in which the decay is ultimately determined by diffusion across a relatively thin cylindrical current sheet, provided that the ratio between the thickness ΔR of the current sheet and the total radius R of the flux tube remains constant during the decay (Gokhale and Zwaan, 1972). Such a radial similarity $\Delta R/R =$ constant fits the radial similarity in the observable sunspot structure. If the diffusion occurs by purely Ohmic dissipation, then the current sheet should be extremely thin: $\Delta R/R$ between 10^{-5} and 10^{-4}.

From a plausible magnetic field configuration $\mathbf{B}(r)$ and the rate of slowest area decay follows a flux decay rate $d\Phi/dt \simeq -6.5 \times 10^{18}$ Mx h^{-1}. This value comes very close to the value $d\Phi/dt \simeq -1.0 \times 10^{19}$ Mx h^{-1} derived by Harvey and Harvey (1973) from magnetic features moving away from two sunspots. If indeed these two spots were in the mode of slowest decay, then it would follow that the diluted magnetic flux crossing the current sheet is carried away in discrete and detectable flux tubes. Certainly rapidly decaying spots often show a fragmentation into magnetic features larger than the socalled moving magnetic features. However, there is no clear picture yet as to what photospheric and chromospheric phenomena around sunspots accompany the decay processes at various rates.

There are some indications that the decay rates of sunspots are related with the structure in the umbrae (Zwaan, 1968). Spots with umbrae showing dark cores without clear structure decay slowly. Umbrae with a dense pattern of bright structures, umbral dots or light bridges or both, seem to indicate rapidly decaying spots. Implications are briefly discussed.

Simultaneous observation of the three aspects of decay, viz.

(i) the photospheric and chromospheric phenomena around the spot (e.g., fragmentation, moving magnetic features);

(ii) the decay rate of sunspot area dA/dt;

R. Grant Athay (ed.), Chromospheric Fine Structure, 233–234. All Rights Reserved.

(iii) the fine structure in the sunspot, would substantially contribute to a better understanding of sunspot structure down in the convection zone and of the mechanisms distributing the magnetic flux over the photospheric surface. When seeing permits the observation of the first aspect then the second and third aspects may be secured without great difficulty.

An account of the investigation of decay rates of sunspots is planned as a paper in the series 'The Structure of Sunspots', to be published in *Solar Physics*.

References

Bumba, V.: 1963, *Bull. Astron. Inst. Czech.* **19**, 91.
Gokhale, M. H. and Zwaan, C.: 1972, *Solar Phys.* **26**, 52.
Harvey, K. and Harvey, J.: 1973, *Solar Phys.* **28**, 61.
Zwaan, C.: 1968, *Ann. Rev. Astron. Astrophys.* **6**, 135.

DISCUSSION

Smith: In your decay phase model you have suggested that you have a very thin current sheet. I imagine that the field of the sunspot is vertical, so I don't see how the current sheets are formed. What kind of field do we have around the sunspot?

Zwaan: Very thin current sheets follow from the assumption of purely ohmic dissipation. If you add extra resistivity, for example, because of turbulence the thickness of the current sheet will grow in proportion. I don't have an idea how the current sheet comes about but it does fit in with the radial similarity in observed sunspot structure and with the lower limit on the sunspot decay rates. A second indication is that the umbra is a rather uniform thing, apart from fine structure that may or may not be present. This suggests that the whole column below the umbra is rather uniform within that column. May I comment on one more thing. I think it is not difficult to observe the decay rates of sunspots accurately from a continuum signal from the magnetograph. The slow decay rate corresponds to one square arcsecond per hour.

Giovanelli: If I understood you correctly, your final diagram showed the umbra of a sunspot in which one portion was dark and another portion contained a lot of little dots, which I assume to be umbral granulation. Your suggestion is that the umbral granulation stays there in more or less the same manner for quite some time. However the lifetime of umbral granules is only of the order of one hour. Isn't there some discrepency here.

Zwaan: I did not say that exactly the same umbral dots are staying there all the time but that a certain area of the umbra shows visible fine structure where the other parts remain dark without observable structure. I think that the irregularity in the appearance of the umbral fine structure indicates that umbral dots are not some convective mode necessary for the physical health of the sunspot.

Bumba: Sunspots that are very regular often dissipate in such a way that granules develop inside the umbra along its border. There are many cases when pieces of umbra or penumbra travel from the sunspot through the intergranular space. This is a very interesting phenomena. I have tried to obtain a movie of it but I have not yet succeeded. My impression is that you have a dark piece of matter pushed out of the sunspot through intergranular space. This dark matter eventually disappears in the intergranular field.

A THEORETICAL MODEL FOR THE CONVECTION OF MAGNETIC FLUX IN AND NEAR SUNSPOTS

F. MEYER and H. U. SCHMIDT

Max-Planck-Institut für Physik und Astrophysik, München, Germany

N. O. WEISS

Dept. of Applied Mathematics and Theoretical Physics, University of Cambridge, England

and

P. R. WILSON

Dept. of Applied Mathematics, University of Sydney, New South Wales, Australia

Abstract. In this paper we investigate the physical processes that lead to the growth and decay of magnetic flux in and near sunspots.

An initial phase of rapid growth is characterized by the emergence of magnetic flux from the deep convection zone. As the flux rope rises through the surface the magnetic field is swept to the junctions of the supergranular network where sunspots are formed. These flux concentrations follow the footpoints of the emergent flux rope as they rapidly move apart.

When all the flux has appeared and the footpoints have become stationary at their maximum extent the second phase sets in. It is characterized by a reversal of the supergranular flow around the spot. During the growth phase the converging super-granular flow descends around the sunspot and cools the region at its base. The spreading penumbral field covers part of the supergranule and reduces the area over which heat is transmitted to the photosphere. As long as the sunspot follows the footpoint of the emerging flux rope and moves rapidly through the supergranular network no steady accumulation of heat in one place occurs. Once the spot has become stationary however, the region around its base is insufficiently cooled. The gas in this region heats up and eventually rises reversing the flow pattern around the sunspot.

This reversal can break up the flux rope and destroys many sunspots within a few days. Some large sunspots however after an initial loss of flux achieve a quasistation-ary configuration. The reversed flow forms an annular cell (the moat) around a concentrated vertical shaft of magnetic flux at the center. Into this concentrated field region the supergranulation flow cannot penetrate. Small scale convection mainly due to the rapid rise of opacity also takes place within the concentrated vertical flux at depths below 2000 km. This provides an eddy diffusivity for the magnetic field.

The interplay between these two regions accounts for the slow decay of the sunspot and the observed motion in the moat around the spot. As the magnetic flux slowly diffuses outwards it enters into the supergranular flow and its photospheric ends are rapidly carried outwards across the surface towards the faculae region. The boundary between the supergranular flow outside and the diffusive region inside will be charac-

R. Grant Athay (ed.), Chromospheric Fine Structure, 235–237. All Rights Reserved.
Copyright © 1974 by the IAU.

terized by a critical magnetic field strength B_c for which the field can prevent penetration of the flow. We show for a cylindrically symmetric model that this leads to a linear decay law for the sunspot magnetic flux in quantitative accord with the observations.

We have tested the assumptions of the model by two-dimensional numerical computations in the Boussinesq approximation. It was found that motion is excluded from the region where the magnetic field strength exceeds some critical value B_c, that the flux decays on a time scale slow compared to the turnover time for the cell, that the decay rate is determined by diffusion throughout the flux rope, and that once fieldlines have diffused into the convecting region they are rapidly swept aside by the motion.

We discuss the implications of this model for the observations of moving magnetic field concentrations around sunspots and for evidence of small scale convective motion within the sunspot.

DISCUSSION

Athay: One aspect of your model that I don't understand is that when you reverse the supergranule flow there does not appear to be any force that holds the sunspot together. Why doesn't the field just disperse then into the surrounding medium?

Meyer: That's an interesting question. Once you have brought the flux together it does not partake further in the supergranular motion. The field is held together by the balance of magnetic pressure and the decrease of gas pressure inside the flux tube. The field just sits there quietly. In the Bussinesq approximation it is just a uniform column. In the non-Bussinesq approximation there would be some reordering of the flux tube with height but in general the column will still be rather verticle.

Athay: Wouldn't this configuration be very unstable however?

Meyer: This is an interesting point. Though there is no dynamic instability of the magnetohydrostatic configuration one might have expected an instability feeding on some thermal-convective effect. However the computations show that this is not the case, and I would like to emphasize that they of course contain the thermal and bouyant effects. The field concentration region excludes the large scale convection pattern, and only as diffusion weakens the field at the outer boundary can it be sheared off. However there will be an instability against large scale supergranular flow when the flux tube is not sufficiently vertical or branches off into different branches at depth. As soon as the flow reverses these will be pulled apart with the supergranular time scale of one or two days. We believe that this is just what happens to many sunspots.

Sturrock: You drew attention to the depression of the sunspot as tending to give it stability. It occurs to me that this might also help to explain the formation of sunspots since if there is a pore on the edge of the penumbra, it would see an inward moving force. In addition, I wish to ask whether there is any way to understand the division of a spot into an umbra and penumbra in terms of your model.

Meyer: That question has been considered by Simon and Weiss who compute what the equilibrium form of the flux tube is by neglecting the gas pressure inside and using the photospheric gas pressure outside. The gas pressure in the photosphere decreases exponentially with height. If there is very little magnetic flux, the bending over of the field lines occurs above $\tau = 1$ and at $\tau = 1$ one sees a pore. If there is enough flux however, the bending over will occur inside of $\tau = 1$. If these field lines are then horizontal and press onto the photosphere, it suggests that you have a penumbra. I am inclined to think that that is a reasonable explanation.

Wilson: I would like to reply to Athay's question. We were worried about the question of stability of a flux tube with motions occurring around the borders of it so I looked into the Kelvin-Helmholz instability in this regard. If you take just a two-dimensional model you obtain the same sort of answer that Chandrasekhar obtained in his book. If you include the cylindrical geometry you can increase the stability against this kind of dissipation quite significantly. May I just briefly show one slide which illustrates this question with possible ways in which the flux tubes move away from the spot. Friedrich has shown two possible cases, the first (the Harveys) is one in which the tube frays away at the top but the rope deep below the spot retains its original concentration. I think it is a fairly important feature of the second possibility that the flux rope starts to fray away from the central rope at some depth below the surface and is then carried towards the surface by the deep supergranule motion. On the first model it would be very difficult to imagine that these loops are produced in a horizontal flux tube solely by the granule motions. The supergranule motions, particularly the deep motions, must play some part in this and give rise to the succession of opposite polarities in the MMFs. However it may be possible to settle this problem by observation, in which case we'd have a much better grasp on what is going on underneath.

Zirin: How is it possible for a homogeneous plage to sit there for 10–14 days if this supergranule motion is sweeping everything to the edges. It's true that the edges of the plage show a supergranule pattern but the large plage that you want to make the sunspot out of is quite permanent and shows no such motion.

Meyer: Surely the supergranule motion exists in the plages?

Zirin: Then why don't we form sunspots all over the place. I don't see that kind of motion.

Meyer: I don't think you do because the foot points of these elements are just going in different direction. This is only a suggestion and I do not want to go into too many details because we have not worked them out. I feel very strongly that the linear decay rate is a very strong argument for the eddy viscosity that we have predicted.

McKenna-Lawlor: I have observed sunspot umbrae to disappear or reduce in size very rapidly following certain major flares. In one case an umbrae with a field of up to 2700 G disappeared in less than 5 h. What mechanism could produce this?

Meyer: I don't really know what I should say to such a case. I do know however that the unipolar flux that you have inside of a sunspot cannot just disappear. It has to either be transported away or opposite flux must move in so it can be submerged under the surface. In the spots you have observed some such effect must have happened. What is the time scale?

McKenna-Lawlor: Unfortunately the cadence of pictures was not better than 5 h. The umbra probably disappeared within a shorter time interval.

Meyer: That seems very short for such a sunspot to disappear by any of the required processes mentioned.

EVOLUTION OF CHROMOSPHERIC FINE STRUCTURES
ON THE DISK

KATSUO TANAKA

Tokyo Astronomical Observatory, Mitaka, Tokyo, Japan

Abstract. Evolution of chromospheric fine structures is discussed from observational points of view. First, various lifetime determinations are reviewed. Then chromospheric oscillation in the interior of supergranule cell is discussed in relation to chromospheric 'grain' structure. We show that the grains are collectively associated with the 180-s oscillations as well as that in K line the grain is identified with local disturbance generated at lower level and propagating to upper levels. Finally the details of evolution of individual mottles are presented. We show some cases which indicate the propagation of a disturbance along the mottle axis. Statistics of evolution patterns of mottles implies a significant tendency for a system of double dark mottles encasing a bright mottle.

1. Introduction

Chromospheric fine structure is different between the supergranule cell and its boundary. In the supergranule cell there are many tiny, roundish features called grains. At the boundaries there are several clusters of elongated mottles as well as small granular features called bright mottles or plagette at the central portion of the cluster. Mottles often extend over the cell interior. Morphological relationship between plagette and mottles is similar to the relation between plage and fibrils in active region. Variety of the fine structure might originate from the difference in strength and orientation of the magnetic field. Studies on the evolution have revealed the difference in the evolution between various fine structures. Present knowledge indicates the dominance of oscillatory changes in the cell interior and in plagette as well as in plage; while mottles and fibrils show non-oscillatory changes. Generally evolution of fine structures in the chromosphere would be closely related to the individual process of heat and mass transfers from photosphere to corona. We may evaluate various modes of heat and mass transfers by studying details of the evolution of individual chromospheric features.

In this paper I first review our knowledge of the lifetimes of various features. Then I discuss chromospheric oscillations associated with the grains in the network cell. Finally I try to discuss the details of evolution of individual mottles.

2. Lifetimes

Lifetimes have been determined most frequently by a method of survival curve. In this method a group of features to be examined are identified at a certain time first and in good quality frames preceding and following this time numbers of mottles that can be identified with the original ones are counted and plotted against time (survival curve). The full width at half maximum of the survival curve might be de-

R. Grant Athay (ed.), Chromospheric Fine Structure, 239–255. All Rights Reserved.
Copyright © 1974 by the IAU.

fined as a lifetime. In Table I those lifetimes obtained by this method are shown with an asterisk mark. Some investigators determined lifetimes by following individual features from their birth to disappearance (Bhavilai, 1965, 1966; Bray, 1968, 1969; Sawyer, 1972). The two methods were compared by Sawyer for the same data. She found for the features showing rapid changes faster than the time resolution of the

TABLE I

Lifetimes

Dark fine mottle			
de Jager (1957)*	16 min	Hα	0.0 Å
Bruzek (1959)*	13		0.0
Beckers (1964)*	18		±0.5
Howard and Harvey (1964)*	9; 15		0.0
Macris and Alissandrakis (1970)*	5.5		−0.6
Sawyer (1972)*	14		+0.3
Beckers (1963)	15.7		±0.5
Bhavilai (1966)	8		+0.75
Bray (1968)	5		±0.3
Rogers (1970)**	7		+0.65
Bhatnagar and Tanaka (1972)	10		−0.5
Bright fine mottle			
Macris and Alissandrakis (1970)*	12 min		−0.6 Å
Sawyer (1972)*	13		+0.3
Bray (1969)	10		±0.3
Grain			
Beckers (1964)*	4 min	K2	
Sawyer (1972)	2.2	Hα	
Bhavilai (1966)	1.3	Hα	
Fibril			
Lippincott (1955)	−45 min		
Foukal (1971)	1–20	Hα	
Supergranule network			
Macris (1956)*	26 h	K	
Janssens (1970)*	21	Hα +0.65 Å	
Simon and Leighton (1964)**	19–21	K	
Rogers (1970)**	25.0±1.6	Hα +0.65 Å	

* survival curve method (see text).
** cross correlation method.

data that the survival curve gives a misleading result. In the following I briefly discuss the results.

Dark fine mottle. Large spread of lifetime from 5 min to 18 min exists in various measurements. There is a slight tendency that a shorter lifetime is obtained for the shorter effective time resolution of the data (Sawyer, 1972). It indicates the existence of rapid change. Bray (1968) found that most of the mottles exhibit well-marked changes after 2.5 min. On the other hand it is shown by the visibility curve analyses (Beckers, 1963; Bhatnagar and Tanaka, 1972) that the visibility of mottles never drops for a time longer than 10 min.

Bright fine mottle. Most of the bright fine mottles are located near the center of dark mottles cluster, thus at a lower height than dark fine mottles (Beckers, 1968; Bray, 1969; Banos and Macris, 1970). There are also bright elongated mottles similar to the dark fine mottle (Bhavilai, 1965). Bray (1969) and Macris and Alissandrakis (1970) found that the lifetime of the bright fine mottle is larger by a factor two than the lifetime of the dark fine mottle. Sawyer (1972), however, found the same lifetime as the dark fine mottle. Sawyer's statistics shows rapid changes with a time scale of 1 min. At the central area of dark mottles cluster the intensity oscillation with a period of about 300 s has been detected by Bhatnagar and Tanaka (1972) in Hα -0.5 Å and by Liu and Sheeley (1972) in K2$_v$. These authors measured the intensity variations directly from movies of Hα filtergrams and K2 spectroheliograms, respectively. Probably the bright fine mottles are responsible for these intensity oscillations.

Grain. From Table I one finds the lifetime of the grain considerably shorter than that of the mottles at the network boundaries. Beckers (1964) found that only 30% of K grains (bright) are recognized after 2.5 min. Similarly, Sawyer (1972) showed that 72% of Hα grains (dark) disappear within 2 min. Bhavilai (1966) also found the bright Hα grains in the cell whose median lifetime is 76 s. These results are consistent with the intensity oscillation of a 180-s period found in the supergranule cell by Bhatnagar and Tanaka (1972) and Liu (1973).

Fibril. Foukal (1971b) finds no essential difference between the lifetimes of the fibril and the mottle. It is pointed out by Foukal (1971a) that the visibility of the fibril oscillates in a few minutes about a 'mean appearance' defined by a characteristic alignment, mean darkness and mean length which is preserved typically for several hours. The latter time scale is interpreted by him as similar to that required for major rearrangement of the magnetic field structure. It is similar to the time scale of the mottle-producing activity in the quiet region (Bhavilai, 1966).

Supergranule network. Both survival curve method and cross correlation method give the same lifetime of about one day. The data analyzed by Janssens (1970) and by Rogers (1970) cover 62 h without an interruption. Time history of the cross correlation Rogers made, indicates a rapid drop of the correlation in the initial 10 min, which corresponds to the lifetime of individual mottles, and more gradual fall of the correlation between 7 min and 4 h. Rogers interprets this stage of gradual fall as the period of readjustment of the coarse dark mottles. It may be suggested that change in magnetic field structure occurs in this time scale.

3. Chromospheric Oscillation in the Network Cell

It has been known from power spectrum analyses that higher frequencies appear in the 5-min oscillations in the chromosphere. Leighton (1963) and Orral (1966) found the broad peak down to 170 s in the power spectrum of K line. Similar results were obtained for Hα by Elliot (1969) and for Hβ by Bhattacharyya (1972). Recently two-dimensional studies have been started by using filtergrams and spectroheliograms. Thus in the central portion of the network cell Bhatnagar and Tanaka (1972) found a well-defined 170-s period for the intensity oscillation in Hα −0.5 Å. The broad peak which appeared in power spectra in one-dimensional analyses might be due to the mixing of two characteristic periods, namely 170 s in the interior of the cell and 300 s at the network boundaries as well as due to the non-oscillatory components of mottles.

The oscillation is easily detected in time-lapse filtergrams of Hα off-band at which overlying elongated mottles are not so visible. One can see the whole area of the interior of the cell oscillate at any time like the waves in the ocean. Obviously the intensity oscillation in Hα wings is accompanied by discrete dark features called grains. However the size of the oscillatory wave is much larger than the size of individual grain (about 1000 km), and occasionally extends over 10000 km. Time-lapse pictures show that the grains tend to appear as a group which takes a form like a chain with a typical length of about 5000 km.

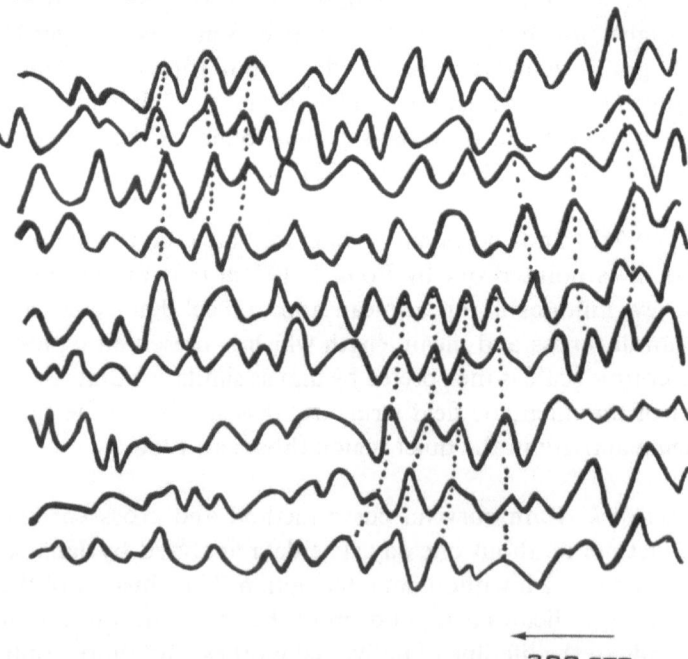

300 SEC

Fig. 1. Sixty-minute records of intensity variation in K filtergram (bandpass 0.3 Å) made at the two-dimensional points in the cell. Dot lines connect the common peaks of oscillatory waves. Each curve corresponds to one place of 5000 km diameter.

In K filtergrams of bandpass 0.3–0.6 Å the oscillations are more evidently detected since no obscuring materials like mottles exist in such wide bandpasses. Besides the bright K grains there seem to be groups of dark points which may correspond to the dark phase of the oscillatory change of the K grain. In Figure 1 intensity oscillations in K filtergram (Bandpass 0.3 Å) are shown for two-dimensional points in the interior of the cell. Each curve represents intensity tracing with time at one 5000 km region. Following records of neighbour points in the ordinate axis one may notice common peaks composing one wave. Time lag between these peaks may lead to apparent horizontal phase velocity of about 100 km s^{-1} within one wave. However, it may not imply actual transverse propagation, but probably represents the difference of arrival time of wavefront at the fixed height of K-line formation. This could happen when the wavefront is originally spherical or when it is modified from plane wave by the effect of magnetic field.

It is interesting to see how the oscillatory wave in the cell interacts with the overlying elongated mottles, whose one end originates at the cell boundary. I have found a few cases in which the grains are formed along the mottle axis as the mottle stretches to the cell interior. They suggest that the top of mottles and grains are not independent phenomena; however their relation on the whole has not been clarified yet.

From a theoretical point of view an attempt has been made to interpret the 180-s oscillation as a trapped oscillation in the chromosphere where the horizontal magnetic fields prevail (Nakagawa, 1973). In such a model, the chromosphere is considered as a continuous medium; discrete features like grains are then secondary phenomena such as local excitation or condensation due to the oscillation. On the other hand there is phenomenological evidence to identify the grain itself as individual propagation of locally confined waves as will be discussed next.

4. Evolution of K and Hα Line Profiles in the Network Cell

Formation of K-line profile, especially the origin of its asymmetry is a still controversial problem. Recently a new aspect has appeared in this problem, namely evolutionary changes of K-line profiles. Wilson and Evans (1971) and Wilson *et al.* (1972) have found various patterns of the evolution as follows:

(1) a single-peak emission evolves into a double-peak emission;

(2) a double-peak emission evolves into a single-peak emission at the blue or red wings;

(3) a double-peak emission with a red (blue) asymmetry changes to a blue (red) asymmetrical emission.

These changes are found to occur with a time scale between 10 and 60 s mainly in the central portion of the cell.

The evolution of K-line profiles in the cell has been studied extensively by Liu (1972, 1973). He has found in several places, not large in number, that brightening moves progressively from the very far wing of K line (±7 Å) towards the line core up to K2 (Figure 2). As the brightening propagates the profile of K-line changes from

pure absorption profiles with enhanced K1 wing to intense single-peak emission at
K2$_v$. The whole phenomenon occurs within about 100 s and tends to repeat with a
period of 180 s. This could be considered as a particularly bright grain.

Most of the places (grains) inside the cell Liu (1973) studied show intensity oscillation with the same period as the above phenomenon, but with relatively small time

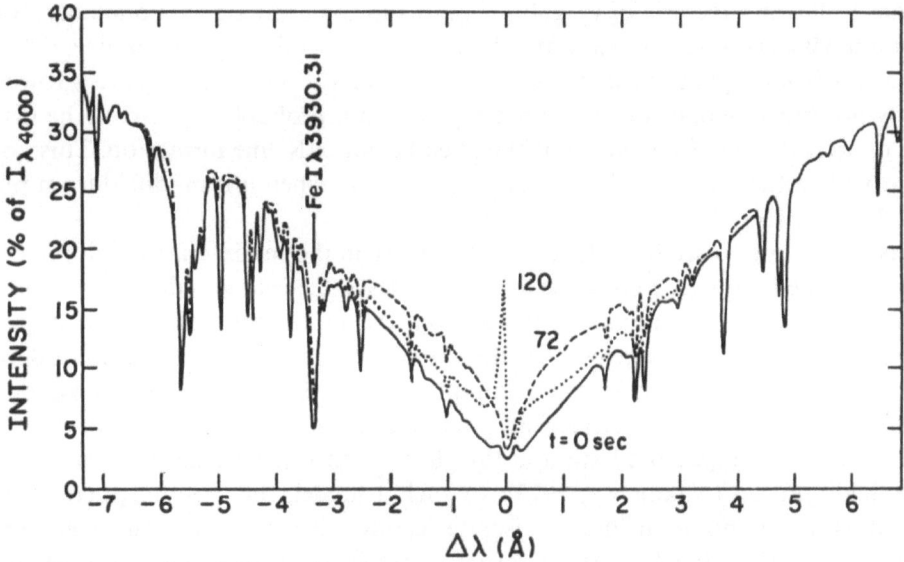

Fig. 2.   Evolution of the K-line profile of the grain which shows particularly large phase lags between
various wavelengths. At $t=0$, the undisturbed profile. At $t=72$ s the disturbance has propagated from
the far wing to the inner wing. At $t=120$ s an intense K2$_v$ peak and the wing starts to recover.
(Courtesy of S. Y. Liu).

lags between various wavelengths. One such example is shown in Figure 3 which exhibits intensity tracings at 7 different wavelengths. One can see conspicuous intensity
oscillation with a 180-s period except K3 core where the amplitude of oscillation is
reduced. There are phase lags between wavelengths in a sense that the brightening
in the wing precedes that near the core. Liu found that 97% of all the 459 peaks show
this tendency. The mean time lag between bright peaks at K1 wing (± 0.44 Å) and
K2$_v$ is found equal to 9.3 s. This value is considerably smaller than the mean period
of 180 s. If we consider also that the intensity enhancements are symmetrical in both
wings beyond K2, it could be concluded that the 180-s oscillation in K line wing is
a pure brightness oscillation.

Liu (1973) interprets these phenomena as the propagation of wave from below the
temperature minimum, assuming that the intensity fluctuations are due to actual enhancement of temperature in an optically deep region. For this explanation he assumes the dissipation due to acoustic wave; if one assumes a height difference of
500 km between the levels forming K2$_v$ and K1 (± 0.44 Å), and takes into account
the mean time lag 9.3 s, one finds a vertical phase velocity $v_p = 54$ km s^{-1} and a group

velocity $v_g = 0.67$ km s^{-1} from the relation $v_p \cdot v_g = c^2$ with $c = 6$ km s^{-1}. Finally using velocity amplitude (1.5 km s^{-1}) obtained from the strong line near K-line center Liu finds mean mechanical flux in the grain equal to 2×10^7 erg cm^{-2} s^{-1}.

Some of this flux is radiated in K line itself; subtracting the flux radiated in K line a net mechanical flux of 7×10^6 erg cm^{-2} s^{-1} is obtained. This value would be re-

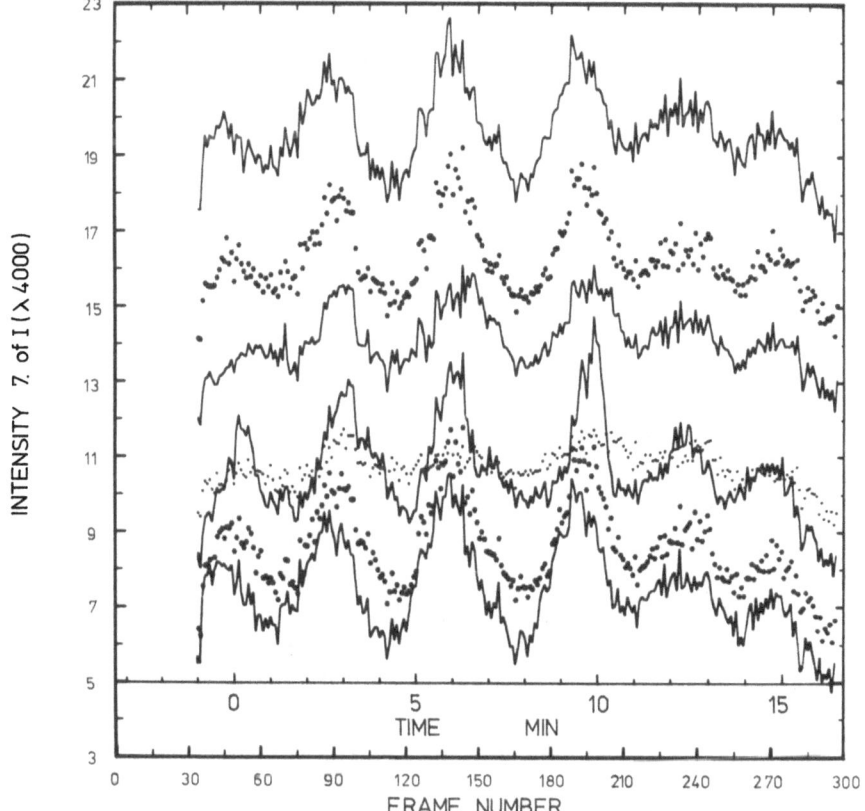

Fig. 3. Example of the intensity oscillation of the grain at various wavelengths (from top to bottom: -0.44 Å, K1$_v$, K2$_v$, K3, K2$_r$, K1$_r$, $+0.44$ Å). Note the phase coherency at various parts of profile. (Courtesy of S. Y. Liu).

duced if we consider the fractional area of grains and radiation loss from other lines. In conclusion, as far as we adopt the hypothesis of acoustic wave, the flux carried by grains seems to be insufficient for the coronal heating, which requires 5×10^6 erg cm^{-2} s^{-1} on the whole surface of the Sun.

At the various phases of oscillation the grain exhibits various K-line profiles. Statistics shows that 50% have double-peak emissions, 30% have single-peak emissions and 20% have pure absorption profiles (Liu and Smith, 1971). In Pasachoff's (1970) result there are more single-peak emissions although he did not specify the position. The single-peak emission appears most frequently at the phase when the bright grain appears at K2, or when the disturbance reaches the highest level. The pure absorption

profile corresponds to the dark phase of the grain or the time when the disturbance
is generated at lower level. It is to be noted that the single-peak emission at K3 is
not observed generally.

The time variation of Hα line profile is qualitatively different from K line. Figure 4
shows the intensity tracings with time at three wavelengths of Hα (0 Å, ±0.5 Å); the
phase differs by about 180 deg in the two wings implying velocity oscillation contrary
to the brightness oscillation in K line. We may note also that this velocity oscillation

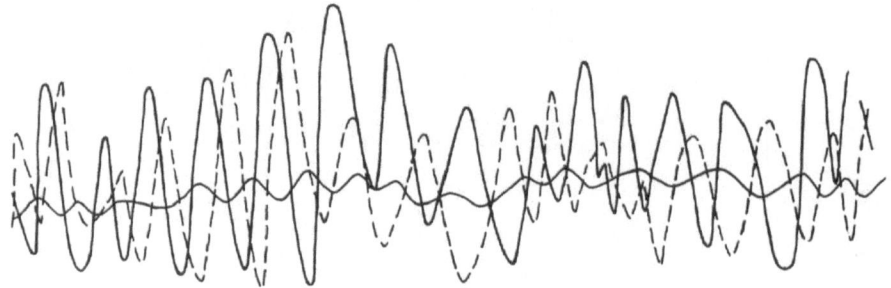

Fig. 4.   Intensity tracings of time-lapse Hα spectrograms made at the central portion of the cell. Time
proceeds from right to left. Note the phase difference between −0.5 Å (thick line) and +0.5 Å (broken
line). The mean interval between intensity peaks is 180 s.

is caused by the Doppler shifts of 3–5 km s^{-1} in the individual dark features, namely
grains. Since this amplitude is well within the width of K3 core, the velocity oscillation apparently missing in K line may well be confined in K3 core, where the brightness oscillation is reduced. It might be suggested that two kinds of oscillations exist
always associated with the grain phenomenon: the brightness change originates at
lower level and propagates upwards and the velocity perturbation is produced at
higher levels, which would be seen in Hα and K3.

While the relation between the two kinds of oscillations remain uncertain, the
interpretation of K-line profiles due to two vertical components (two clouds) looks
interesting. In this model the dark feature absorbs various parts of K3 core of the
relatively wide emission originated at slightly lower level than the dark feature; thus
double or single-peak profile can be produced as the phase of the velocity oscillation
of the dark feature changes. Independent of this dynamical model Suemoto (1971)
has examined the detailed characteristics of H and K lines profiles by two vertical
components model taking their velocity distributions into account and succeeded in
fitting the observed mean and rms profiles by adjusting optical thicknesses of the two
components and number of elements superposed in the line of sight. The optical
thicknesses in K line center equal to 1000 and 10 are obtained for the bright feature
and the upperlying absorption feature respectively.

Another interpretation of time-dependent K-line profiles has been proposed by
Cram (1973), who considers the height variation of source function in a single structure. In this case transfer equation in presence of velocity field would represent the
problem.

One characteristic feature in the Hα spectra of the grains is the tilt of profile to the direction of dispersion. There has been observed a reversal of this tilt against the dispersion direction in a time of about 1 min. For the tilt of emission line observed outside the limb Beckers (1972) suggested two explanations: (1) merging of two or more unresolved features; (2) rotation or differential motion in the structure. In case of the grain the tilt is evidently associated with the velocity oscillation of two neighbouring features in favor of the first explanation.

5. Evolution of Individual Mottles at the Network Boundary

Evolution of elongated mottles located at the supergranule boundaries are very complex since not only brightness but size and shape changes continuously. Howard and Harvey (1964) called these changes seething motion. When we look at the intensity variation of one dark mottle, continuous fluctuations are found with a time scale from 1 min to 5 min. Most of them are by no means periodic. The basic problem in the evolution of mottles would be to determine the fundamental mode of these changes. Particularly in relation to spicule it would be important to know the direction of the fundamental motions, whether it occurs along the axis of mottle or perpendicular to it.

Beckers (1963, 1964) studied time variations of mottles visibilities in the blue and red wings (± 0.5 Å) of Hα. Out of 112 mottles he finds 41% show the visibility maximum first in the blue wing and then in the red wing with a mean time lag of 7.0 min, 35% show periodic visibilities and 24% show no distinct variation during 20 min. Beckers interprets the mottles in the first type as rising and falling spicules. Generally the velocity determination by using filtergrams is not straightforward (Athay, 1970). However, so far as a mottle can be considered as a cloud isolated from any other medium as has been shown by Grossmann-Doerth and von Uexküll (1971), it would be correct to interpret a blue sentive mottle as rising.

Bhavilai (1965) and Dunn (1971) have shown that some mottles are visible only in blue wing, while others appear only in red wing. In a cluster of mottles the positions of blue and red mottles are mixed. Bhavilai (1965) considered a flow of material in a loop to explain this feature. We have no clear evidence at present whether the mottles occur in an open field or in a closed field such as the loop although faint connection is occasionally seen between two centers of the clusters of mottles suggesting the loop structure (see Figure 7(e), (i) for example).

Rising and falling velocities as inferred from Hα spectra of mottles are considerably smaller (4 km s^{-1}) compared to the rms velocity of the spicule (15 km s^{-1} according to Pasachoff et al., 1968). Grossmann-Doerth and von Uexküll (1973) attempted to explain this discrepancy due to the decrease of contrast by seeing; it is found that the observed velocity is reduced to 10–25% of the true velocity provided that the size of mottle is about the same as the spatial resolution. Under the ideal seeing condition we might expect 15–40 km s^{-1} for the Doppler shift on the disk.

Direct evidence of upward motion of mottles can be obtained by examining evolution of an individual mottle. I have found that the top of the mottle extends with an apparent velocity from 20 km s^{-1} to 200 km s^{-1}. The mean velocity is 55 km s^{-1}. The frequency of elongation at one fixed position is 0.17 min^{-1}. We can see that a mottle starts to stretch suddenly from its original length or from the center of cluster and continues to stretch for a duration 20 s to 2 min. Although it indicates some transfer from below it is uncertain as yet whether it is associated with actual mass motion or wave propagation.

Under the high spatial resolution it is common to see a system of the double dark mottles with a bright gap between them. Bhavilai (1965) found this feature which he called double chains. Tanaka (1972) found about 30% of all the dark mottles are double or can be resolved into double at any one instance. It consists of the two parallel, straight or curved dark mottles with a separation of 1″ or less and a gap between them which is brighter than the background. In Hα center line picture the lower part of the bright gap is often identified with the bright elongated mottle; it loses its contrast in the wing picture, thus only the double mottles are visible there. Figure 5 illustrates double mottles taken at Hα −0.5 Å.

It has been found that any mottle shows a stage of 'double mottles' in its evolution; statistically it occurs at one position in the cluster of mottles every 4.4 min al-

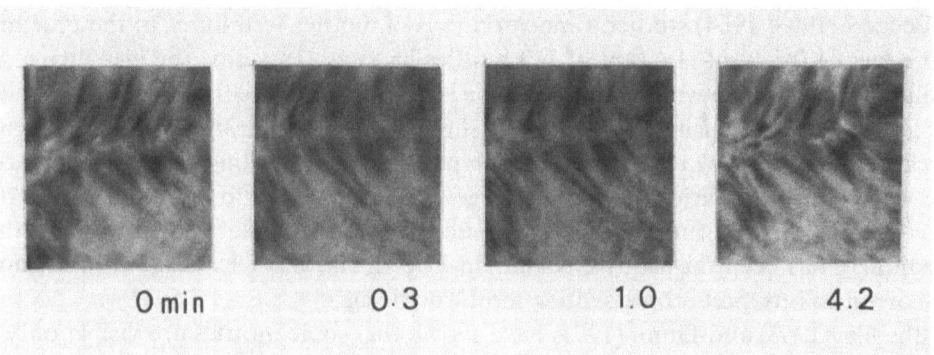

0 min   0.3   1.0   4.2

Fig. 5.   Example showing the double mottles at Hα −0.5 Å. Note the stretch and shrink with time. One side of each picture corresponds to a scale 25″.

though the occurrence is not always periodic. The lifetime of this system ranges from 10 s to 15 min with the average of 40 s. Their evolutions are shown in Figures 6 and 7.

In the final part of this paper I try to classify patterns of the evolution of mottles in the quiet region. For data I used filtergrams taken at Hα −0.3 Å every 10 s for 40-min duration. Table II summarizes distribution of numbers of various phases which can be found at the position of a mottle; the numbers are the averages of 70 mottles out of 10 clusters. Phases of a mottle include (1) double mottles system, (2) multistreaks or several faint dark streaks parallel to each other, (3) one or more roundish knots, (4) diffuse cloud-like stage, (5) short dark mottle or a group of short segmentary mottles with various orientations. Among these various phases the double

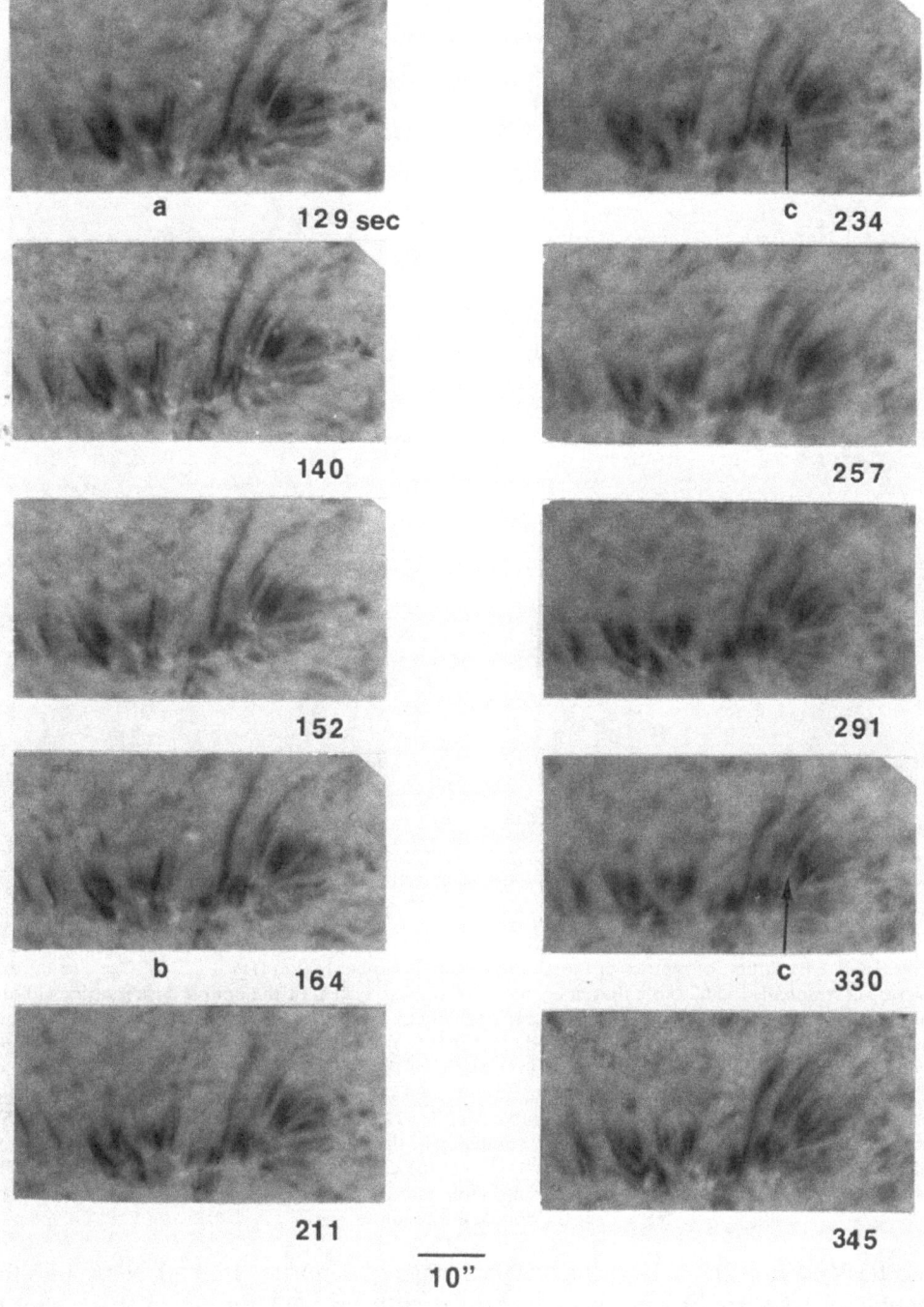

Fig. 6.   Evolution of mottles in semi-active region (Hα −0.7 Å). (a) double mottles, (b) widening of the double mottles, (c) the double mottles (234 s) splits into the narrower double mottles (330 s). This may be a newly emerging double mottles system. See text for discussion. See many double mottles at the frame 345 s.

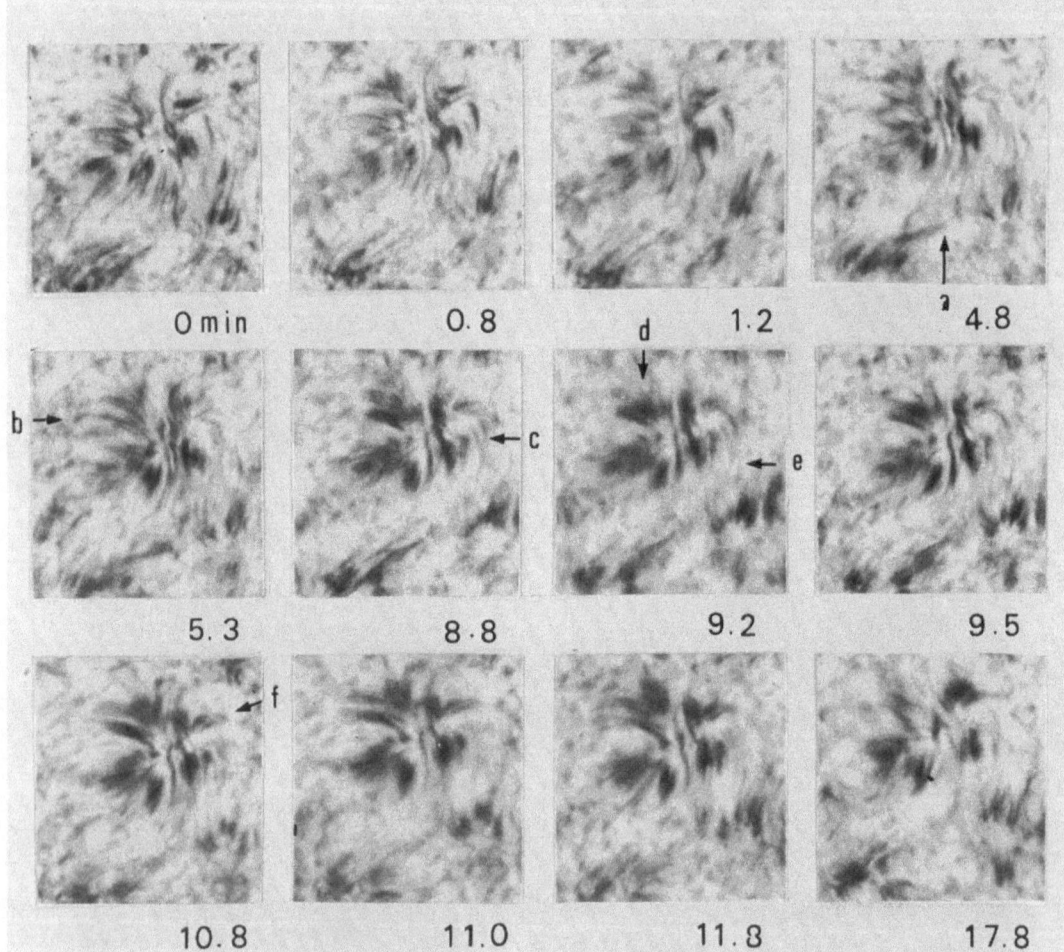

Figs. 7a–b. Evolution of mottles in the cluster (rosette) in quiet region (Hα −0.3 Å). One side of each picture corresponds to 30″. Note that many mottles exhibit a system of the double dark mottles with the bright gap (bright elongated mottle); for example (a), (b), (j), (k), (l) etc. (a) typical example of bright elongated mottle stretching through the double dark mottles, (b) three sets of the double mottles, (c) a bright point (8.8 min) elongates slightly (9.5 min), (d) transfer of the diffuse matter between neighbour mottles (8.8 to 9.5 min), (e) and (i) faint connection between two centers of the clusters, (f) connection of the broken mottles (10.8 min) to the curved double mottles (11.0 min), (g) stretching of the curved double mottle, (h) transient short double mottles, (j) stretching of the double mottles (20.0 min) from the broken stage of mottles (19.2 min), (k) apparent lateral shift of the double dark mottles (22.5–23.7 min), (l) bright point grows at 25.0 min, extends through the double mottles (26.3–27.0 min). Note also the stretch of the double dark mottles.

mottles system is the only well-developed stage of a mottle although sometimes the double mottles are hardly resolved. Also it matches with the phase that a mottle stretches and becomes dark. Stages from (2) to (5) occur in general in the intermediate phases between the double mottles stage.

On the whole there do exist some features corresponding to (1) to (5) when one

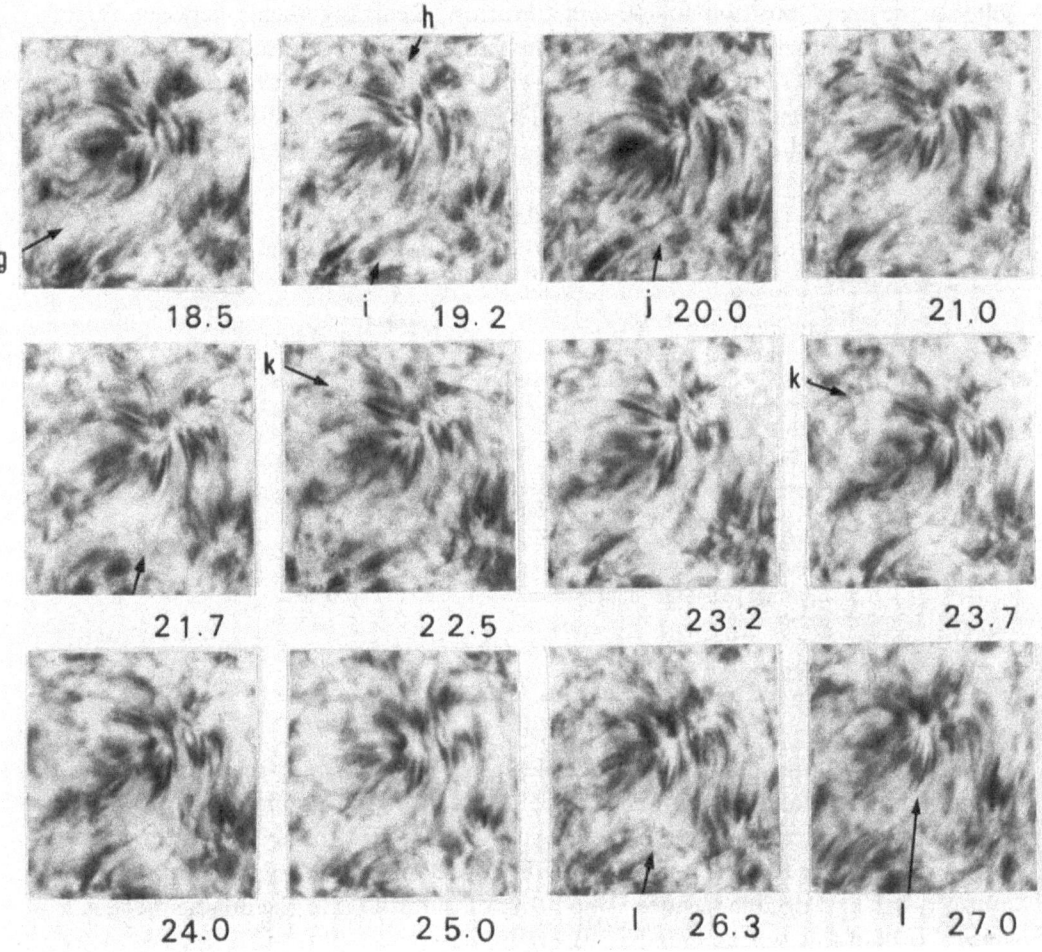

Fig. 7b.

TABLE II

Mean frequency of various phases of a mottle at one position during 40 min

	Double system	Multi-streaks	Knots	Diffuse	Short
Quiet	9.8	1.7	1.4	4.4	3.2
Semi-active	21.9	5.7	3.2	7.2	7.7

	Slow stretch[a]	Darkening
Quiet	6.2	3.1
Semi-active	7.3	3.8

[a] Visually recognizable stretch with velocities smaller than 500 km s^{-1}. See text for the definitions of various phases.

follows the fixed position for 40-min duration. Rearrangements between various phases occur almost continuously with the shortest time scale equal to the time resolution namely 10 s. Table III shows the classifications of the evolutionary patterns to and from the double mottles system. The pattern which occur most frequently is the formation of the double mottles from pre-existing streaks, knots or small segments

TABLE III

Phases before evolving to the elongated mottles

A.	Multi-streaks	20.5%
B.	Multi-knots	18.5
C.	Short scattered mottles	17
D.	Bright point → Bright elongated mottle with double mottles	15.7
E.	Dark point at the root	14.5
F.	Diffuse cloud	9.6
G.	Diffuse knot	4.8

Phases to which elongated dark mottle evolves

A.	Diffuse cloud	29%
B.	Break-up	25
C.	Splitting to another double mottles	19
D.	Shortening	11
E.	Multi-streaks	10
F.	Knots	7

(A, B, C, see Figures 6 and 7 for examples); in this case small, scattered mottles or knots suddenly become straightened or connected into the elongated mottles (double mottles). In some cases dark mottles are formed adjacent to the pre-existing mottles as though they are transferred to lateral direction. When this occurs one finds always a bright gap and double mottles. A splitting of a mottle into the double mottles is a special case of this type of change (see Figure 6).

For mechanism of sudden change of mottles visibility we may suggest (a) rapid change of the Doppler shift due to the oscillatory wave, (b) rapid condensation from pre-existing matter, (c) wave propagation along the axis of mottles. The last explanation is analogous with the evolution described below although the velocity of propagation larger than 500 km s^{-1} is required to explain the rapid formation such as restricted by the time resolution.

An important pattern of the evolution is the stretching of the double mottles from the central part of the cluster of mottles (D, E). Type D would be particularly interesting as it suggests some heat transfer from lower chromosphere to corona along the magnetic field line. A bright point appears first at the central portion of the cluster, increases its size and intensity and finally elongates through the double dark mottles, which develop together with the bright mottle elongation (see examples in Figure 7a, c, j, l and Figure 8). Occasionally one can recognize the top of the bright elongated mottle moving outwards with a velocity 20–40 km s^{-1}. Then it loses its contrast near the top of the double dark mottles as if it becomes too hot to be visible in Hα. The time scale from the brightening to the break-through ranges between 40 and 100 s. In the red wing picture we see a diffuse bright mottle appear at the intermediate position of mottles at 3.5 min after the break-through occurred in the blue wing (see Figure 8). This amount of the time lag is very common for the morphological similarity between the blue and red wings.

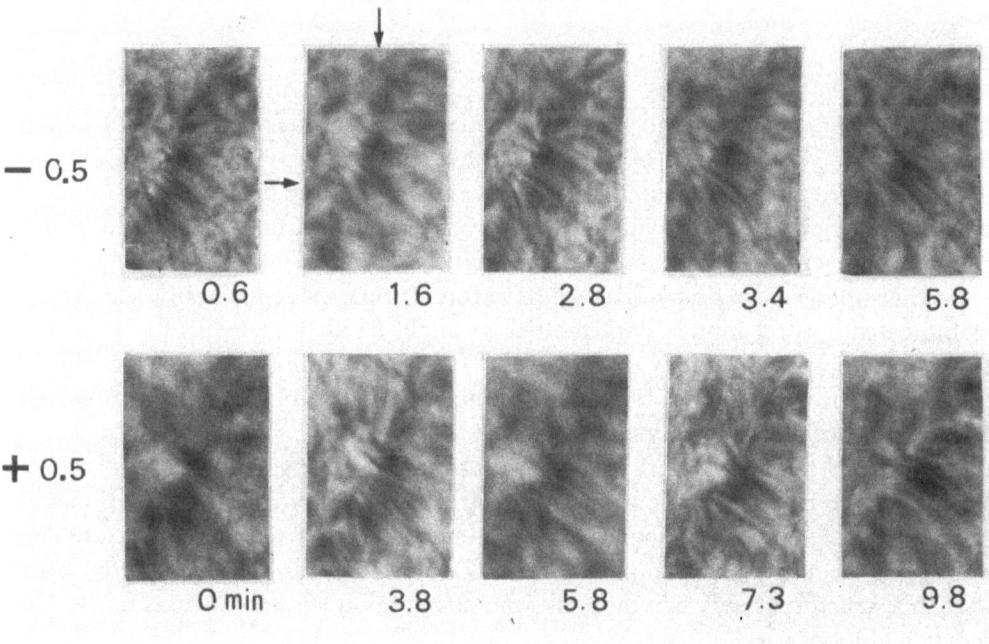

Fig. 8. Evolution of mottles seen in Hα −0.5 Å (top) and Hα +0.5 Å (bottom). In Hα −0.5 Å a bright point (arrow at 1.6 min) extends through double dark mottles from 2.8 min to 3.4 min. In Hα +0.5 Å bright feature appears at 5.8 min at intermediate position between the root and top of the diffuse dark mottle at 5.8 min. The double mottles with a bright gap are seen after 7.3 min and gradually fall towards the root. There is a time lag of 4 min between the appearances of the double mottles in Hα −0.5 Å and +0.5 Å.

The outgoing velocity of this phenomenon is similar to the typical velocity of the spicule. We may note that spicule spectra of Hα show the double structures like the double mottles although they appear usually in one wing. It is suggested that this phenomenon or bright mottle elongation through double dark mottles might be considered as a physically meaningful spicule phenomenon. I assume that a bright elongated mottle or bright gap represents a hot core of the spicule and the double mottles are a cold cylindrical envelope or two discrete features formed at the boundaries of the hot core.

Summarizing the evolution of mottles 30% are formed due to the stretching from shorter features and 50% are reformed from pre-existing materials. However there is no way to reject a possibility that the latter case denotes very fast stretch independent of the pre-existing materials. It is to be added that we often see apparent falling of the dark fine mottles in the movies taken near the limb. But this case has not been detected so often in this analysis made at disk center.

6. Concluding Remarks

We have seen that the mode of evolution is different between the cell interior and the supergranule boundaries. In the cell photospheric oscillation extends to the chromosphere with the period being modified. It accompanies individual feature called grains. The character of the grain, however, seems to be dual: (1) grains are associated with a large scale oscillatory wave and show group behavior, (2) in spectral evolution the grain is an individual event of vertical propagation of disturbance. The relation between these two images is still vague.

At the supergranule boundaries we have found some evidence of upward motion as indicated by stretching of dark mottles. It might be identified with the spicule although the observed velocity of 55 km s^{-1} seems too large for the spicule. Half of the mottles show rapid formation which cannot be explained by a velocity lower than 500 km s^{-1}. Alfvén wave may be invoked.

A system of the double dark mottles with the bright elongated mottle in between them seems to be physically one structure. Its evolution indicates close relation with the heat transfer to the corona. The higher temperature parts of this structure may be seen in the high resolution pictures of EUV and XUV ranges. It should be noted that this structure is very common also in active region features such as fibrils, filaments and loops.

Evolutionary studies of K and Hα line profiles are important for understanding the dynamical behaviors of the network cell. Such study is also necessary for the supergranule boundaries.

Studies on the evolution of individual fine structures at various levels would clarify the individual process of the generation of the disturbance in the granule, its propagation and development in the chromosphere and heat transfer to the corona. In particular we need to study the height between the granule and the mottle as well as the transition region between the chromosphere and the corona.

I wish to thank Dr H. Zirin for providing the data for this analysis and Drs Z. Suemoto, T. Hirayama, Y. Uchida, and E. Hiei for valuable discussions.

References

Athay, R. G.: 1970, *Solar Phys.* **12**, 175.
Banos, G. J. and Macris, C. J.: 1970, *Solar Phys.* **12**, 106.
Beckers, J. M.: 1963, *Astrophys. J.* **138**, 648.
Beckers, J. M.: 1964, Thesis, Utrecht; AFCRL Env. Res. Papers No. 49.
Beckers, J. M.: 1968, *Solar Phys.* **3**, 367.

Beckers, J. M.: 1972, *Ann. Rev. Astron. Astrophys.* **10**, 73.

Bhatnagar, A. and Tanaka, K.: 1972, *Solar Phys.* **24**, 87.

Bhattacharyya, J. C.: 1972, *Solar Phys.* **24**, 274.

Bhavilai, R.: 1965, *Monthly Notices Roy. Astron. Soc.* **130**, 411.

Bhavilai, R.: 1966, in K. O. Kiepenheuer (ed.), *Fine Structure of the Solar Atmosphere*, Franz Steiner Verlag, Wiesbaden, p. 96.

Bhavilai, R.: 1968, *Solar Phys.* **5**, 471.

Bray, R. J.: 1968, *Solar Phys.* **5**, 323.

Bray, R. J.: 1969, *Solar Phys.* **10**, 63.

Bruzek, A.: 1959, *Z. Astrophys.* **47**, 191.

Cram, L. E.: 1973, this symposium, p. 51.

de Jager, C.: 1957, *Bull. Astron. Inst. Neth.* **13**, 133.

de Jager, C.: 1959, *Handbuch der Physik* **52**, 80.

Dunn, R. B.: 1971, private communication.

Elliot, I.: 1969, *Solar Phys.* **6**, 28.

Foukal, P.: 1971a, *Solar Phys.* **19**, 59.

Foukal, P.: 1971b, *Solar Phys.* **20**, 298.

Grossmann-Doerth, U. and von Uexküll, M.: 1971, *Solar Phys.* **20**, 31.

Grossmann-Doerth, U. and von Uexküll, M.: 1973, *Solar Phys.* **28**, 319.

Howard, R. and Harvey, J.: 1964, *Astrophys. J.* **139**, 1328.

Janssens, T. J.: 1970, *Solar Phys.* **11**, 222.

Leighton, R. B.: 1963, *Ann. Rev. Astron. Astrophys.* **1**, 19.

Liu, S. Y.: 1972, Thesis, Maryland University.

Liu, S. Y.: 1973, Preprint (HAO).

Liu, S. Y. and Sheeley, N. R.: 1971, *Solar Phys.* **20**, 282.

Liu, S. Y. and Smith, E. V. P.: 1972, *Solar Phys.* **24**, 301.

Macris, C.: 1957, *Ann. Astrophys.* **5**, 179.

Macris, C. and Alissandrakis, C. E.: 1970, *Solar Phys.* **11**, 59.

Nakagawa, Y.: 1973, this symposium, p. 157.

Orrall, F. Q.: 1966, *Astrophys. J.* **143**, 917.

Pasachoff, J. M.: 1970, *Solar Phys.* **12**, 202.

Pasachoff, J. M., Noyes, R. W., and Beckers, J. M.: 1968, *Solar Phys.* **5**, 131.

Rogers, E. H.: 1970, *Solar Phys.* **13**, 57.

Sawyer, C.: 1972, *Solar Phys.* **24**, 79.

Simon, G. and Leighton, R. B.: 1964, *Astrophys. J.* **140**, 1120.

Suemoto, Z.: 1971, private communication.

Tanaka, K.: 1972, Report of Big Bear Solar Observatory, No. 125.

Wilson, P. R. and Evans, C. D.: 1971, *Solar Phys.* **18**, 29.

Wilson, P. R., Rees, D. E., Beckers, J. M., and Brown, D. R.: 1972, *Solar Phys.* **25**, 86.

DISCUSSION

Deubner: I would like to comment on the horizontal phase velocities of 100 km s^{-1} that you mentioned in your talk. It has been shown that these velocities are probably not real velocities and not even real phase velocities on the solar surface but can be explained by random phase relationships of individual elements, which are uncorrelated and pulsating on their own frequency and own phase.

Meyer: I would like to come back to the remark you made about the 180-s oscillation in the chromosphere and about the question of whether these are propagating or not. I recall when some years ago Dr Schmidt and I investigated the oscillations of the atmosphere we found one particular mode that is a little bit curious. It is a surface wave at the interface between the corona and the chromosphere. Such a wave decreases in amplitude rapidly with increasing depth but stays in phase and has a relatively high frequency. I wonder if there are observations that would fit to this type of mode.

Tanaka: What is the frequency of that mode?

Meyer: The range near 180 s would be acceptable. Waves of this type would be seen most prominently in the upper chromosphere near the corona interface. Perhaps Stix has a comment.

Stix: I do not believe that the 180-s oscillations are surface gravity waves because the wavelength of such waves is too short to fit with Dr Tanaka's observations.

EVOLUTION OF STRUCTURES IN
THE BRIGHT Hα NETWORK

DAINIS DRAVINS

Lund Observatory, Lund, Sweden

Abstract. We present off-band Hα filtergrams showing the bright Hα network (solar filigree) and its evolution, with subarcsecond resolution. In quiet regions the total area covered by elements in the network is constant over periods of hours but changes in the geometry are observed over 10 min (the timescale of granule and dark mottle lifetimes), that can be interpreted as due to (macro)turbulent horizontal velocities of about 1 km s^{-1}. The network can be resolved into strings of tiny bright dots; on the order of 100 per supergranule cell boundary. These dots may mark the location of magnetic filaments. The contrast of the features is higher in the blue than in the red wing which may indicate that in the lower chromosphere, the downdraft at supergranule boundaries does not occur in the bright crinkles themselves but rather adjacent to them.

1. Introduction

The bright photospheric network is cospatial with quiet region magnetic fields, at least when viewed with a resolution of about 2″ or worse (Chapman and Sheeley, 1968a, b; Sheeley and Engvold, 1970). The evolution of this network, as seen in spectroheliograms in the CN bandhead, has been studied by Sheeley (1969).

In off-band Hα photos the bright network can be seen, provided the seeing is good. Its properties when seen under very high resolution were first studied by Dunn and Zirker (1973) and following them we adopt the terminology 'solar filigree' for the bright network and 'crinkle' for the individual elements in the network. At Hα − 1.0 Å a typical width of the crinkles is about 0.5″ and further out from line center, like 2 Å, the widths of the structures are about 0.3″, which makes them very difficult to resolve. At about Hα − 1.0 Å they are coarse enough to be observable with moderate size telescopes, yet retain their identity in quiet regions where adjacent crinkles are typically separated by more than their widths. We are using a 25-cm telescope at Big Bear Solar Observatory with a Zeiss 0.25 Å bandpass filter. Since the structures are separated by distances much larger than their intrinsic widths, the image smearing in the telescope is determined (apart from seeing) by the point spread function (rather than the MTF). For a 25-cm aperture at Hα its full width at half-maximum (50% modulation) is half an arcsecond while the widths at 30% and 20% modulation are 0.4″ and 0.3″, respectively.

2. Observations

A sequence spanning 26 h of evolution is in Figure 1. This is a quiet region well removed from all sunspots and here we can recognize some larger structures from one day to the next; some but not all. The structures in the lower left seem to be quite similar on the 14th and on the 15th, while the complex in the upper part is

R. Grant Athay (ed.), Chromospheric Fine Structure, 257–261. *All Rights Reserved.*

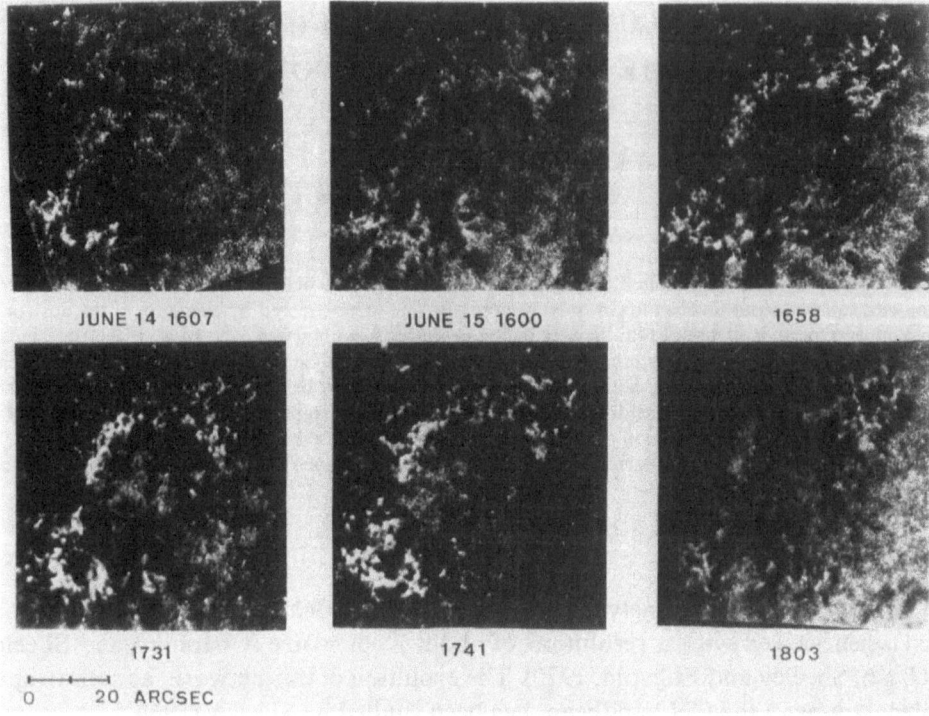

JUNE 14 1607   JUNE 15 1600   1658

1731   1741   1803

0   20 ARCSEC

Fig. 1.   Evolution in a field June 14–15, 1972. Filtergrams at Hα −1.0 Å (June 14) and Hα −0.9 Å (June 15). Numbers show times in UT. Scale is in arcseconds; 1″ = 735 km on the Sun. Other classes of features have been suppressed in the printing process.

only visible on the 15th. On the two frames taken during the best seeing (1731 and 1741 UT) we can see numerous changes. Figure 2 shows a small part of the region still better.

It is seen that:

(1) Changes have taken place.

(2) The total area taken up by the bright network has not changed significantly and the bright structures, although changed, still occupy the same regions in the frame.

(3) Some distinct features, like the 1000 km diameter ring-type structure in the lower part, have not changed very much and can be identified on both frames.

(4) There is no coherent flow pattern in the changes.

From this we may conclude that:

(a) For features in the subarcsecond network, changes take place on a timescale shorter, but not much shorter, than 10 min. This is similar to the timescales of granule and dark mottle lifetimes.

(b) That the total area covered by bright crinkles does not change, suggests that the features are the same ones that have been pushed around by some (macro)turbulent flow. An estimate of the average displacement of discrete features between the

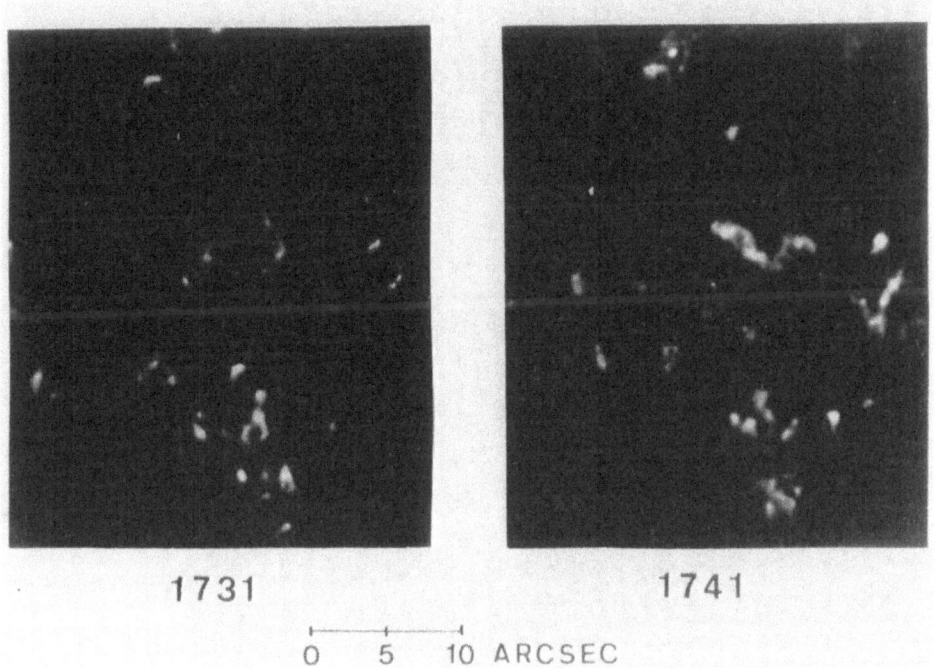

Fig. 2. Evolution in a small field on June 15. Filtergrams at Hα −0.9 Å.

frames in Figure 2 is about $0.8'' = 600$ km. Averaged over the time of 10 min this means an average horizontal velocity of about 1 km s^{-1}.

In contrast to this, the enhanced network around a growing sunspot shows a more complex evolution with brightenings and appearence of new features.

The filigree are more prominent in the blue wing than in the red, as can be directly seen by comparing filtergrams taken in different line wings. This, using a naive Doppler interpretation, is contrary to what would be expected if the downdraft at the supergranule boundaries took place in the bright crinkles in the network. One can interpret this as the downdraft taking place immediately adjacent to the crinkles. That dark chromospheric material will be redshifted and the crinkle will be more obscured by the downfalling material in the red wing, making it appearently brighter in the blue wing.

3. Intensity Structure Within Network Elements

In a negative print (Figure 3), the network structures appear dark, i.e. positively exposed, and this makes it easier to study the intensity fine structure within the network. In Figure 3 we see how patchy the bright network is when viewed under high resolution; indeed in many areas it is resolved into strings of discrete dots. The diameters of these dots are below 1″ or so. Along the borders of one supergranule cell we

Fig. 3. Negative print of an area surrounding a few very young sunspot pores near center, 1607 UT, 1972, June 14. Filtergram at Hα −1.0 Å, filter bandpass 0.25 Å. Area = 6 × 4.5′. This print shows the intensity structure within the network, much of which can be resolved into dots. The arrows point to some obvious cases.

estimate that there are on the order of 100 such dots. There is no obvious correlation between individual dots and the overlaying dark mottles.

It is tempting to identify the bright areas as locations of strong magnetic fields and an extrapolation to the dots would then identify them as discrete magnetic filaments. From studies of Zeeman-sensitive spectral line profiles, it has been argued that magnetic fields in quiet regions are channeled through narrow filaments of subarcsecond diameter (Frazier and Stenflo, 1972; Stenflo, 1973), which would be consistent with our interpretation. Also Sawyer (1971) observed small size dots on averaged filter magnetograms.

4. Possible Sources of Error

While the seeing as far as image sharpness is concerned, is good in some of our frames, this does not mean that image motion and geometrical distortions necessarily are negligable. Inspection of sequences including the present frames shows that there is an image motion present of some arcsecond also in times of good image sharpness. Does this affect our conclusions? A study of the frames in Figure 2 shows that it would not be possible to obtain one frame from the other by simple geometrical distortion, why we retain our conclusion on significant evolution there over 10 min.

A more difficult factor to estimate the effect of is that of mottle obscuration. If some dark chromospheric material moves into the line of sight to a bright element, that will be obscured and some previously obscured element may be displayed.

Acknowledgements

The observational data were obtained at Big Bear Solar Observatory (Hale Observatories; California Institute of Technology, Carnegie Institution of Washington) while the author was a visiting ESRO/NASA fellow there. I thank the staff there and also Dr J. O. Stenflo of Lund Observatory for several discussions.

References

Chapman, G. A. and Sheeley, N. R.: 1968a, *Solar Phys.* **5**, 442.
Chapman, G. A. and Sheeley, N. R.: 1968b, in K. O. Kiepenheuer (ed.) 'Structure and Development of Solar Active Regions', *IAU Symp.* **35**, 161.
Dunn, R. and Zirker, J. B.: 1973, *Solar Phys.* **33**, 281.
Frazier, E. N. and Stenflo, J. O.: 1972, *Solar Phys.* **27**, 330.
Sawyer, C.: 1971, in R. Howard (ed.) 'Solar Magnetic Fields', *IAU Symp.* **43**, 316.
Sheeley, N. R.: 1969, *Solar Phys.* **9**, 347.
Sheeley, N. R. and Engvold, O.: 1970, *Solar Phys.* **12**, 69.
Stenflo, J. O.: 1973, *Solar Phys.* **32**, 41.

DISCUSSION

Pecker: I think one should be very careful in assigning a velocity in the downward direction to a blue shift. It is by no means obvious.

Dravins: Yes, of course, it is difficult to infer such velocities, but I should point out that we observed far out in the line wings and are spatially resolving the features.

Bhavilai: I would like to know whether these small features have been observed in the line center for Hα.

Dravins: I haven't observed them.

Bhavilai: My impression is that the bright mottles at line center overlie the filigree.

Dravins: Yes these are bright regions in the line center. The filigree structure (with the resolution I have been using) becomes visible at about 0.7 Å from line center.

Hα FINE STRUCTURE AND THE DYNAMICS OF THE SOLAR ATMOSPHERE

FRANZ-LUDWIG DEUBNER

Fraunhofer Institut, Freiburg, G.F.R.

Abstract. Series of simultaneous spectra in different spectral regions, representing all 'visible' layers of the solar atmosphere, of high spatial and temporal resolution are used to compute power spectra and diagnostic diagrams of intensity and velocity fluctuations in order to study the dynamic response of the chromosphere to disturbances from below.

Although these spectra provide the highest spatial resolution presently available, most of the oscillatory power appears to be concentrated towards relatively low wavenumbers corresponding to wavelengthes of 12″. According to the periods observed, these waves must be evanescent ones. Therefore, internal gravity waves can be ruled out as a model for the 300-s oscillations, whereas the observations fit well with the diagnostic data predicted by Ulrich, Leibacher and, more recently, Wolff on the basis of their models of trapped subphotospheric sound waves leaking into the photosphere.

Nevertheless, on closer inspection of the raw, un-transformed spectral data (cf. Figure 1) one finds sufficiently well isolated events, which exhibit in detail the different features predicted by the 'local disturbance' theories: acoustic modes in a 'nearby zone' about 2″ in diameter, and internal gravity waves spreading further away from the center of disturbance. The energy involved in these events leaves no detectable traces in the diagnostic diagrams. Whether it merges into the larger scale 300-s oscillation energy remains unclear.

The observational material being completed by simultaneous slit jaw pictures (K-line and white light), it is possible to study the evolution of structures under the slit to some extent in three dimensional space. It is shown how a disturbance, starting as a 'granular ring' in the photosphere, propagates upward (cf. Figure 1 and 2) to the top of the chromosphere, where a 'spicule' is formed.

The non-uniform structure of the magnetic field plays an important role in this process.

At the borders of a supergranule, where the magnetic fields are usually strong and vertical, the rising material is guided upward and kept together or even concentrated by the surrounding field lines. Away from the supergranular network the chromospheric magnetic field is either bent in different directions (as the fine structure elements become more and more elongated) or generally too weak to govern the fluid motions.

In the first case the disturbance will be deflected, maybe still visible as a mottle, but in most cases not visible in the plane of the spectragraph slit. In the second case the disturbance will, from the photosphere, propagate freely in all directions (instead of one) and will, therefore, be not powerful enough to eject a spicule.

R. Grant Athay (ed.), Chromospheric Fine Structure, 263–266. All Rights Reserved.

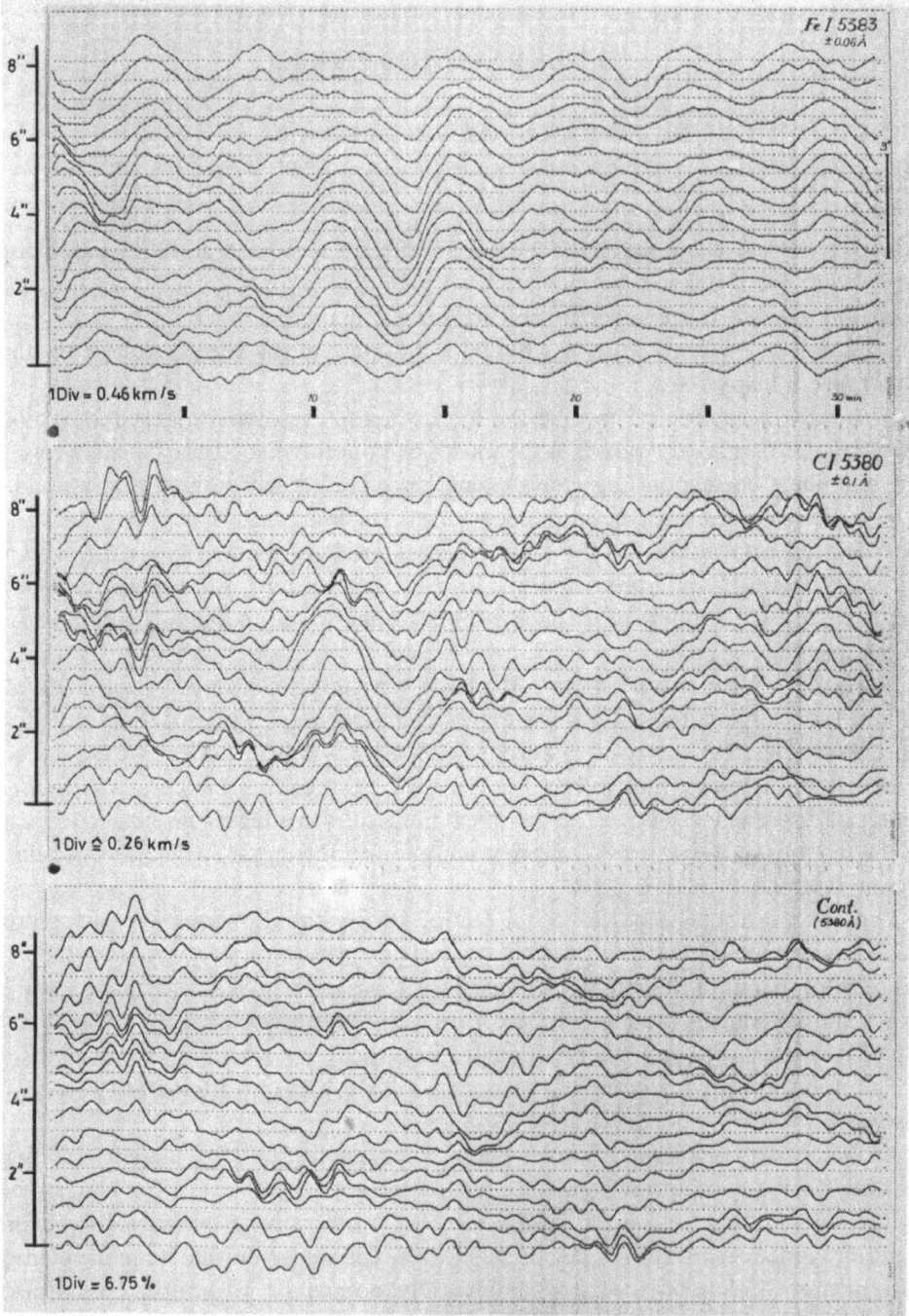

Fig. 1.

Fig. 1–2. Continuum intensity and velocities derived from line shift measurements in photospheric and chromospheric lines of increasing height of formation. The relevant scales are given in the figure. A photospheric disturbance is shown, which starts ∼9 min after the beginning of the observations, propagates through all layers of the atmosphere exciting various oscillating modes in the lower chromosphere and finally produces in Hα an abruptly rising spicular structure.

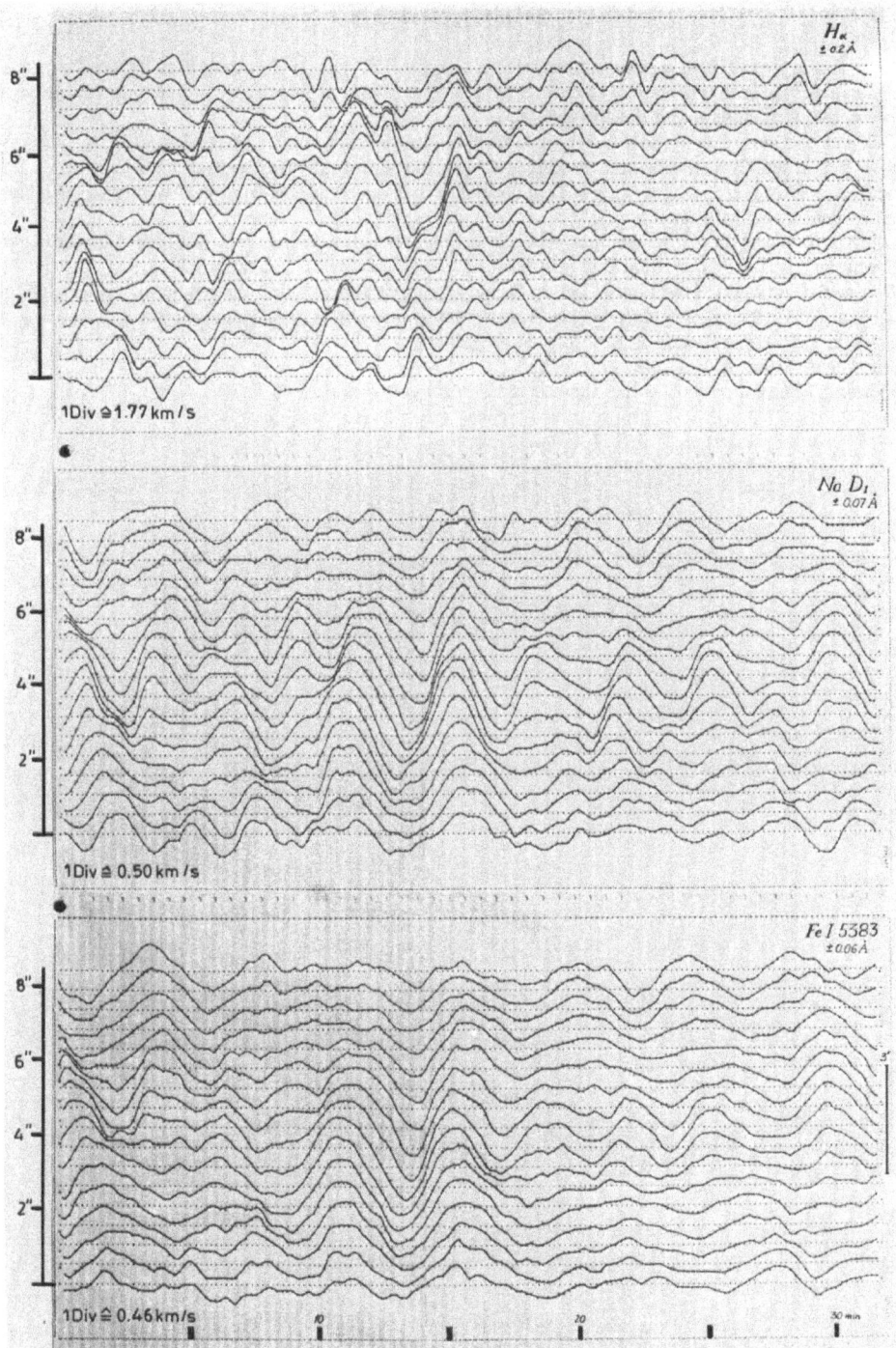

Fig. 2.

DISCUSSION

Athay: Exactly what is the mechanism that drives the spicule – what is the connection between the granule and the disturbance that drives the spicule? There is, after all, six orders of magnitude difference in density between the photosphere and the spicules.

Deubner: I think it is a shock wave that develops in the high layers. In the lowest layers it is just a spraying of a density variation that produces a wave that eventually, in the higher layers, evolves into a shock wave. However, perhaps it is not even necessary that we have a shock wave – we just get an enhancement of density at the higher layers.

Giovanelli: Can you tell me whether your slit jaw pictures show that the slit was indeed on the network and whether the particular event that you have described was in part of the network.

Deubner: The region I was investigating was a very quiet region so on the whole slit jaw picture there was hardly any network structure visible. There were photospheric gaps visible on the spectra, however. The feature shown happened close to one of the gaps.

Giovanelli: What do you mean by a gap?

Deubner: The filling in of a line that occurs in a magnetic knot.

PART VII

ENERGY BALANCE, HEAT TRANSFER AND HEATING MECHANISMS IN CHROMOSPHERIC FINE STRUCTURES

THE CHROMOSPHERIC ENERGY BALANCE

J. H. PIDDINGTON

CSIRO Division of Physics, National Standards Laboratory, Sydney, Australia 2008

Abstract. The problem of the solar atmospheric heat and energy balances is here divided into three sections.

(i) Basically the problem is an outward temperature gradient involving additional radiation losses whose estimate involves a model atmosphere. Proposed models described are based on optical and radio emissions together with various theoretical considerations such as thermal conductivity. The models are useful in providing estimates of the required flux of mechanical energy from below. However, they are limited by the use of averaging processes, and the failure to take full account of the many structural features which make up the chromosphere.

(ii) It has long been accepted that the main energy input is by acoustic waves generated in the 'small-bubble' Vitense model convection zone. We review recent observations of chromospheric magnetic and velocity fields and their interpretation which strongly suggest that this model is not valid and that the acoustic theory of heating is without basis. At the same time the new data remove the objections of Osterbrock and others to a theory of heating based on Alfvén and/or slow-mode hydromagnetic waves.

(iii) Observations of a variety of individual magnetic-plasma chromospheric structures, together with the adoption of heating by waves which follow the field lines, suggest a new approach to the whole problem of the heat balance. We discuss the heating problem in a number of these magnetic-plasma structures including the emission network, spicules and related disk features, arch filament systems and flares.

1. Introduction

Following the discovery more than three decades ago that the solar atmosphere is hotter than its surface, a clear objective was to determine its temperature and density distributions. Such a model had interest in its own right and also allowed estimates of heat losses and the energy input. Several models were developed using optical and radio data combined with a variety of theoretical considerations, but all were necessarily crude because they were based on measurements of emissions integrated along long lines of sight. Nevertheless they were, and still are, useful in providing averaged estimates of emissions and hence of heat requirements.

Great impetus was given to these studies with the availability of the extreme ultraviolet spectrum where the continuum is orders of magnitude weaker than in the visual spectrum, and the permitted lines correspondingly outstanding. Lines from the upper chromosphere and the transition region are now visible on the disk, and so much more detailed models are available. However, some integration along the line of sight and over a given area is still involved, and in addition there is an internal system of energy transfer in the new models involving thermal conduction and mass gas motions. Thus the new chromospheric models, although much more refined, are still very uncertain.

Meanwhile studies continued of the mechanism of heat input, long attributed to mechanical waves which originate in and below the photosphere. Following the classical study of Osterbrock (1961) acoustic waves have been favoured, and it is now widely accepted that the acoustic theory of solar heating has been established.

Some reservations about this theory stem from the very short wave period (< 30 s) which seems to be required, and the very strong concentration of heating observed in magnetic regions. However, it now appears likely that the whole basis of the theory, the small-bubble Vitense model of the convection zone, is in doubt. The discovery of supergranulation has led to the development of quite different convection-zone models, which are unlikely to provide the acoustic flux. The high degree of regularity of the polarization of surface magnetic fields appears inconsistent with tangled fields in the convection zone as required by the acoustic theory. On the other hand, the observations of strongly concentrated surface magnetic fields removes the previous objections to heating by an Alfvén-wave flux and clears the way for such a theory.

All of the above considerations suggest the necessity for an entirely new approach to the problem of the chromospheric heat balance. It is evident that magnetic fields divide the chromosphere into a number of largely isolated magnetic-plasma structures, and it now seems likely that each structure is heated by waves which travel up along the very field lines which define the structure. Some structures which invite study are the network of enhanced line emission, the spicules and their disk counterparts, arch filament systems and, finally and outstandingly, flares.

2. Chromospheric Models

Some two decades ago a popular pastime among solar physicists was the construction of models of the chromosphere. These were based on a variety of observational data and theoretical features, together with unrealistic simplifying assumptions. The most popular, and indeed necessary, assumption was that the chromosphere is a plane stratified atmosphere or, on the larger scale, spherically symmetrical; this implies that magnetic fields are absent or everywhere radial. Pictures of the chromosphere in the continuum and line emissions show that it is highly *irregular*, and the simplifying assumption has been attacked accordingly. In defence one might claim that the models give some sort of statistical average and provide useful estimates of the energy requirement, that the present-day attempts to evaluate the transition zone are based on much the same assumption, and finally that the structure is so complex that one must either adopt this averaging process or else study individual elements one by one. The last method is clearly our ultimate objective, but meanwhile averaged models serve a useful prupose.

Some of the first horizontally stratified models were those of Schatzman (1949) and Giovanelli (1949). In the former, energy was provided by the dissipation of shock waves and was removed by radiation, all other effects being ignored. The latter was based on an assumed constant conductive flux of 6×10^5 erg cm^{-2} s^{-1} through a chromosphere-corona transition region, all other possible sources and sinks of energy being ignored. The conductive model yielded a temperature (T) against height (h) gradient $dT/dh \approx 10^{-1}$ K cm^{-1} (near $T = 10^5$ K), and the shock model 10^{-2} K cm^{-1}. A model (Piddington, 1954) based on the interpretation of the observed radio spectrum of the whole Sun, together with some eclipse data, yielded a gradient of only $\approx 10^{-3}$ K cm^{-1}. The extreme case was provided by a model based on the assumption of hydrostatic equilibrium (Pottasch, 1964), for which $dT/dh \approx 10^{-4}$ K cm^{-1}. Thus the gradients determined were spread throughout the range 10^{-4}–10^{-1} K cm^{-1}. However, in none of these models were the simplifying assumptions clearly justified, nor was it evident just what was implied by the avaraging processes involved.

The next generation of chromospheric models was based mainly on the extreme ultraviolet (EUV) spectral line fluxes observed on the solar disk, and has been very adequately reviewed by Goldberg (1967), Athay (1971) and Noyes (1971). These data are both sensitive to the nature of the transition region and easier to interpret, because they do not involve very long line-of-sight integrations as do the optical limb observations and the disk radio observations. Because they yield large temperature gradients and consequently narrow transition regions with roughly isothermal regions above and below, the term 'model chromosphere' has been replaced by 'model chromosphere-corona transition region'.

2.1. SPHERICALLY SYMMETRIC SHELL MODELS

The classical picture of the transition region is that of a thin, spherically symmetric shell; implicit in this model is the assumption that the magnetic field is either vanishingly small or everywhere vertical. The development of this model from EUV data has been reviewed by Athay (1971), the region being defined by a gradient $dT/dh \approx 0.1$ K cm^{-1} and a product of electron density (n_e) and temperature $n_e T = 6 \times 10^{14}$ k cm^{-3} throughout. It is interesting to note that the model agrees with the simple conductivity model proposed by Giovanelli (1949). Athay found that the conductivity flux F_c is large compared with the radiation loss, and that this may explain the near constancy of F_c. The reason is that if F_c varied appreciably then some additional source or sink of energy within the transition zone would be necessary; no such source or sink was then apparent.

More recent investigations (Lantos, 1972; Piddington, 1972b) suggest that even as a first approximation this model may be inadequate, and that there *is* a simple mechanism for varying the conductivity flux as much as desired. This is a continuous upward expansion and flow of thermal energy carried by (a mass motion of) gas which has been heated by the downward diffusion of thermal energy F_c. The equation of energy for a static transition interface is $E_d = E_r + E_c$, where E_d represents mechanical wave dissipation and E_r and E_c radiative and conductivity losses (all per unit volume). If the assumption of a static atmosphere is abandoned we then have

$$E_d = E_r + E_c + E_f, \tag{1}$$

where $E_f = \nabla \cdot \mathbf{F}_f$ represent the upward flow of thermal energy carried by the expanding gas. The model of Lantos (1972) is based on electron densities given by radio measurements. That of Piddington (1972b) uses physical arguments and more clearly reveals the processes involved, and so is described here.

The thermal energy density of a fully ionized gas with n ions cm^{-3} is $3nkT$, and so for an upward expansion velocity $u_f(h)$ we have

$$E_f = \frac{dF_f}{dh} = \frac{d}{dh}(3nku_f T). \tag{2}$$

In the steady state, so that $nu_f = \text{const.}$,

$$F_f = 3nku_f T. \tag{3}$$

At the base of the corona $T = 10^6$ K, $n = 6 \times 10^8$ cm^{-3}, and if F_f balances a downward thermal flux $F_c = -3 \times 10^5$ erg cm^{-2} s^{-1}, then $u_f \approx 12$ km s^{-1} which seems a reasonable flow.

It may be of interest to examine the temperature profile of this model transition region within the limits where we might reasonably assume E_d and E_r unimportant. We then equate F_f to be downward flow of thermal energy, so that

$$3nku_f T - \sigma T^{5/2} \frac{dT}{dh} = 0, \tag{4}$$

where σ is a constant. Integrating again and rearranging we have

$$h - h_0 = \alpha T^{5/2}, \quad \alpha = \frac{2\sigma}{15nku_f}. \tag{5}$$

A reasonable value of σ might be 10^{-6} (see Noyes, 1971), in which case $\alpha \approx 1.3 \times 10^{-6}$. The corresponding parameters of the transition region are given in Table I.

TABLE I

A model corona-chromosphere transition region

T(K)	$h - h_0$(km)	$F_f = F_c$ (erg cm^{-2} s^{-1})	u_f(km s^{-1})
3×10^4	-2.2	10^4	0.4
10^5	0	3×10^4	1.2
3×10^5	640	10^5	3.6
10^6	1.3×10^4	3×10^5	12

These velocities might be varied a little to take account of the energy expended in lifting the corona against gravity. They must, of course, be changed if E_d or E_r are important. As they stand, the temperature gradients are 0.3 K cm^{-1} below the level of 10^5 K and 3×10^{-3} K cm^{-1} above that level. The former is in fair agreement with the EUV value, but the latter is much smaller, and close to the early radio model.

A rather obvious feature of this 'expanding' model is that it cannot be both steady-state and spherically symmetrical. The quiet solar wind represents a continuous mass flow from the base of the corona of $\approx 10^{13}$ atoms cm^{-2} s^{-1} (Kuperus, 1969) while the above flow is some 70 times larger. Presumably, then, most of the gas which moves up across the transition region must return again at another time or place. There seems to be no obvious objection to such irregular motions and, indeed, spicules themselves may represent an upward flow of $\gtrsim 10^{16}$ atoms cm^{-2} s^{-1} averaged over the Sun (Athay, 1971) which is an order of magnitude larger than the above requirement. The model simply provides an alternative mechanism for heat disposal to that of Kuperus and Athay (1967) and Kopp and Kuperus (1968) which invokes intermittent eruptions of gas instead of the stedy flow.

2.2. MAGNETIC FIELD EFFECTS

The more-or-less spherically symmetric chromospheric models discussed above are useful in providing semi-quantiative estimates of the various physical factors involved. However, observations reveal that the chromosphere is highly irregular and that the irregularities are caused mainly by magnetic fields which protrude through the solar surface. These fields introduce new density gradients, horizontal as well as vertical, they introduce new temperature gradients partly because the thermal conductivity remains appreciable only along the field lines, and they introduce a complex velocity field. In addition to all of these effects, magnetic fields may play a major role in transporting and guiding the mechanical-wave energy from the convection zone into the chromosphere.

On the basis of these considerations it seems likely that the chromospheric structure and energy balance may be described only in terms of numerous models of individual magnetic-plasma structures. Meanwhile, in reasonably quiet regions we may adopt an averaging process based on assumed vertical fields in the photosphere and the known supergranule motions which push the fields into the boundary regions.

The resulting magnetic structure is shown schematically in Figure 1 where the photospheric fields in the supergranule boundary regions (SBRs) have strengths of many hundreds of Gauss (Livingston and Harvey, 1969). Before reaching the corona, the field lines diverge to provide a uniform vertical coronal field of a few gauss. As the diameter of a supergranule cell is $\approx 3 \times 10^4$ km, the transition region is a relatively thin

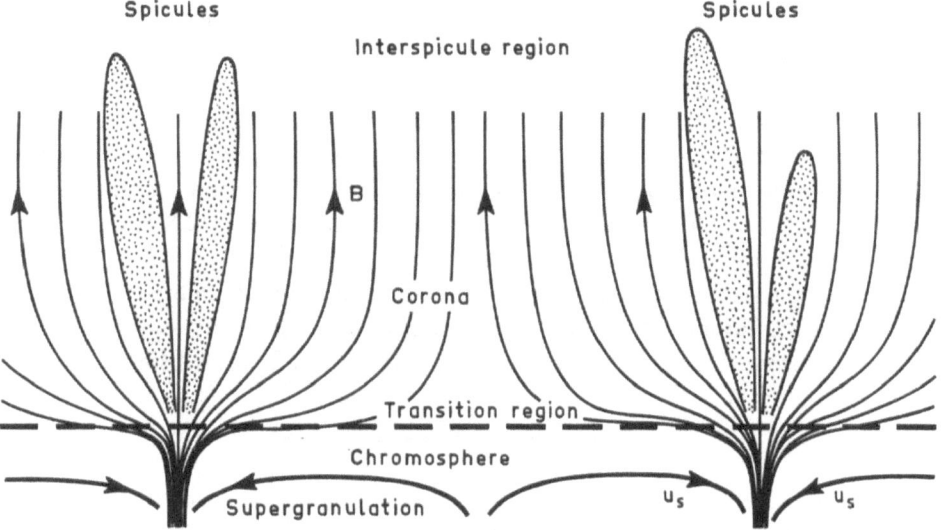

Fig. 1.   Schematic cross section of the solar atmosphere over a supergranule cell, showing the fluid flow lines u_s, the consequent bunching of the field lines at photospheric levels, and the locations of the chromospheric-coronal interface and the spicules.

sheet as shown by the dashed line; field lines traverse this sheet at a variety of angles. Spicules are confined to the SBRs as shown; theories of their origin are discussed in Section 4.

Kopp and Kuperus (1968) have used this magnetic model to determine a model transition region based on the simplifying assumptions of negligible wave dissipation and radiative heat loss. However, they also neglect the possibility of upward expansion of the plasma (E_f in our Equation (1)), and in addition their model provides a very thick transition region (average $\approx 10^4$ km) with most of the emission from regions above the centres of the supergranule cells, which is contrary to observational evidence (Tousey, 1971). For these reasons it seems that this model and that of Dubov (1971) must be rejected.

More recently Kopp (1972) has developed a quite different model based on the observational evidence (Tousey, 1971) that emission from the transition region is concentrated into a network which is the upward projection of the Ca emission network. If emission is confined to a fraction q of the disk then the transition region should be correspondingly thicker in order to provide the same total emission over the disk. This means that the temperature gradient must also be reduced and the thermal conductivity flux reduced by a factor q^2 to 5×10^4 erg cm^{-2} s^{-1}. Kopp finds a radiation loss which considerably exceeds this conductivity flux and concludes that the dissipation of mechanical wave energy must play a dominent role in the local energy balance. He invokes slow-mode hydromagnetic waves which have degenerated into shocks.

In summary, we consider that these various one- and two-component statistically averaged models have served a very useful purpose in indicating roughly the energy radiated, and hence the input required. However, as far as internal energy transfers are concerned, there are too many unknowns to make the models very useful. For example, if upward or downward gas drifts are possible, as seems likely, then they may settle the thickness of the transition zone. As seen in Table I, the temperature gradient below the 10^5 K level may be large, as in some of the earliest models. On the other hand, if the transition region proves to be very thin, then at any point on this thin sheet the direction of the magnetic field is the same at the top and bottom surfaces; there is no convergence of conductive heat flux as envisaged in the models of Kopp and Kuperus (1968) and Kopp (1972).

We consider that future study of this problem might best be aimed at *individual* magnetic-plasma structures, each of which must have different temperature and density structures. This approach is discussed and used in Section 4 below.

3. The Source of Thermal Energy

The only major source of the chromospheric heat input which is worth consideration is the dissipation of mechanical waves generated in and below the photosphere. There are three basic wave types, acoustic, Alfvén and gravitational, but the last two have been mainly rejected and it is now usually taken as established that the primary source of heating of the solar atmosphere is acoustic energy (for a review, see Kuperus, 1969), and that we are 'reasonably certain that there is a mechanical flux of at least 2×10^6 erg cm^{-2} s^{-1} associated with acoustic noise from the granulation' (Goldberg, 1971). The theory was developed by Lighthill (1952, 1954, 1967), Proudman (1952), Stein (1967, 1968), and others using the convection zone model of Vitense (1953) and Böhm-Vitense (1958). In this model hot bubbles of dimensions equal to a scale height move upwards continuously while cool gas moves downwards. The acoustic power so generated has been computed by Osterbrock (1961), de Jager and Kuperus (1961) and Kuperus (1969, where other references are given). The upward propagation of the sound waves and their dissipation has been discussed by Osterbrock, de Jager and Kuperus, and more recently by Ulmschneider (1967) and Kuperus (1969) who agree on a flux of mechanical energy of about 3×10^7 erg cm^{-2} s^{-1}. The energy required to replace radiative and other losses has been computed by Osterbrock (1961), de Jager and Kuperus (1961), Athay (1966), Ulmschneider (1970, 1971) and others with more recent estimates in the vicinity of 5×10^6 erg cm^{-2} s^{-1}, which is comfortably smaller than the acoustic flux available.

In the above theory, magnetic fields play important, but strictly secondary, roles. The observed enhancement of heating in regions of stronger magnetic fields is attributed to the tangling and amplification of the fields in the convection zone, and to the consequent increased efficiency of sound production (Kulsrud, 1955; Osterbrock, 1961; Kuperus, 1969). The theory also takes account of the conversion of sound waves to the fast hydromagnetic mode in those parts of the atmosphere where magnetic forces equal or exceed compressional forces.

The basis of the above acoustic theory is the Vitense convection model with rms bubble velocities ranging up to 2.3 km s^{-1} (see Osterbrock, 1961, Table 3) and assumed magnetic fields of strength $\lesssim 50$ G. More recent measurements of the surface magnetic and velocity fields seem incompatible with both of these features and suggest the necessity of a complete review of the theory of the origin of the chromospheric heat input.

3.1. VELOCITY AND MAGNETIC FIELDS IN THE CONVECTION ZONE

The acoustic theory of solar heating depends on gas bubbles moving upwards with velocity u; they have dimensions equal to a scale height H and so produce a spectrum of sound waves peaking at frequency $v = u/H$ (Osterbrock, 1961). Magnetic fields are 'passive' in the sense that they are too weak to influence gas motions, and so become tangled and thereby increase acoustic output by an order of magnitude in the regions of strong fields.

The above 'small-bubble' model of the convection zone was developed before the discovery of the supergranule motions and appears incompatible with those motions. There is now little doubt that the true convective motions extend over a number of scale heights, with consequent increase in efficiency (Simon and Weiss, 1968; Wilson, 1972b, who also reviews other models). These large-scale, much slower motions would not produce an acoustic flux comparable with that of the small, fast-moving bubble model. Thus the theoretical basis of the acoustic theory of solar heating is removed.

Observational evidence against the Vitense model and the acoustic theory is provided by the observed surface magnetic fields. The criterion for equipartition of magnetic and kinetic energy within the convection zone is

$$4\pi\varrho \langle u^2 \rangle = B^2, \tag{6}$$

where ϱ is the gas density, $\langle u^2 \rangle$ the mean square convective velocity and B the field strength. This yields $B = 450$ G at a depth of a few hundred kilometres (Osterbrock, 1961), so that weaker fields are tangled and fields up to $\gtrsim 1000$ G should be strongly twisted. Also, since the observed surface fields have scales ranging from ≈ 1000 km up to $> 10^5$ km, they must have the same pattern as fields only $\gtrsim 500$ km below the surface.

The observed surface fields are completely incompatible with the convection model and the acoustic theory. Most of the flux is concentrated in more-or-less vertical flux tubes with dimensions ≈ 1000 km and fields $\gtrsim 1000$ G (Harvey, 1971, Vrabec, 1971 and others). To some extent these are controlled by the supergranule motions, but there is no evidence that they are tangled or even strongly tilted by any smaller-scale motions. One might argue that there are weaker fields, not observable, that are tangled; this is not possible, because after tangling they would be squeezed and amplified by the supergranule motions and so become visible. Finally, we refer to the largest-scale surface magnetic feature, a field of a single polarity extending for $\gtrsim 10^5$ km (Vrabec, 1971, Figure 4).

These results appear to rule out completely the assumption made by Kulsrud (1955), Osterbrock (1961), Parker (1971) and many others that the fields in the convection zone are passive and are tangled by the gas motions. It also invalidates the claims of Parker (1971) and Weiss (1966) that after the fields are tangled, the small-scale components are eliminated by magnetic annihilation across neutral sheets. Even the earliest stage of tangling would be observable in the surface fields at some period or other in the life of an active region. Also, as shown earlier (Piddington, 1972a, 1973a), only components of scale $\gtrsim 100$ km can be eliminated by magnetic diffusion, which leaves all larger fields to be eliminated by the ubiquitous neutral sheet – a highly improbable assumption.

As far as the acoustic theory of solar heating is concerned, there remains the possibility that it may be modified by basing it on the observed granule motions rather than the Vitense model. However, this suggestion meets two major difficulties: it still leaves unexplained the observed close correlation of heating with magnetic fields, and it is quantitatively inadequate. The observed velocities ($\gtrsim 1$ km s^{-1}), cell sizes ($\gtrsim 1000$ km) and periods (≈ 300 s) yield acoustic waves which are too weak by a factor of > 100 (Piddington, 1973c) and with much too long a period to suit Ulmschneider's (1970, 1971) requirement of $\gtrsim 30$ s.

3.2. THE SOURCE OF CHROMOSPHERIC HEAT

From the above arguments it appears that the theoretical basis of the acoustic theory of chromospheric heating no longer exists, and that the theory is also incompatible with a great deal of observational evidence. Accordingly we consider the possibility of heating by one of the other wave modes (Piddington, 1973c).

Gravitational waves may only develop in stably stratified regions, which excludes the convection zone (Lighthill, 1967). It is possible that some convective motions project locally into the overlying stable regions as 'tongues of turbulence' and that these generate gravity waves. However, as these tongues have densities orders of magnitude less than the convection zone gases, and velocities equal to or less than those of the Vitense model, it seems unlikely that the necessary power would be provided. A further argument against gravity waves is the observed close relationship between heating and magnetic fields. This has no obvious explanation in terms of gravity waves alone.

This brings us to the Alfvén mode which was suggested some years ago (Piddington, 1956) as a source of chromospheric heating. The mode was rejected because it was generally accepted that magnetic fields at the solar surface are generally weaker than 50 G in which case the waves would be mainly absorbed (Osterbrock, 1961). In addition, the Alfvén speed in the convection zone is $\gtrsim 0.2$ km s^{-1}, or 0.1 of the Vitense turbulent speed, in which case the predicted Alfvén energy flux is negligible (Lighthill, 1967). It is now known that the first of these objections is not valid and that the second is most doubtful, so that Alfvén waves (and slow-mode waves) should be reconsidered.

A satisfactory theory based on these modes requires an established convection zone model and this is not yet available. This deficiency may be partially rectified by the use of observed surface motions which have been used to provide two sub-models of chromospheric heating. The first, described below, is based on observed granule motions; the second (Section 4.2) is based on observed supergranule motions. In addition to a velocity field, the heating model requires a magnetic field structure, and here we are on firmer ground. It is now known that almost all surface flux is concentrated into small ($\gtrsim 700$ km), more-or-less vertical flux tubes with field strength of order 1000 G (Chapman and Sheeley, 1968; Livingston and Harvey, 1967; Frazier and Stenflo, 1972, and others).

The sub-model based on granule motions is illustrated in Figure 2a where a single granule is shown in vertical section. The vortex motion V_g distorts an originally straight, vertical flux tube B as shown and an Alfvén wave must propagate upwards. Values of V_g have been discussed by Bray and Loughead (1967) who point out that the measurements give at best only a rough indication of order of magnitude of V_g. For the purpose of illustration we assume an rms velocity of 1 km s^{-1} in the low photosphere ($\tau_{5000} = 1$) in the presence of a field of 500 G. With density $\varrho = 2 \times 10^{-7}$ g cm^{-3} we have $V_A \approx 3.2$ km s^{-1} and an energy flux $F = \varrho V_g^2 \, V_A \approx 6.4 \times 10^8$ erg cm^{-2} s^{-1}. This flux is some 200 times the heat requirement of the average chromosphere, and so takes account of the fact that fields of 500 G occupy only a corresponding fraction of the surface.

The dissipation of Alfvén waves provides two distinct problems. In and below the chromosphere the flux tubes which transport the energy are of relatively small cross section, and are separated by regions of much larger extent. The flux tubes move like taut wires vibrating in a compressible fluid, and so are likely to cause strong localized heating outside the tubes. In the corona the field lines spread out and merge to form a more-or-less uniform medium as shown in Figure 1, and strong dissipation is likely only after the formation of shocks (Osterbrock, 1961).

Different aspects of the vibrating-wire mechanism are illustrated in Figure 2b

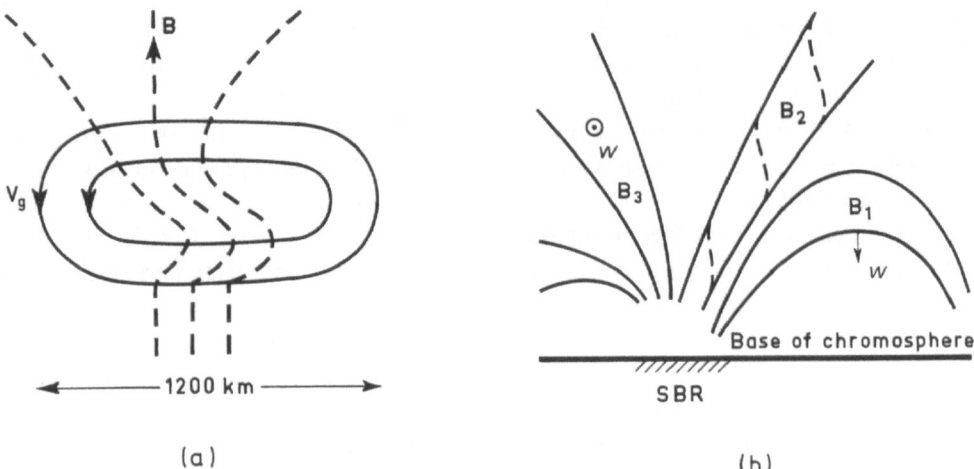

Fig. 2.   Solar heating by Alfvén waves. (a) Generation of waves in mainly vertical magnetic field lines B by the rotation of a granule cell. – (b) Several flux tubes emerging from a supergranule boundary region (SBR), and providing optical emission patterns which depend on the direction of the perturbation velocity u.

where three flux tubes are shown emerging from the photosphere. It may be assumed that they have been concentrated so by the supergranule motions. The tube B_1 loops back into the photosphere and at the moment illustrated the upper part of this loop is moving downwards with velocity u. The gas below the tube will be compressed and will provide enhanced emission; later, when the tube moves upwards, the gas above the tube will emit more radiation. The tube B_2 illustrates the possibility of a helical twist, strongly suggested by the forms of some chromospheric features, and capable of energy transmission by a twist Alfvén wave. The tube B_3 is moving into the paper and the enhanced emission will show a red Doppler shift; during the second half of the cycle this will change to a blue shift. Other possible effects include the formation of a shock on one side of a tube when u exceeds the sound velocity, the collision of adjacent tubes and, finally, different gas densities inside and outside a tube. Each of these effects may show up as more, or less, complicating factors in the interpretation of line profiles.

4. Structural Components of the Chromosphere

The evolving picture of the chromosphere is that of many individual structural components, each of which is undergoing continuous changes. It follows that the chromospheric models discussed in Section 2 are statistically averaged models whose physical nature is not clear. A true physical picture of the chromosphere and a knowledge of its energy balance thus requires satisfactory models of each of the various structural components.

The basic structural components seen on the disk have a variety of names including bright and dark, fine and coarse mottles, fibrils, filaments and threads, and there is not always agreement about just what object is referred to. Without entering this controversy, we suggest that the vibrating wire model described above is capable of providing a variety of bright and dark structural components either elongated if the field is orthogonal to the line of sight or small blobs if along the line of sight. Collectively, these components define the supergranule boundary regions (SBRs) and are termed the emission network. The energy balance of the network is discussed in Section 4.1.

The most notable structural component seen on the limb is the spicule and this, with its related phenomena, is discussed in Section 4.2.

Turning to active regions, we again find that the various structures correspond with and are presumably determined by magnetic fields. Again we find the various network components (mottles, fibrils and so on), but with sufficient magnetic flux these may fill supergranule cells instead of being mainly confined to the SBRs. Another difference is that while the magnetic ffields in quiet regions are predominantly vertical, those in active regions may have substantial horizontal components. The latter form closed loops within the chromosphere and low corona, often referred to as arch filament systems and discussed in Section 4.3.

Finally, in Section 4.4 we discuss the two most notable of all structural features: sunspots and flares. The sunspot energy deficit and the flare energy requirement constitute the two largest components of the energy flux.

4.1. THE EMISSION NETWORK

The fact that the lower chromosphere, as observed for instance in the H and K lines of Ca II, exhibits a complex network of enhanced emission has been known for many decades. On the other hand, the extension of this network into the chromosphere-corona transition region has been observationally established only recently by high-resolution EUV spectroheliograms (Tousey, 1971). From such data it appears that the network persists at least to the 2×10^5 K level, and an analysis of OSO-4 data by Reeves and Parkinson (1972) suggests that it may extend into the low corona where $T \gtrsim 2 \times 10^6$ K. Confirmatory evidence of coronal enhancement at a much lower level of resolution is provided by the earlier results of Hansen et al. (1971) who measured the K corona above plages.

These results are extremely important in the theory of the heat balance because they prove that the energy emerging from the photosphere is dissipated mainly in the SBRs and the regions of strong, more-or-less vertical magnetic fields (Figure 1). This result is entirely consistent with the main conclusion of the last section that the energy concerned is carried by Alfvén and/or slow-mode hydromagnetic waves, both of which tend to follow the field lines.

The energy requirement of the network must exceed that of the average chromo-sphere by some factor, and using He 10830 spectroheliograms de Jager and Loore (1970) found a factor of 7. As this seems to agree with the ratio of the areas concerned it is acceptable and gives an energy requirement for the network of roughly 3×10^7 erg cm^{-2} s^{-1}.

However, this requirement is insignificant in comparison with that of the usually neglected photospheric or white-light faculae. Their requirement was recently con-firmed by Wilson (1971) as about 10^{10} erg cm^{-2} s^{-1} or some 3000 times that of the higher levels, so that a theory of heat balance may be dominated by these faculae. It is unlikely that they are a direct result of enhanced convection because, as is well known, gas motions in the network are generally downwards. If they are caused by mechanical waves then they pose a new problem in the determination of the energy balance. The energy flux required is some 300 times that estimated for acoustic waves (Kuperus, 1969), and so provides further evidence against that theory. However, as shown in the following subsection, this flux is not beyond the capabilities of a magnetic flux tube with field strength of order 1000 G.

Wilson (1972a) has already proposed that some of these faculae may be provided by an Alfvén energy flux through sunspots. This will not account for all of them because some are present in active regions before and after the spots, and some occur in polar regions where there are no spots.

4.2. Spicules and related phenomena

The solar disk features discussed above are likely to have equivalent limb features, and of these the spicules are the most notable. Beckers (1968, 1972) has provided a comprehensive review of their properties, the first being an upward growth rate of ≈ 30 km s^{-1} and Doppler shifts with rms values $\gtrsim 20$ km s^{-1}. This suggests that spicules are upward jets and may be an important factor in the energy balance, either directly by the upward transport of kinetic energy or indirectly through control of the plasma mass balance. As seen in Section 2, there appear to be upward plasma streams representing ≈ 100 times the solar wind loss and so there must be downward streams representing an almost equal flow. These streams are also likely to be very important in the heat balance.

Models of spicules have been reviewed by Beckers (1968, 1972) who rejects those not depending on magnetic fields because of the compelling association. All of the models attempt to describe an upward-moving supersonic gas jet, but such a model meets difficulties when spicules are identified with disk features. There is little doubt that during different phases spicules correspond with both dark and bright fine mottles (Beckers, 1968), so that one should see blue Doppler shifts of $\gtrsim 20$ km s^{-1} from many of these features. Observational results have been reviewed by Beckers and further analysis made by Grossmann-Doerth and von Uexküll, 1971) which show line-of-sight velocities in mottles of only ≈ 4 km s^{-1}, the average being slightly red-shifted.

A possible explanation of this discrepancy may lie in the Doppler-shifted line profiles (Beckers, 1968, Table XII). The large rms values found are largely due to the high-velocity tails to the distribution. If we exclude the measurements made in active regions, because these may involve a somewhat different phenomenon, the *mean* velocities lie in the range 2.5–8.0 km s^{-1}. Now the disk measurements of mottles must refer to the lower, denser part of the spicule, and this moves up or down with mean velocity ≈ 4 km s^{-1}. We suggest that the high-velocity tail seen at the limb relates to a higher level, where the gas is moving upwards or downwards with velocities $\gtrsim 25$ km s^{-1}. The actual direction depends on whether the spicule is tilted away from or towards the observer and is not known. Accordingly, we have suggested (Piddington, 1972b) that spicules involve an upward gas motion averaging 2×10^{15} atoms cm^{-2} s^{-1} averaged over the solar surface (200 times the solar wind) followed by an equivalent *downward* flow. The latter comprises a flow of ≈ 20 km s^{-1} in the upper parts of a spicule (density 10^{11} atoms cm^{-3}) falling to 2 km s^{-1} in the lower parts. There is also evidence of associated wave motions propagating upwards with speeds of 400 km s^{-1} (Beckers, 1968).

These and other considerations have led to a model of spicules, polar plumes and quiet-region faculae (Piddington, 1972b) illustrated in Figure 3. The flux tube B is

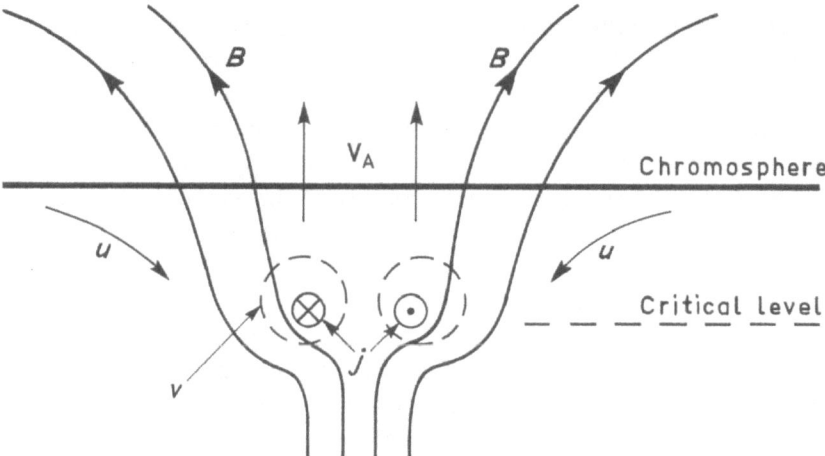

Fig. 3.   A magnetic field configuration B following hydromagnetic collapse caused by converging gravitational-acoustic waves u. Above a critical level the field resists the plasma pressure, but a few scale heights lower it is strongly compressed. An inner current loop j introduces a Lorentz force with a vertical component. As shown by Altschuler *et al.* (1968), a vortex ring v moves upwards with the Alfvén velocity. It is invoked to explain spicules, polar plumes and heating of the quiet chromosphere and corona.

located in an SBR and is a detail of the field of Figure 1; the likely field strength indicated by observations is $\lesssim 1000$ G. The velocity field has a variety of components including the supergranule cell motion, the 300-s motion (Frazier, 1968), and the granule motions which may combine to provide localized velocities with peak values equal to the sum of the components, perhaps as much as 3 km s^{-1}. The vertical flux tube will easily withstand these motions in the chromosphere, but below a critical level given by Equation (6) the motions shown will compress the tube. For a field of say 500 G the critical level is at $\varrho \approx 2 \times 10^{-7}$ g cm^{-3} a little below that of unit optical depth. Since the scale height is only ≈ 200 km and the horizontal scale much larger, the transition layer is relatively thin and the field lines are bent sharply.

The hydromagnetic configuration becomes non-linear and extremely complicated so that only a qualitative discussion is attempted. However, there appear to be some interesting effects involved. First, there is the nature of the gas compression. which depends on the radiative relaxation times. At the lowest levels concerned, these times are less than 1 min (Ulmschneider, 1971) so that compression is nearly isothermal and the ratio of magnetic to thermal energy density increases in proportion to B. The relaxation times increase rapidly upward, so that above the critical level (Equation (6)) compression is adiabatic and the gas resists compression more effectively. This further narrows the region of transition from gas to magnetic domination.

Gas flowing into the SBR is blocked above the transition region but continues onward at lower levels. The likely result is that some gas flows downwards from the upper layers through the transition layer. For a fall of only 200 km the gravitational energy gained per unit volume of the original gas (mass 2×10^{-7} g) is 10^5 erg which is that of a field $B = 1600$ G. The fall may well be several hundred kilometres, particularly in view of the fact that the original supergranule motion is mainly downward near the boundaries. Thus fields of several thousand Gauss seem possible, particularly as we envisage inflow of gas from all sides to compress a cylindrical force tube.

This hydromagnetic collapse can neither continue indefinitely nor attain a stable state. The deformation implies the introduction of a current loop j shown in section in Figure 3, and when combined with the local, mainly horizontal, field the result is a mechanical force of density $f = j \times B$ which is mainly upwards. In that direction the inertial force opposing expansion is provided by a layer of thickness only one scale neight, and so an eruption appears likely.

A similar non-linear hydromagnetic disturbance has been analysed by Altschuler *et al.* (1968) as a model of surges. The model is necessarily simplified by the assumption of incompressibility, and a numerical solution is obtained over the period taken for the transfer of magnetic to kinetic energy until equipartition is achieved. They show that the magnetic perturbation propagates upwards at approximately the Alfvén velocity V_A. The whole perturbation corresponds to two current rings, one of which is shown in our Figure 3. This ring moves upwards and at the same time contracts towards the axis of symmetry. The second current ring is not shown here, but it is obvious that it must lie below and outside the ring marked j, so as to account for the lower bends in the field lines. This ring moves away from the axis as the field lines straighten, and needs no further discussion here.

At the base of the photosphere ($\varrho \approx 2 \times 10^{-7}$ g cm^{-3}) a field of 500 G provides upward velocities of $V_A \approx 3$ km s^{-1}. For corresponding gas mass motions, $u = 3$ km s^{-1}, the upward energy flux is $F_d = \varrho u^2 V_A \approx 5 \times 10^9$ erg cm^{-2} s^{-1}. This is roughly the requirement of the photospheric faculae, which may absorb most of this energy flux. A small residual is ample to account for the estimated spicule kinetic energy of $\gtrsim 10^7$

erg cm $^{-2}$ s^{-1} (Athay, 1971) and the remainder of the network radiation and other losses.

4.3. MAGNETIC-PLASMA STRUCTURES IN ACTIVE REGIONS

Observational results have been summarized and extended by Frazier and Stenflo (1972) to show that almost all of the non-spot magnetic flux in active regions is concentrated into thin filaments of extent $\lesssim 1000$ km and strength $\lesssim 1000$ G. When the field strength is substantially greater, perhaps 1500 G, pores and spots form.

The extension of this pattern to three dimensions is effected by measuring both longitudinal and transverse magnetic fields (Severny, 1965) and by observing the evolution of an active region from its first hours (Weart, 1970; Harvey, 1971; Frazier, 1972 and others). It appears that flux tubes erupt through the photosphere over a period of a day or so, and organize themselves into an east-west bipolar configuration. Each flux tube forms an arch filament with two feet held firmly by the gas in and below the photosphere. The supergranule motions move the feet about, and if they separate further then the filament grows accordingly; their dimensions range threfore from zero up to the size of the whole active region. Thus the arch fflament system (AFS) appears to be the main structural feature of a spot-free active region, and so is likely to be the controlling factor in the energy balance.

A fully developed arch filament system constitutes a dome-shaped magnetic field which projects into the chromosphere and corona. As the strands emerge from the photosphere they lift cool plasma through the chromosphere, but at the same time plasma appears to flow down the field lines (Frazier, 1972) and presumably some collects at the feet of the arch. A second feature of such a group of AFSs is that some are more luminous than the surrounding chromosphere (Weart, 1970) and so may represent concentrations of hotter or more dense gas. A third feature of the dome field is provided by radio observations of the slowly-varying thermal component (Kundu, 1965) which indicate localized concentrations of very hot, dense gas. Such gas must be contained by a magnetic field, presumably of the dome form. Finally, such dome fields are often the sites of flares which develop from the pre-existing plage; these are discussed in the following subsection.

It would appear therefore that an arch filament system is likely to have its velocity, density and temperature distributions determined mainly by its magnetic field. The heat balance in such a system may have no relationship to that of the surrounding chromosphere, and must be studied quite separately.

4.4. THE HEAT FLUX OF SUNSPOTS

Apart from the general optical emission, the two surface phenomena which involve the largest heat fluxes are *flares* and the *sunspot energy deficit*. During a major flare, energy of unknown form is converted to thermal and mass kinetic forms at a rate of $\approx 3 \times 10^{29}$ erg s^{-1}. Because of its lower temperature a sunspot radiates less than the photosphere, the deficit for a large spot being about the requirement of a major flare and, incidentally, of the entire quiet atmosphere.

The flare energy source is usually assumed to be a store of energy located in the atmosphere before the flare commences, but a recent investigation (Piddington, 1972c) suggests that corresponding flare models are unsatisfactory. Accordingly an old suggestion (Piddington, 1958) is revived: that a flare is the result of an enhanced flux of Alfvén-wave energy through the solar surface during the flare (Piddington, 1973b).

Several writers (Danielson and Savage, 1968; Musman, 1967; Savage, 1969) have shown that 'overstable' oscillations are likely to occur in spot flux ropes. Savage has shown how their upward escape may be prevented, and Wilson (1972a) how their leakage may account for umbral dots and other effects.

We suggest that they may normally escape *down* the flux rope, but that this energy sink is sometimes blocked for some minutes to reverse the flow and cause a flare. The blockage may result from the collision of the spot flux ropes following the reversal of spot polarity; it is well known (Zirin, 1970; Sakurai, 1972) that such reversal is very effective in flare production.

References

Altschuler, M. D., Lilliequist, C. G., and Nakagawa, Y.: 1968, *Solar Phys.* **5**, 366.
Athay, R. G.: 1966, *Astrophys. J.* **146**, 223.
Athay, R. G.: 1971, in C. J. Macris (ed.), *Physics of the Solar Corona*, Reidel, Dordrecht, Holland, p. 36.
Beckers, J. M.: 1968, *Solar Phys.* **9**, 51.
Beckers, J. M.: 1972, *Ann. Rev. Astron. Astrophys.* **10**, 73.
Böhm-Vitense, E.: 1968, *Z. Astrophys.* **46**, 108.
Bray, R. J. and Loughhead, R. E.: 1967, *The Solar Granulation*, Chapman and Hall, London.
Chapman, G. A. and Sheeley, N. R.: 1968, *Solar Phys.* **5**, 422.
Danielson, R. E. and Savage, B. D.: 1968, in K. O. Kiepenheuer (ed.), 'Structure and Development of Solar Active Regions', *IAU Symp.* **35**, 112.
de Jager, C. and Kuperus, M.: 1961, *Bull. Astron. Inst. Neth.* **16**, 71.
de Jager, C. and Loore, C.: 1970, *Solar Phys.* **13**, 126.
Dubov, E. E.: 1971, *Solar Phys.* **18**, 43.
Frazier, E. N.: 1968, *Astrophys. J.* **152**, 557.
Frazier, E. N.: 1972, *Solar Phys.* **26**, 130.
Frazier, E. N. and Stenflo, J. O.: 1972, *Solar Phys.* **27**, 330.
Giovanelli, R. G.: 1949, *Monthly Notices Roy. Astron. Soc.* **109**, 372.
Goldberg, L.: 1967, *Ann. Rev. Astron. Astrophys.* **5**, 279.
Goldberg, L.: 1971, in C. J. Macris (ed.), *Physics of the Solar Corona*, Reidel, Dordrecht, Holland, p. 333.
Grossmann-Doerth, U. and von Uexküll, M.: 1971, *Solar Phys.* **8**, 690.
Hansen, R. T., Hansen, S. F., Garcia, C. J., and Trotter, D. E.: 1971, *Solar Phys.* **18**, 271.
Harvey, J.: 1971, *Publ. Astron. Soc. Pacific* **83**, 539.
Kopp, R. A.: 1972, *Solar Phys.* **27**, 373.
Kopp, R. A. and Kuperus, M.: 1968, *Solar Phys.* **4**, 212.
Kulsrud, R. M.: 1955, *Astrophys. J.* **121**, 461.
Kundu, M. R.: 1965, *Solar Radio Astronomy*, Wiley, New York.
Kuperus, M.: 1969, *Space Sci. Rev.* **9**, 713.
Kuperus, M. and Athay, R. G.: 1967, *Solar Phys.* **1**, 361.
Lantos, P.: 1972, *Solar Phys.* **22**, 387.
Lighthill, M. J.: 1952, *Proc. Roy. Soc. London A* **211**, 564.
Lighthill, M. J.: 1954, *Proc. Roy. Soc. London A* **222**, 1.
Lighthill, M. J.: 1967, in R. N. Thomas (ed.), 'Aerodynamical Phenomena in Stellar Atmospheres', *IAU Symp.* **28**, 429.
Livingston, W. and Harvey, J.: 1969, *Solar Phys.* **10**, 294.
Musman, S.: 1967, *Astrophys. J.* **149**, 201.

Noyes, R. W.: 1971, *Ann. Rev. Astron. Astrophys.* **9**, 209.
Osterbrock, D. E.: 1961, *Astrophys. J.* **134**, 347.
Parker, E. N.: 1971, *Astrophys. J.* **163**, 279.
Piddington, J. H.: 1954, *Astrophys. J.* **119**, 531.
Piddington, J. H.: 1956, *Monthly Notices Roy. Astron. Soc.* **116**, 314.
Piddington, J. H.: 1958, in B. Lehnert (ed.), 'Electromagnetic Phenomena in Cosmical Physics', *IAU Symp.* **6**, 141.
Piddington, J. H.: 1972a, *Solar Phys.* **22**, 3.
Piddington, J. H.: 1972b, *Solar Phys.* **27**, 402.
Piddington, J. H.: 1972c, CSIRO Div. Physics. Rep. No. 610.
Piddington, J. H.: 1973a, *Astrophys. Space Sci.* **24**, 259.
Piddington, J. H.: 1973b, *Solar Phys.* **31**, 229.
Piddington, J. H.: 1973c, CSIRO Div. Physics. Rep. No. 603.
Pottasch, S. R.: 1964, *Space Sci. Rev.* **3**, 816.
Proudman, I.: 1952, *Proc. Roy. Soc. London A* **214**, 119.
Reeves, E. M. and Parkinson, W. H.: 1972, *Solar Phys.* **24**, 113.
Sakurai, K.: 1972, *Solar Phys.* **23**, 142.
Savage, B. D.: 1969, *Astrophys. J.* **156**, 707.
Schatzman, E.: 1949, *Ann. Astrophys.* **12**, 203.
Severny, A. B.: 1965, *Publ. Crim. Astron. Obs.* **33**, 34.
Simon, G. W. and Weiss, N. O.: 1968, *Z. Astrophys.* **69**, 435.
Stein, R. F.: 1967, *Solar Phys.* **2**, 385.
Stein, R. F.: 1968, *Astrophys. J.* **154**, 297.
Tousey, R.: 1971, *Phil. Trans. Roy. Soc. London A* **270**, 59.
Ulmschneider, P.: 1967, *Z. Astrophys.* **67**, 193.
Ulmschneider, P.: 1970, *Solar Phys.* **12**, 403.
Ulmschneider, P.: 1971, *Astron. Astrophys.* **12**, 297.
Vitense, E.: 1953, *Z. Astrophys.* **32**, 135.
Vrabec, D.: 1971, in R. Howard (ed.), 'Solar Magnetic Fields', *IAU Symp.* **43**, 329.
Weart, S. R.: 1970, *Astrophys. J.* **162**, 987.
Weiss, N. O.: 1966, *Proc. Roy. Soc. London A* **293**, 310.
Wilson, P. R.: 1971, *Solar Phys.* **21**, 101.
Wilson, P. R.: 1972a, *Solar Phys.* **22**, 434.
Wilson, P. R.: 1972b, *Solar Phys.* **27**, 363.
Zirin, H.: 1970, *Solar Phys.* **14**, 328.

MAGNETIC FIELDS IN THE CONVECTION ZONE

Addendum by J. H. Piddington

The main problems of the heat and energy balance in the solar atmosphere might be listed as follows.

(1) The nature and energy flux of the mechanical waves responsible.

(2) Radiation losses and other energy transfers, up and down.

(3) A model of the atmosphere – which divides into three sections as follows.

(4) Models of small structural features – spicules, mottles, etc.

(5) Models of intermediate features – arch filaments, X-ray knots, flares, etc.

(6) Models of large features – active regions, quiescent prominences (filaments).

There is no doubt that magnetic fields play a major part in all of these problems, and that lack of an understanding of the fields *below* as well as above the surface has been the main cause of lack of progress. The major theories relevant to fields and heating may also be listed as follows.

(1) The dynamo theory of the origin and form of the magnetic field and of the 22-yr cycle of activity.

(2) The acoustic theory of heating.

(3) The theories of the concentration and strengthening of subsurface and surface fields by the various convective motions.

(4) Magnetic field annihilation across neutral sheets as a cause of flares, etc.

These four basic theories are rather closely related in that they invoke the control of fields by subsurface gas, which results in varying degrees of twisting and tangling of field lines. The theories have largely dominated solar physics for a decade or so. I believe that they are all incompatible with observations and invalid – I believe that collectively they have provided a major blockage to the advance of solar physics.

I will suggest an alternative picture of large-scale, enduring, subsurface fields and give a brief description – more details will be published elsewhere.

Following many earlier workers, let us assume a substantial flux rope which has been created by differential solar rotation *below* the level of convection. Let us assume that for an unknown region this rope develops an inverted U section which projects upwards and approaches the surface. The rope is not affected by convective motions and remains intact until its upper, horizontal surface nears the photosphere. The rope might have a diameter of 2×10^4 km which is about 100 scale heights at that level; its total flux might be 10^{21}–10^{22} Mx.

When the upper surface reaches a critical level where the gas pressure outside the rope equals the magnetic and gas pressure inside the rope, drastic effects must occur. The *upper* surface fields of the rope cannot be retained and must billow upwards and become weaker; fields at lower level are still intact. Small flux tubes lift from the upper surface of the rope and are then lifted through the photosphere by the super-granule motions. These are observed as arch filaments and as the main rope rises further, innumerable such filaments provide the arch filament system. Because the horizontal section of the rope spans several supergranule cells, the arch filaments are rather disordered, but a degree of order is imposed by their connections to the rope.

Finally, when the horizontal section of the flux rope emerges through the critical level there must be another rather drastic change in the surface pattern. The feet of all of the arch filaments will move towards one or other of the *vertical* flux rope sections and the whole pattern of the arch filament system will become ordered. The two vertical flux ropes will cause two spots to form, and if any of the arch filament magnetic fields are strong enough to provide spots or pores, then these must all flow into the main spots. This curious effect is entirely consistent with observations.

Spots endure sometimes for periods of weeks, and so must have some coherence to resist the convective motions. A likely explanation is the presence of a helical twist which will provide cohesive magnetic stresses. When the rope untwists it is likely to 'fray' by the loss of 'strands' or small magnetic tubes. This is also observed, because a spot may divide into smaller spots or may decay by the loss of small magnetic elements which are carried away by the supergranule motions. The original flux rope will fray to steadily increasing depths but it endures for months or years. The individual flux tubes may separate into smaller tubes as a result of untwisting, but there seems to be no reason why tubes of diameter $\gtrsim 100$ km (in the convection zone) should not last for a few years. They will be spread by differential rotation and by supergranule motions and eventually provide large 'preceding' and 'following' unipolar magnetic regions.

Each such region is under the control ultimately of the original subsurface flux rope, and the whole provides a magnetic structure which may extend below to $\approx 10^5$ km and across the surface for a few times 10^5 km. Such large structures appear to be essential to account for the observed stability and long duration of large quiescent prominences and of the large decaying magnetic regions with a single polarity observed over enormous areas.

Such a picture is, of course, quite incompatible with the dynamo theory of solar fields.

DISCUSSION

Sturrock: I want to comment on the comparison of the acoustic wave hypothesis of heating to the Alfvén wave hypothesis. I don't think it is fair to quote Osterbrock, who used the Lighthill-Proudman estimates for the generation of energy, because their formulation is based on the hypothesis that the wavelengths of the waves being generated are small compared to the scale height of the atmosphere. This simply isn't so.

Piddington: I think it's all right. I can quote 175 km for a certain depth, and that is exactly the scale height. I am fairly sure he took the scale height.

Sturrock: But the wavelengths of the waves being generated have wavelengths much larger than that, so the use of the Lighthill-Proudman formula is entirely inappropriate. One should use formulas that are based on the assumption that the scale height is in fact less than the wavelength. Under those conditions you will get much more efficient conversion of photospheric motion into propagating acoustic energy. It seems to me that one of the difficulties with the assumption that the energy transfer is due to Alfvén waves is that one would then expect a very strong dependence of the heating of the solar atmosphere on the magnetic field strength. In fact one would then get an energy flux depending upon the cube of the magnetic field strength. That means that if you went from 1 G to 10 G you would go up by a factor of 1000 in the energy flux, and 100 G would give you a factor of one million. It is my impression that the temperature of the corona goes up with the average field strength of the photosphere but not all that rapidly. It goes from one or two million to three or four million. This is not a tremendous variation and argues against the Alfvén wave hypothesis.

The other point I wish to comment on is the flare hypothesis, which is due to the conversion of energy that should come out of sunspots. The point here is that you have simply taken one fact about flares, namely the energy conversion, per unit area per unit time. Lots of other facts about flares against which you could compare your theory and our theories exist. I think it is very unfair to compare flare theories on the basis of only one factor.

Piddington: In answer to the first comment, I agree that attacking other people's theories is not very rewarding but I think that it is necessary if one is going to put forward an alternative theory, particularly with such a strongly entrenched theory as the acoustic theory. Even as late as 1967, Ulmschneider used the Böhm-Vitense model and as far as I can see much the same theory to develop his acoustic theory. Since then it seems that he has more or less postulated the existence of the waves needed to account for the heating that is observed. That is not a very sound basis and it leaves a critic in a very difficult position. What do you criticise about such a theory?

To your second question, I would answer that perhaps the energy does go up as the cube of the field strength, but that doesn't mean that it is all dissipated. In these loop structures, which are very common, the Alfvén waves could just propagate back and forth. A second answer to that – and there might be others – is that very strong fields will tend to suppress the motions that generate Alfvén waves, so it's not entirely clear-cut. Finally, if we invoke the kind of model that I have suggested, within flux tubes, the cubic dependence is not at all obvious because a very strong flux tube simply resists the flow of gas around it and the development of Alfvén waves then is very rare.

In connection with flares, naturally I did not base my theory upon mere coincidences with the sunspot energy deficit. Perhaps I should not even have introduced flares in this review. I think perhaps the strongest point in favour of the Alfvén wave theory arises from the difficulties met by the magnetic storage theories. This comes down to examining the magnetic configuration in many, many flares, in particular in the simplest, smallest flares. I agree with Kiepenheuer and de Jager that these occur as very, very small domes, which seems to exclude the neutral sheet. I don't think the Alfvén wave theory is more than a hypothesis but it is an interesting one. I hope the observers will look for motions of a few kilometers per second in sunspots.

Thomas: Could you give just a rough sketch of the energy dissipation as a function of height? It seems to me that you have concentrated mainly on the corona so far as the energy input is concerned and one worries very much about having all the energy put in way above the region where the conduction comes in. Could you sketch on the board some idea of how the dissipation varies with height?

Piddington: No. I think that's far too detailed a question. After all, you have a flux tube that you might imagine as a taut string without changing its cross section. This is vibrating in an atmosphere with a density scale height that is small compared to the height of the flux tube, and it's a very, very difficult question.

Thomas: But if you are going to criticise other people's theories you ought to at least make some kind of a rough idea as to the rate of energy dissipation as a function of height. That's the one thing that we have some kind of an observational handle on, even though it may not be very accurate.

Piddington: I think it's very inaccurate. I think it's also sufficient that with a little bit of very rough arithmetic I can show that the input from granulation with a photospheric speed of 1 km s^{-1} is more than 10^8 erg cm^{-2} s^{-1}.

Thomas: It seems to be a characteristic of theories for heating the chromosphere that at least six mechanisms can be shown to have potentially the amount of energy that one needs. The question is, do you really get the energy where you need it. That is the big point.

Piddington: You get it where the magnetic fields are. That seems to be overwhelmingly important.

Schmidt: I have a number of small comments. First your comment that the follow-up work of Parker has made much out of a factor of 10. Basically Parker showed that there is a parametrical change. The exponent in the mach number of the convection which comes into the sound production goes up by two, not by a factor of ten but by an addition of two. But this happens to be a parametrical change and can be very large, depending upon the details of the convection in the sunspot which we do not know. We have seen in this symposium some evidence presented by Giovanelli which seems to indicate that we might just have in the umbra of sunspots mechanical acoustic flux of several times 10^8 erg cm^{-2} s^{-1}. I admit that there are certainly difficulties with the production of acoustic flux and difficulties with the production of Alfvén waves. For the latter case I would only mention one. You really have to get the magnetic flux into the granules, and this is a clear-cut question of observation. It is terribly important to learn whether the average magnetic flux enters the granule or not. If it does then certainly your proposal has a very good chance. If it does not, it has none at all.

Piddington: Let me answer your comment on Parker's work. I would not have raised that point at all had it not been for Peter Sturrock. I raised it because I was afraid that he would if I didn't. I don't take much stock of it myself.

Schmidt: There is still a good chance that we might have much acoustic production in the model using strong fields, which might be sufficient, and I think Parker's argument is well taken that the acoustic flux wouldn't grow so terribly and would grow in proportion more or less to what we observe, that is by a factor of ten or so in active regions.

Newkirk: In connection with waves in the corona that are responsible for heating, I would like to remind people that the observations so far are completely inconclusive. Observers at HAO, Sacramento Peak Observatory and the people at Max-Planck Institut have collaborated to look for both compressional waves and Alfvén waves in the corona in both the green line and in white light. So far these efforts have shown no positive evidence. We do have the possibility in the Skylab experiments and of two successful experiments directed towards these ends carried out during the last eclipse; also Dr Liebenberg has reported that he sees changes over a period of several hours, so this point may be resolved perhaps during the next week.

Gabriel: I have a question about your rising flux tubes and the critical line that you have drawn. Would you not expect the pressure balance to be in a quasi-steady state with the external gas pressure balancing the magnetic pressure in the flux rope with the tube expanding so that you would always maintain equilibrium? In that case what is your critical line?

Piddington: Yes, I agree with that, but the scale height is only a few hundred kilometers and the thickness of the tube is perhaps 10 000 km. It projects through several scale heights. The disparity becomes very acute and I mentioned that it should be a big expansion. Assuming that the field is weak enough, the supergranulation motion can take control.

Zirin: I like the idea of some lower influence on the magnetic field because I think that any model of the supergranulation must take account of the fact that the lifetimes of the magnetic structures in the quiet region are of the order of one day, whereas the lifetimes of plages are several weeks, with little apparent change. It is only on the edge of the plages where the magnetic fields are somewhat weaker that one sees evidence of the supergranulation and even there the lifetime is of the order of three or four days. We still don't understand how the large unipolar regions arise. We only see fields originating in the form of sunspots and arch filaments. Although you have drawn a very large flux tube you haven't shown how the rest of the group might come up in this weak field configuration. Secondly, do you have any idea about the asymmetry between preceding and following spots? I am referring to the well-known fact that preceding spots are bigger and live longer than following spots.

Piddington: So far as the arch filament is concerned, I have drawn the three supergranule cells at the top. I imagine that the flux loop at the top expands and weakens considerably so that the supergranules can get hold of part of the magnetic field and give a collection of arch filament systems that will tend to be oriented with the flux rope but also controlled considerably by the supergranule circulation. I would expect a rather random collection.

Zirin: That's great for sunspots but one does not observe a similar phenomenon contributing to the large unipolar regions. That is what bothers me with your picture.

Piddington: Surely that would be in the form of coronal arches or very enlarged coronal structures?

Beckers: When I look at your model of the upward-moving flux tubes and the similar one presented by Vrabec yesterday, and when I put in numbers for time-scales and distances, I get an upward flow of about $1\frac{1}{2}$ km s^{-1}. I expect that the Evershed effect might result from this upflow, resulting from the upward motions of the flux tubes. However a general upward motion of $1\frac{1}{2}$ km s^{-1} is well within observational capabilities and I am not aware of any observations that indicate such motions. Do you have any estimates as to what we might expect to observe?

Piddington: The flux rope breaks through the photosphere with the first arch filament and from then on it is continually shedding arch filaments. When they have all moved through the photosphere the flux rope is more or less vertical, and at that stage I cannot see why there should be any particular flow of matter. Every arch filament goes upward but it sheds its excess downward.

Bracewell: Could you make it clear to me why the flux loop rises in the first place. That is a difficult point. Schatten has mentioned that he is trying to revive Alfvén's old theory in which a toroidal loop propagates up parallel to the main magnetic field. This would fit in with the theory that I have. However I can't see a field of 3000 G propagating up a field of 1 G. I simply do not understand this and I don't think anyone understands. Buoyancy seems to be negligible since the gas pressure is so enormous and the magnetic pressure is negligible.

Wilson: Let me comment in answer to Bracewell. If you believe in giant cells – and I believe this requires a certain amount of belief – then it is natural to think of the flux ropes emerging in the region where the upflow in the giant cells is concentrated most, and one could develop a sunspot theory also related to the motion in the giant cells.

Piddington: But you do have to believe that the giant cells exist in the first place.

Bracewell: I wonder whether negligible buoyancy is a correct concept. If you have something deep down in water at immense pressures and the density is a little bit less than that of water, it will come up even though the buoyancy is very small. So if we had some way of explaining how the magnetic pressure could form a little below the equilibrium value, that would be enough. Perhaps it could occur that the flux rope would develop some instability that would cause it to be locally curved. Then if I use physical intuition with a little hand-waving, there is a longitudinal tension in the rope in the region of the bends. If there is any friction then you would expect a slightly smaller flux density in the region where you develop a loop, so the thing would expand a little with the same flux but a little less flux density. It would seem, therefore, that a random variation would lead to an instability and cause perturbations to grow. Perhaps by the time you have a situation where the flux rope was actually vertical, and especially if there is another bend in the flux rope below similar to that you have shown at the top, I can see very well that the tension in the lines of force would not be able to maintain the same flux density around the corners, and perhaps there is an increasing tendency for the rope to become more buoyant.

Piddington: If you limit the field to 5000 G, at some depth where the flux rope exists the gas pressure will exceed the magnetic pressure by 1000 times and the buoyancy factor is 1 part in 1000. If you squash the field and make it stronger and stronger, you can of course get any buoyancy you like, but I hate to admit fields of 100 000 G because these have to be generated.

Smith: You have said that you think flares occurring low down cannot be due to magnetic field annihilation. Yet if you think of a flux rope coming up it has to interact, as Hal Zirin showed yesterday, with the flux that is already sitting there. Very possibly the flux coming up could indeed cause reconnection. Is there anything wrong with the idea that reconnection occurs low down by the inclusion of flux from below coming up to the flux that is already there?

Piddington: No, there is nothing wrong with that, but some of the pictures we have seen here as well as in the past few days showed a very marked neutral line separating regions of rather uniform polarity, so flux loops of opposite polarity would be very unusual and I cannot image them carrying very much energy.

Brown: In the absence of reconnection, how do you suggest getting such a large proportion of the energy into energetic particles?

Piddington: Are you referring to type III bursts?

Brown: No, I am referring to hard X-ray bursts. I don't know that the X-ray particles are accelerated as fast as the type III's, but the type III's pose a problem of very rapid accelerations and I am willing to give the type III's to the magnetic neutral sheet picture. The time-scale I am talking of is of the order of 1–10 s.

Piddington: That is a very difficult thing to describe.

Brown: The reason I mention the X-rays rather than the type III's is that they involve very much more energy.

Athay: I hate to bring up the subject again but I would like to make a comment about spicules in relation to the comment that Dr Schmidt made on Monday. We have all been guilty of ignoring the work done by the spicules in overcoming the gravitational energy. In making an estimate of the amount of work done in this process, I find for a typical spicule density of 6×10^{10} particles cm^{-3} and a vertical velocity of 25 km s^{-1}, the rate of doing work against the gravitational field is about 5×10^8 erg cm^{-2} s^{-1}. This is a very large amount of energy. If you average it over the solar surface by assuming that the spicules cover 1% of the surface, you still find an average energy flux of 5×10^6 erg cm^{-2} s^{-1}. This is an order of magnitude more than is required to heat the corona and is even somewhat larger than the energy required to heat the chromosphere. If this estimate is correct, then the major part of the mechanical energy coming into the atmosphere from below is going into the spicule motions and is used to overcome the gravitational energy.

Piddington: I would agree with that but I would put it in a different way. The spicules are driven by hydromagnetic forces in nonlinear Alfvén waves.

Schmidt: May I comment on this last discussion. I agree with the estimate; however part of the energy is regained reversibly. The spicules are basically a recurrent phenomenon and the energy put into them when they go up, to a large degree, but certainly not fully, comes back into the magnetic field and the matter falls down again.

Giovanelli: I would like to return to the emerging flux region. It seems that if you have an emerging flux region which is going to give rise to ropes that come out, then gather themselves together by some handwaving, as Piddington describes, these have to be twisted, and indeed that is what has been shown on the board. I am going to assume that a wreath of twisted rope can have little pieces that will shred off, but if this is the case, they will emerge with a particular orientation. When the top of that ropes comes up, the orientation will be in one particular direction, so we should see emerging flux regions with a typical orientation which reverses or at least twists around as the bottom of the flux rope comes through. This is something that should be observable, and I would like to ask the observers who study these flux regions whether they have found such an effect.

Frazier: That is exactly what I have observed. In a *Solar Physics* article several months ago I pointed out that one sees a sequence of arch filament systems over several days and one can trace the plane of rotation as successive loops come up. The top loop is rotated, but when you get down to the bottom it is indeed oriented at the proper angle.

Martin: We have observed the arch filament systems to rotate as much as 120° during the first day or two of the development of the region.

Zirin: I understand that when people study laboratory magnetic fields there are something like 132 instabilities that are always occurring. This doesn't seem to happen on the Sun. Except for flares, things seem to be surprisingly stable. Is there any reason for this?

Schmidt: Nature looks for the stable solution.

Zirin: Yes, but one should see it going from the unstable towards the stable. Do you think it has had enough time to relax by the time that we observe it?

Piddington: I think it is very strongly relaxed. There is really no difficulty here, it just provides co-herence.

RAY TRAPPING IN STELLAR ENVELOPES, PULSATIONAL INSTABILITIES AND HEATING OF EXTERNAL LAYERS

PIERRE SOUFFRIN

Observatoire de Nice, France

Abstract. The relation between local criteria for hydrodynamic instabilities and stellar stability is reconsidered. This leads to a discussion of the classical problem of stellar pulsation with a special emphasis on the pressure modes of high order spherical harmonics. It is shown that within the frame of the geometrical approximation, acoustic rays exist which are trapped inside any specified slab of a stellar envelope. The pressure modes associated with such rays are conjectured to give rise to small scale instabilities whenever a region exists in an envelope where any *local* criterium indicates pulsational instability. The model is expected to support and somewhat extend the recent developments concerning the role of pressure non-radial modes on the dynamics of the solar atmosphere and the heating of external layers.

Preliminary report available from Observatoire de Nice.

DISCUSSION

Wilson: Would you identify then the oscillations like the 5-min oscillations with the *T*-modes that are trapped below the surface.

Souffrin: Yes, that has been suggested by several people. I can give some numbers to make it more specific. We may ask what wavelengths and frequencies are expected in the Sun within the framework of observations. It turns out that periods may be expected in the range 250–750 s. This gives wavelengths of 2500 to 25000 km. The numbers that are best suited to the ray approximation give a period of 350 s and λ about 8000 km. This is just right and I am happy with that.

Stix: Did you compute growth rates for these oscillating modes? We have computed some but I think we still need more growth rates.

Souffrin: The growth rates computed by Ulrich and Wolff are maybe good to order of magnitude, but this is really a technically difficult point and I do not have too much confidence in the calculations. In the Ulrich calculations the boundary conditions are not well known. In Wolff's calculations he uses the variational principle in the specific case where it is not really applicable. I understand from Bob Stein that he is doing this kind of computation, and in Nice, George Conczi is also doing the calculations. This will come out very soon.

Uchida: Would you expect regions of high and low temperature in the photosphere associated with the *T*-mode?

Souffrin: Well, yes, it is a pressure mode, but I have not looked into this question.

Delache: There are two wavelengths in your results, the wavelength of the ray and the wavelength of the acoustical wave. Which wavelength have you given? And can you give the other one?

Souffrin: I am speaking of the acoustical wave. I have not calculated the wavelength of the ray.

R. Grant Athay (ed.), Chromospheric Fine Structure, 293. *All Rights Reserved.*

A MAGNETIC MODEL OF THE CHROMOSPHERE-CORONA TRANSITION REGION

A. H. GABRIEL

Astrophysics Research Division of the Appleton Laboratory, Culham Laboratory,
Abingdon, Berkshire, England

Abstract. The structure of the quiet solar atmosphere is dominated by effects due to the magnetic field distributions produced by field concentrations at the super-granual boundary regions. This effect has been studied by previous workers (Kopp and Kuperus, 1968, Kopp, 1972), and is also evident in the enhanced intensity at the SBR's of spectral lines formed in the upper chromosphere and transition region. An attempt is being made to evaluate the field distributions more precisely, taking into account the physical processes involved in the interactions between the field and plasma, and thereby to predict the intensity distributions expected for the EUV spectra. This work is not yet complete, but some important results can be foreseen.

The field distribution resulting from a preliminary model is shown in Figure 1, in

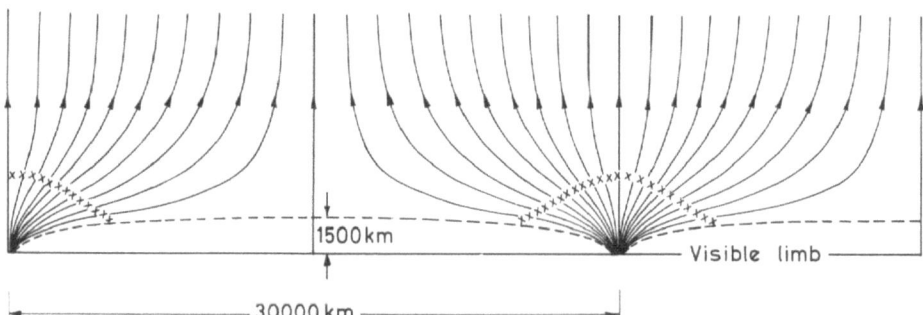

Fig. 1.   A preliminary model for the supergranule structure, showing the primary (× 's) and secondary (dashed) transition regions.

which a mean field of 1 G has been assumed. The lower boundary is the photospheric surface, and the dashed line is the upper limit to the velocity field characterising the super-granule convection flow. Two quite separate transition regions are produced. The first, indicated in the figure by ×'s is physically similar to that predicted by spherical solar models. It will be rather thicker than that arising in the spherical models, and similar to that proposed by Kopp (1972). It is from this region that the majority of the EUV radiation for transition ions will be emitted and it is therefore termed the 'primary transition region'. The dashed surface covering the network centre is also a transition region in the sense that it separates coronal from chromo-spheric material. However, it is closely parallel to the magnetic field, and in this simple model will have an infinite temperature gradient, i.e. negligible material at

R. Grant Athay (ed.), Chromospheric Fine Structure, 295–298. All Rights Reserved.
Copyright © 1974 by the IAU.

transition temperatures. This is referred to as the 'secondary transition region'. Obviously such a gradient will suffer a range of instabilities, which will result in a small mixing, and therefore a limited emission from a very thin transition layer. All these features are broadly consistent with the present EUV observations. One important consequence of the model shown in Figure 1 is that the secondary transition region will be at a higher pressure (and density) than the primary region. This is the opposite to the situation found in other cases of inhomogeneities, such as quiet to active region comparisons, in which the higher densities are normally associated with higher brightness regions. Spectroscopic observations of the network using density sensitive line ratios could therefore prove an important test for the validity of this model.

References

Kopp, R. A.: 1972, *Solar Phys.* **27**, 373.
Kopp, R. A. and Kuperus, M.: 1968, *Solar Phys.* **4**, 212.

DISCUSSION

Athay: It seems to be generally assumed that the supergranulation circulation extends through the chromosphere. The only actual observation of the supergranule motion however is restricted to the photosphere. The question I would like to ask is whether the observers that are here see any prospect of measuring the supergranulation circulation in the chromosphere?

Deubner: Actually the dependence of supergranule circulation has been measured in several spectral lines. It shows a slight decrease from photospheric lines up to lines such as the sodium D lines, but I am not sure whether it has been measured in Hα.

Gabriel: If the chromosphere is not circulating at these heights, then the field will spread out much more rapidly, and on this model you would expect radiation from the transition region lines coming right down to the photosphere.

Sturrock: Have you considered the effect of the field lines on the acoustic wave propagation and excitation as well as on the heat conduction?

Gabriel: Only to the extent that the heat conduction will be channelled along the magnetic field lines.

Sturrock: If acoustic waves are being produced in a magnetic field that is strongly inclined to the vertical, then the cut-off frequency is reduced and this could mean that much more of the turbulent energy is coupled into the acoustic waves as a result of an inclined field.

Giovanelli: I am still worried about this particular model because I do not think that it lines up with the things we see in the chromosphere. We do see fibrils that leave the network and turn over towards the horizontal, and these represent bundles that are discrete; they are not touching one another, at least not always. It doesn't seem to me that this particular model in which the field lines end up going vertically, is quite compatible with the picture we get from the calcium and hydrogen chromospheres. There the tubes seem to be two, three or even four thousand kilometers wide. I think that these need to be contained by an external gas pressure in which the field is lower. Your model shows all of the field lines swinging up and becoming vertical so that if you get to 10000 km or less, or even at 3000 or 4000 km, the field scarcely knows what is down below. The model one gets from the chromosphere suggests that as the flux tubes come up they are kept separate.

. *Gabriel:* One limitation of this model is that it is two-dimensional, and I have assumed that the boundary of the network has a uniform field distribution whereas my understanding from the observations is that the field is fairly well clumped into bundles. That is the sort of thing that could provide the limitation you are describing.

Schmidt: I want to comment on this question. I think one is dealing with the quiet chromosphere where all of the flux is leaving the Sun, most probably to a very high percentage, is connected to the solar wind. I do not believe that fibrils are to be seen in regions of the Sun where the flux is connected to another supergranular boundary or another distant field concentration. If the fibrils connect to foot points then it is not a matter of the dimensionality of the model, but the model which we have for the fibrils embedded in that medium described by Dr Gabriel at greater heights. The height of the fibrils is probably something like 4000 km and they should be embedded in coronal matter which has field lines connected to the source regions described in this paper to the corona above the transition layer. This model is provided by Pikel'ner and I think is a beautiful non-LTE model.

Athay: I understand from talking with Jack Harvey at Kitt Peak Observatory that although the network is statistically embedded in a region of uniform polarity, it is also a common feature of the network to have small fields of opposite polarity adjacent to the main fields of the network. In other words the uniform field polarity of the network is a statistical result and does not hold in detail, but rather there is a substantial complement of opposite polarity present on a fine scale. Could you comment on the effect that this would have on the model that you were talking about?

Gabriel: I am not familiar with the work by Harvey but it would clearly introduce additional closed field lines. Isn't this very much in line with what Professor Schmidt just said?

Beckers: When we study the variation of spicule diameters with height, we find that spicules actually grow smaller when they go away from the limb. If spicules occurred in a divergent magnetic field configuration as you have drawn, I would expect the spicule diameter to increase with height, but that is not the case. Maybe this is related to the bundling that Giovanelli has mentioned, or maybe it is related to the possibility that spicules occur in voids in the magnetic field where the field lines have been pushed apart by the granulation, to form a region of avoidance that converges back to uniform field at greater heights. I didn't mention this yesterday but there is some evidence in the work I have done that this indeed may be the case – that the spicules are not on top of the filigree but open spaces between the filigree.

Vrabec: I would like to reply to Athay's comment concerning Jack Harvey's observations. I am not familiar with this work but our experience at Aerospace in looking at our K-line spectroheliograms is that for very extended regions of the network outside of active regions, the polarities seem to be very strictly of one magnetic polarity. The only place we observe a mixture of polarities is where these extended regions interface or in the active regions themselves.

Schmidt: May I just at this point stress the terrible importance of such observations, of stating just that there are no opposite polarities intermingled, or there are. This deserves the utmost effort. As soon as there are opposite polarities within an average quiet chromosphere cell, even though it may be only on a very small scale, the models of Pikel'ner and Uchida for spicules become very efficient and could easily describe what we see. But please don't find spurious opposite polarities.

Acton: At those locations where we do see two polarities in otherwise quiet regions, this is precisely where in the corona we see X-ray bright points. Up to this point there is a one-to-one correlation that wherever you see a bipolar structure, no matter how small, there is a bright point in the corona. When you look at high-resolution photographs from the corona compared to chromospheric or transition region structure, the corona is generally very diffuse, except at these small bipolar features. Clearly when the bipolar features are present the model Dr Gabriel has been discussing breaks down and we have an entirely different state of affairs.

Brueckner: Dr Gabriel referred to the observations of NRL, and I would like to show slides in order to make sure that everyone understands just what has been observed. These are mixed slides from Skylab and recent rocket observations. The network in helium 584 has a good resemblance to the H and K network. The network structure disappears at the polar caps as can be seen in the slide. However, bright points are still visible in the polar cap region and we identify these with bipolar regions. The network in He II, 304 is identical to the network in He I. It is a carbon copy, including the absence of network in the polar caps. In O IV your first impression is that it is similar to the helium network, but on closer inspection there is more and more concentration into bright points and there is no region of avoidance in the polar caps, at either pole. In O VI there is even more and more of a concentration into very bright points, and I am sure they are below the resolution of the instrument in size. In Ne VII almost everything is concentrated in the very bright points that are below the resolution limits of the instrument in size. The polar cap is visible but very difficult to make out. The last slide shows a comparison of a helium image with an image of the corona. It is obvious from the X-ray picture of the corona that the coronal holes are very depressed. You find a close correlation between the absence of the helium network and the corona, but you also find that the bright points that still exist can be traced back into the corona. You also find that at the boundary of the suppressed region where you see coronal rays coming out, you can see that the helium emission is enhanced.

Vrabec: I think we have an important question for discussion. Dr Acton has just mentioned that wherever they see bipolar regions they see bright points in the corona. Now Brueckner has shown us pictures indicating that there are bright points in the polar cap regions and there is the interference that these overlie mixed-polarity regions. But the rest of the Sun is completely dotted with more bright points. Is the inference that all of these bright points are bipolar? That appears to be a real problem.

Jordan: Some years ago Tony Hearn and I did calculations where we allowed matter to flow out from the Sun as observed in the solar wind, and we calculated what effect this would have on the line emissions. The ions moved through the transition region with velocities consistent with the solar wind. The interesting result that we found, which we never got around to publishing, is that the He II line is very dependent upon the velocity because of the high excitation potential of the ions compared to the surrounding temperature. If you concentrate the solar wind motion into one-thirtieth of the solar area, you can raise the intensity of the 304 line by a factor of ten over the value you get for the static Sun. This could mean that the absence of network in He II indicates an absence of outflow in the helium material. This effect would not be seen in any of the lines except the He II lines and the He I lines. It is not present in Ly-α because of the transfer problems.

EXCESS HEATING OF CORONA AND CHROMOSPHERE ABOVE MAGNETIC REGIONS BY NON-LINEAR ALFVÉN WAVES

YUTAKA UCHIDA

Tokyo Astronomical Observatory* and High Altitude Observatory**

and

OSAMU KABURAKI

Dept. of Astronomy, University of Tokyo, Tokyo

Abstract. Excess heating of the active solar atmosphere is interpreted as the decay of MHD slow-mode waves produced in the corona through the non-linear coupling of Alfvén waves supplied from subphotospheric layers. It is stressed that the Alfvén-mode waves may be very efficiently generated directly in the convection layer under the photosphere in magnetic regions, and that such magnetic regions, at the same time, provide the 'transparent windows' for Alfvén waves in regard to the Joule and frictional dissipations in the photospheric and subphotospheric layers. Though the Alfvén waves suffer considerable reflection in the chromosphere and in the transition layer, a certain fraction is propagated out to the corona. A large velocity amplitude, exceeding the local Alfvén velocity, is attained during the propagation along the magnetic tubes of force into a region of lower density and weaker magnetic field. The otherwise divergence-free velocity field in Alfvén waves, in such a case, couples with a compressional component (slow-mode waves) which again is of considerable velocity amplitude relative to the local acoustic velocity when estimated by using the formula for non-linear coupling between MHD wave modes derived by Kaburaki and Uchida (1971). Therefore, the compressional waves thus produced through the non-linear coupling of Alfvén waves are eventually thermalized to provide a heat source. The introduction of this non-linear coupling process and the subsequent thermalization of the slow-mode waves may provide a means of converting the otherwise dissipation-free Alfvén mode energy into heat in the corona. The liberated heat will readily be redistributed by conduction along the magnetic lines of force, and thus the loop-like structure of the coronal condensations (or probably also the thread-like feature in the general corona) is explained in a natural fashion. Matter density in the arches is expected to be higher than the ambient density because of the higher temperature and, consequently, greater scale height in the arches.

(Full paper is to be submitted to Solar Physics.)

Reference

Kaburaki, O. and Uchida, Y.: 1971, *Publ. Astron. Soc. Japan* **23**, 405.

* Tokyo Astronomical Observatory, University of Tokyo, Mitaka, Tokyo.
** High Altitude Observatory, National Center for Atmospheric Research, Boulder, Colorado, is sponsored by the National Science Foundation.

THE STABILITY OF A MAGNETIC FLUX ROPE AND ITS RELATION TO SUNSPOTS, FACULAE AND FLARES

P. R. WILSON

Dept. of Applied Mathematics, University of Sydney, Sydney, Australia

Abstract. The stability of a magnetic flux rope is investigated in the presence of an Alfvén wave flux. It is shown that instabilities will arise when the transverse perturbation velocity exceeds the Alfvén velocity for the flux rope. With this restriction the maximum permitted Alfvén energy fluxes have been calculated under a variety of conditions typical of faculae, flares and sunspots. It is shown that the power requirements for faculae are very close to the theoretical limit and that flares may or may not exhibit a flash phase depending on whether the stability criterion is exceeded or not. The Alfvén fluxes which have been suggested for sunspots are, however, generally less than the limit based on this criterion.

DISCUSSION

Schmidt: If I am not mistaken, I think that dissipation of an Alfvén wave effectively sets in at the limit which you describe as an instability, so would you object to an interpretation where this is an onset of dissipation of a wave rather than the onset of an instability, or do you really mean that it is the same thing?

Wilson: Much the same thing. I think that Piddington suggested this morning a mechanism whereby an Alfvén wave may dissipate through pushing the neighboring material around, but this would be a fairly steady dissipation, and you could imagine such a process continuing for quite a long while and providing the heating for fairly stable structures. Once you exceed this limit the dissipation will be sudden and rapid. You may call it either dissipation or instability, I am quite happy either way.

R. Grant Athay (ed.), Chromospheric Fine Structure, 301 All Rights Reserved.
Copyright © 1974 by the IAU.

GENERAL DISCUSSION

Newkirk: The observations I will now show from Skylab are relevant to the question raised somewhat earlier about the connection between the chromosphere, the transition region, and the overlying corona. The slide is provided by the Harvard College Observatory, far ultraviolet experiment. It shows an area of the Sun with images in Ly-α, C III and Mg X. Bright points are visible in all three of the images. However there is a portion of a coronal hole which you can see. Within the coronal hole, the bright points appear to be suppressed somewhat in the coronal line of Mg X. One interpretation that might be placed on this is that the coronal holes are regions where the magnetic field is open to the interplanetary medium, so that what we are seeing in the coronal hole is a region in which the high effectiveness of conductivity outward into space is able to cause a depletion of the corona even though the bright points continue to show from the chromosphere through the transition region into the corona. This has relevance to the type of model suggested by Dr Gabriel.

Schmidt: Did I understand correctly that you stated that the connection to interplanetary space is restricted to these coronal holes; if so, how about the magnetic field balance in the interplanetary space?

Newkirk: I am not sure I understood your question, but we may consider the polar corona as being coronal holes and there we clearly do not have to worry about any flux balance in interplanetary space.

Schmidt: But are you implying that only the coronal holes are connected to interplanetary space, and no other regions on the Sun?

Newkirk: I think that would be hazardous.

Schmidt: You only say that these regions are connected.

Newkirk: These regions are connected to interplanetary space in a much more effective way. The solar wind has to do much less work in plowing its way out of the Sun. Where the field is more or less unipolar in character, the solar wind does not have to change the magnetic field configuration of the Sun. In a region of more complex polar structure, this is not true.

Schmidt: What fraction of the solar area would you ascribe to these coronal hole regions?

Newkirk: By looking at the X-ray pictures that have been published and in which the coronal holes are easily identifiable, I would guess 20–30%, excluding the poles.

Schmidt: Isn't it true that one needs an average field strength for the whole Sun of about 2 G to be in balance with the interplanetary magnetic field? Don't you need therefore to connect to some 80% of the quiet Sun – or, if you connect primarily only to the coronal holes, you need to connect some reasonable percentage of the field lines to high field areas, and this latter picture seems a little improbable.

R. Grant Athay (ed.), Chromospheric Fine Structure, 303–307. All Rights Reserved.
Copyright © 1974 by the IAU.

Meyer: I think this is an important point that should be discussed.

Gabriel: Do you consider that the solar wind will have a higher velocity in these coronal hole regions?

Newkirk: The magnetic field allows the conductivity to carry the heat away more effectively. The scale height is higher in the coronal holes than it is in the rest of the regions where there is a restriction to the field. Even though the base temperature in the coronal holes may start out lower it decays less rapidly. Your question was, do I expect the velocity to be higher in the holes than elsewhere. Jerry Pneuman should answer that question but he is not here.

Smith: The answer to the question is yes, and that is consistent with observations. Art Hundhausen has shown that in fact the high-velocity winds observed in the solar stream are very well correlated with the coronal holes.

Gabriel: This seems to be contrary to the suggestion that Carole Jordan just made that the helium enhancement is consistent with higher solar wind velocities.

Jordan: Can I ask Dr Brueckner a question? Do you see the helium emission suppressed in the coronal holes?

Brueckner: Yes, the helium network is suppressed in the coronal holes, in the same way that it is over the poles.

Delache: I would direct a question to the one raised by Alan Gabriel. If the velocity of the wind is going to be larger in the coronal holes, what about the matter flux – is it larger or lower than in other regions?

Newkirk: It depends upon how much you ascribe to other regions. There is some question as to whether or not we have ever sampled any of the high-density regions that we see, in coronal streamers.

Delache: Let me put it in another way. If you observe a correlation between coronal holes and high velocities of the solar wind, what is the enhancement or the decrease of the mass flux?

Brandt: Mass flux is almost independent of speed, and statistically the high-velocity streams are associated with low density. The high-velocity streams generally have low density and high temperatures. The subject of the solar wind and coronal holes came up before in this symposium and I will make the statement again that I think it is premature to identify the coronal holes as the major course of the solar wind. There is simply too much evidence that demands a flow from a rather large fraction of the Sun.

Brueckner: I would like to raise a question about the identification of the coronal holes with high-velocity solar wind. On my last slide I showed the base of the coronal streamers as having intense helium network and not the suppressed helium network that we saw in the slides, at the location of the coronal holes.

Jordan: You see these little bright spots just at the bottom of the polar plumes as well, don't you?

Brueckner: This is still debatable, but always where we see a coronal streamer coming out we have an enhancement of the helium network.

Schmidt: I propose that we now discuss a little bit the question that has come up in a number of talks: how much of the magnetic flux coming out of the photosphere is going into the chromosphere and staying there, and how much goes on into the corona? There seems to be strong advocates for a large percentage of the flux staying in the chromosphere. The fibrils discussed by Dr Zirin and Dr Giovanelli and the MMF's discussed by Vrabec give independent evidences that there is a sizable percentage of flux staying in the chromosphere. On the other hand there is a serious problem with the Lorentz forces. If there is a sizable flux staying in the chromosphere, what we need is a flat-iron holding it down. This flat-iron has to have an exact weight which can be calculated. It comes out to be roughly 100 to 1000 times the total weight of the atmosphere available. We should discuss the possibilities for finding a solution to this problem. My own suggestion is that we consider a situation in which we have only a slight percentage – and I mean only 1 to 4% of the flux staying in the chromosphere and the rest going into the corona. This would be nearly a force-free situation. Certainly the fields of force are not force-free but they are nearly so. This raises the question as to how this can be reconciled with the evidence for field remaining in the chromosphere. The evidence boils down to a discussion of the radius of curvature of the fine structure in the chromosphere. This is very good evidence but I simply ask, is it not evidence that comes from a single very thin layer near 4000 km. Therefore I am tending to conclude that the radius vector of these curved features lies in the layer itself and not perpendicular to it. That is to say, we are dealing with spiral structures that lie at a level of 4000 km and a little bit deeper down. If so, we still have the question of what is the reason for the curvature. I would suggest that it is the very fact that about 60% of the flux in the regions where we see the spiral structure, that is near to the sources of the field, is leaving the region more nearly to the vertical than the angle subtended by the 4000 km height from half the diameter of the supergranule cell, which is a very small angle. That 60% of the flux which comes out, if you assume that the field is nearly force-free, has to go out somewhere between the fibrils that are so conspicuous. If so, I don't think that we have a serious problem with the spiral structure, because then we balance Lorentz forces against Lorentz forces. We do not have to balance them against the waves which are not there. We simply have flux lines which go out to large heights in the corona, and maybe to interplanetary space, which are as strong in their field strength as the surrounding fields intrinsic to the fibrils, but they have different curvature.

Zirin: I have two suggestions in regard to this very important question. In the filaments we have an example of a case where flux is confined very near the surface for very long distances, which in a potential type theory must – as Dr Schmidt has suggested – rise high above the surface. It is clear from observations by Foukal and others as inferred fields and by direct measurements of the magnetic field in prominences at the limb, that filaments exist wherever there is a magnetic boundary between one polarity and the other, where for long distances magnetic fields run parallel to the surface. In fact there we see a flat-iron, namely the material that occupies the filaments. This seems to be a very stable situation that lasts for a long time. There is no doubt that these lines of force in the filament are running parallel to the surface.

We are not talking much about the sharp curvature, but wouldn't you agree once again that you would not expect it to run parallel to the surface without having a flat-iron to hold it down.

The second point is that I think it is very important to determine just how high lines of force in substantial active regions go. It is my impression from studying my photographs – and I realize it is difficult to express this quantitatively – that the stronger the general connection between an active region with distant fields, the more sharply suppressed is the magnetic field underneath. In this case, the flat-iron seems to be the strong magnetic field between the active regions and the distant regions, which apparently keeps other fields from rising forward. It seems clear that the stronger the field, the more sharply the turning to the horizontal by the lines of force.

Schmidt: I accept your flat-iron with respect to any filament, be it active or quiescent. Certainly this is a small region visible on the disk where we have sufficient mass at a high elevation, suspended in the fields where exactly the magnetic tensions are balanced by the weight of the filaments. These cover a certain per cent of the active region, and I leave it to the observers to estimate that. I do not accept at all the other flat-iron you suggest. I did not mean that there has always to be an integral of the mass per cm^2 that balances the magnetic stresses. On the other hand, this does imply that at almost every point of the Sun the field is almost exactly vertical. There has to be some flux that stays within a scale height or two of 4000 km for a long distance, so that we can have correlated features visible in Hα at that level. But we need only a tiny bit of the solar flux to account for that. Someone needs to look into this problem.

Dravins: I believe it is difficult to have only one or two per cent of the flux being in the chromosphere. This is because one can determine the magnetic field at different heights in plages and sunspots, and one can make an estimate as to how rapidly the magnetic structures in the network diverge with height. This gives consistent values in the sense that there is a distinct weakening of the longitudinal field with height which is because the field is bending over in the first few thousand kilometers, and on the inside of the network one sees the fibril type of structure. One doesn't see them down at the very bottom of the photosphere. If we believe that these structures follow the magnetic field, we can trace the field with height consistent with the observation that the field must have bent over quite sharply. It is difficult to estimate how much field will continue to remain at low heights, but if the fibril field does not remain at low heights then one must imply an additional kink in the field going through another bending upwards. Another comment – there is a problem with Lorentz forces coming from the magnetic kink in the chromosphere near active regions. It was mentioned yesterday that these moving magnetic features might be associated with magnetic field lines going up and down in the photosphere, and doing this in just a few thousand kilometers, in very strong fields. I think that in that type of field geometry the implied Lorentz forces are orders of magnitude stronger than the fields suggested by the Harveys.

Kiepenheuer: The fact that you observe the 4000 km altitude to consist of a very flat buttress for these fibrils, etc., does not mean that the field lines there are really at rest. As soon as these structures are lifted a little bit they will empty out and disappear by emptying out, so that there still could be flux above this level.

Deubner: My remark is similar to Dr Kiepenheuer's. So far as observations are concerned, I believe it is clear that the chromospheric fibrils tend to follow the average of the field direction but nobody so far has shown that indeed they follow the inclination of the field lines.

Vrabec: I suggest that a question of possible relevance is the interconnection of different active regions by magnetic flux tubes. Here perhaps the radio astronomers and X-ray astronomers can shed some light on the problems.

Miss Shahinay Yousef: I would like to ask Dr Newkirk if he thinks that the coronal holes are free from magnetic arches connecting distant active regions, and are the coronal holes related to the magnetic sector structure in the interplanetary magnetic field, or do they have something to do with the active regions?

Newkirk: I would ask Dr Altschuler to answer the first part of the question.

Altschuler: We generally find that coronal holes are associated with diverging fields. On the average the field is unipolar diverging outwards, but this is only in average conditions.

Schmidt: I wonder whether, at some future symposium, we might discuss the subject of possible dissipation in the chromosphere. At times we have talked about annihilation, at times about diamagnetic acceleration, at times we have talked about downward flow, or the need for it, and at other times we have talked about the transport of streams of mass into the corona by spicules and its return back to the chromosphere again. On the other hand we have both heard and seen beautiful evidence for dissipation in the chromosphere. We have seen brightenings at the borderlines of arch filament systems or eruptive flux regions; we have seen bright points of several kinds; and we have seen filigree. It would be very helpful if we would be able to ascribe mechanisms to the different observed phenomena.

INDEX OF SUBJECTS